AF389611

Characterization of Bioactive Components in Edible Algae

Characterization of Bioactive
Components in Edible Algae

Special Issue Editor

Leonel Pereira

MDPI • Basel • Beijing • Wuhan • Barcelona • Belgrade • Manchester • Tokyo • Cluj • Tianjin

Special Issue Editor
Leonel Pereira
MARE-Marine and Environmental Sciences Centre
University of Coimbra
Portugal

Editorial Office
MDPI
St. Alban-Anlage 66
4052 Basel, Switzerland

This is a reprint of articles from the Special Issue published online in the open access journal *Marine Drugs* (ISSN 1660-3397) (available at: https://www.mdpi.com/si/marinedrugs/edible_algae).

For citation purposes, cite each article independently as indicated on the article page online and as indicated below:

LastName, A.A.; LastName, B.B.; LastName, C.C. Article Title. *Journal Name* **Year**, *Article Number*, Page Range.

ISBN 978-3-03928-560-0 (Pbk)
ISBN 978-3-03928-561-7 (PDF)

Cover image courtesy of Leonel Pereira.

Contents

About the Special Issue Editor

Leonel Pereira (Ph.D.) A. Biography

- Birthplace and date: Lisbon, Portugal; 13 February, 1966.
- Personal information: Married, one daughter, mobile phone: 00-351-960-039-727
- Institutional address: Department of Life Sciences, Faculty of Science and Technology, 3000-456 Coimbra, Portugal. Tel. (+351)-239-240-782 leonel.pereira@uc.pt
- Research Center: MARE, University of Coimbra
- Research Group: Coastal Systems and Ocean
- Research Lines: Biodiversity and Ecosystem Functioning, Environmental Risk, Biotechnology and Resource Enhancement
- ORCID ID: http://orcid.org/0000-0002-6819-0619
- Scopus Author ID: http://www.scopus.com/authid/detail.uri?authorId=7201962433
- Researcher ID: http://www.researcherid.com/rid/M-3527-2013
- ResearchGate: https://www.researchgate.net/profile/Leonel_Pereira
- GoogleScholar: https://scholar.google.com/citations?user=YrGbzsEAAAAJ&hl=en&oi=ao
- Portuguese Seaweeds Website: http://macoi.ci.uc.pt/ B. EDUCATION
- High School: Escola Sec. José Falcão, 1985, classification: 15 out of 20.
- Graduation: Biology, University of Coimbra, 1991, classification: 15 out of 20.
- University Diploma in Coastal Biology - option: Marine Algae - University of Caen (France), 1992, classification: good.
- Scientific and Pedagogic Capacity Test in Biology (Phycology), University of Coimbra, 1997, classification: Very Good.
- Ph.D.: Biology (Cellular Biology - Phycology), University of Coimbra, 2005, classification: Very Good. C. PROFESSIONAL CAREER
- Trainee Assistant, Department of Botany, University of Coimbra, 1992–1997
- Assistant, Department of Botany, University of Coimbra, 1997–2005
- Assistant Professor, Department of Botany, University of Coimbra, 2005–2009
- Professor, Department of Life Sciences, University of Coimbra, 2009–Current. D. TAUGHT CLASSES
- Graduation: Algae and Fungi, Bioinformatics, Field Biology
- Post-graduation: Algae Biotechnology, Marine Ecology, Marine Phycology, Economic Marine Botany E. AREAS OF SPECIALIZATION
- Marine and Biotechnology Phycology
- Phycocolloids Biochemistry - FTIR, FT-Raman, MNR
- Marine Botany - Ecology, Biodiversity, Economic Marine BotanyF. Membership in Scientific Societies
- 1995/Actuality - Member of the Executive Board of APAA (Portuguese Association of Applied Algology)
- 2006/Actuality - Member of Sociedade Broteriana.
- 2007/2009: Member of WSEAS (World Scientific and Engineering Association and Society)
- 2008–Current: Permanent Forum for Maritime Affairs.

Preface to "Characterization of Bioactive Components in Edible Algae"

Numerous factors affect the quality of modern life, so the population must be aware of the importance of foods that contain substances that help to promote health by improving their nutritional status. The incidence of death from cancer, cardiovascular accidents, stroke, atherosclerosis, and liver disease, among others, can be minimized through good eating habits. Functional and nutraceutical foods have been commonly considered synonymous; however, functional foods should be in the form of common foods, consumed as part of the diet and produce specific health benefits, such as reducing the risk of various diseases and maintaining good physical and mental health. Biologically active substances found in functional foods can be classified as probiotics and prebiotics, nitrogenous foods, pigments and vitamins, phenolic compounds, polyunsaturated fatty acids, and fibers. The marine environment provides a huge source of many healthy foods, including algae (green, red, and brown macroalgae), which are an example of a marine product that is part of the diet in several countries around the world. In addition, marine samples are sources of an abundance of chemicals, many of them with biological properties so are therefore called bioactive compounds. These chemicals can be extracted and incorporated into various food matrices, showing potential as new functional foods. With little fat, seaweed has polysaccharides that behave, most of the time, as non-caloric fibers. Therefore, algae appear to be the best way to correct current global nutritional deficiencies (in developed, emerging, and/or underdeveloped countries) due to their varied range of essential constituents including minerals (iron and calcium), proteins (with all essential amino acids)), and vitamins and fibers; these nutrients absolutely necessary for primary human metabolism.

Leonel Pereira
Special Issue Editor

 *marine drugs*

Editorial

Characterization of Bioactive Components in Edible Algae

Leonel Pereira

Marine and Environmental Sciences Centre (MARE), Department of Life Sciences, Faculty of Sciences and Technology, University of Coimbra, 3004-517 Coimbra, Portugal; leonel.pereira@uc.pt

Received: 13 January 2020; Accepted: 16 January 2020; Published: 19 January 2020

Introduction

From the origin of our planet, about 4.6 billion years ago, 1.400 million years had to pass in order for an event of maximum transcendence to develop in the evolutionary history of life on Earth: the appearance of the first photosynthetic organisms, with which the biochemical machinery that maintains ecological dynamics was launched, as we know it today, both in the aquatic environment, where it was born, as well as terrestrially [1].

Photosynthesis allowed them to synthesize organic matter, the basis of the feeding of every living being, from inorganic compounds, at the same time that it was an oxygen contribution, which at that time generated a radical change in the composition of that primitive atmosphere and conditioned the next steps of evolution. Among these pioneers are the most primitive algae, with a precarious unicellular constitution that lacks a nucleus (prokaryotic cell organization). They are known as Cyanobacteria or blue-green algae, although today there is a discrepancy between systematics when considering them as bacteria or algae [2].

The next step in the evolutionary process was the formation of a complete cellular system with a true nucleus (eukaryotic cellular organization), shared by unicellular and multicellular algae, which happened just 1.4 billion years ago, with an atmosphere that was already rich in oxygen. However, 700 million years had to pass before the great algae and the first invertebrate animals were developed, and 300 million more so that, from the green algae, the first vascular plants colonized the terrestrial environment, free of living beings until that moment. The last stage in plant evolution was reached with the appearance of flowering plants (phanerogams) only 100 million years ago, which, if we compare the age of the Earth with the average life expectancy of human beings, is equivalent to two years in the life of a person [1].

Algae as Food

Larger seaweeds, macroalgae, are a source of food ignored in the western world, while constituting an essential part of, for example, the Japanese diet. Macroalgae are, in many respects, optimal for human nutrition since they contain a prodigality of minerals, trace elements, vitamins, proteins, iodine, and essential polyunsaturated fatty acids, as well as an abundance of both soluble and insoluble dietary fibers, with few calories [3].

In the same way, macroalgae have an enormous and unexplored potential for gastronomic innovation. They can be used practically in any dish of the daily kitchen as well as haute cuisine. Considering that there are about 10,000 species worldwide of edible algae and that the genetic differences between any of these species are as large as those between plants and animals, it is clear that this splendid abundance provided by the oceans of the planet deserves greater attention from scientists and the inhabitants of the western world [4,5].

Today, the societies of western countries, so-called developed, live immersed in an illusory abundance and diversity of food. We are driven to consumption without rules or dietary care and to fast food, high in calories and unsaturated fats. This appears to be the miraculously adequate response to the hectic pace of urban life—so much so that we have even adopted the designation of ready food, or fast food, as a style and misperception of a reality, where food is seen merely as organic fuel doses to meet our most immediate energy needs. The consequences of such a diet (antagonistic to traditional slow food, more carefully prepared), where the lack of essential nutrients is evident, are obesity (and collateral diseases) as well as other diseases related to excessive intake of sugars (diabetes) and fats (arteriosclerosis), among others [3,6].

On the other hand, this illusion does not express itself with the same impact in underdeveloped or transitional countries, as seen from an economic perspective, or on those emerging economies; although in the latter, the tendency is towards their imposed consolidation rather than towards its eradication. Countries such as Brazil, with a considerable coast, face the same dilemma and have before them the path that other countries can take: where food practices can and should be adapted to local resources. Less-developed countries with an appreciable coastline, such as Angola and Mozambique, may adopt new food strategies to overcome the strong shortages still felt [7].

The question that arises at this point of consciousness is simple—what contribution or benefits can seaweed bring to the human diet in terms of food, gastronomy, or diet? The answer seems simple in the light of current knowledge—it is the exact opposite of the concept of fast food: a natural, yet wild and abundant food (with a growth rate capable of sustaining an intensive culture), capable of providing high nutritional value but at reduced caloric value. Poor in fat, seaweed has polysaccharides that behave, for the most part, as non-calorie fibers. Therefore, algae seem to be the best way to correct not only the lack of food for ingestion but also the nutritional deficiencies of the current diet worldwide (in developed, emerging, and/or underdeveloped countries) due to their wide range of essential constituents—minerals (iron and calcium), proteins (with all essential amino acids), vitamins and fiber [3]; nutrients that are absolutely necessary for human primary metabolism. They are, therefore, a guarantee of survival, which the human being, sooner or later, will resort to, now more out of whim and curiosity (thanks to some pioneering work and investments that are starting to pay off) and later, of course, to meet the demands of an explosively-growing human population that already numbers more than 7 billion people, increasingly concentrated in Asia and Africa [8].

The marine environment provides a huge source of many healthy foods including algae (seaweed–green, red, and brown macroalgae), an example of a marine product that has been part of the diet in several countries around the world. Furthermore, marine specimens are also sources of an abundance of chemicals, many of them with biological properties and therefore called bioactive compounds. These chemicals can be extracted and incorporated in several food matrices leading to new potential functional foods [6,9].

Some of the edible species referred to in this special issue have been the subject of studies related to some of their bioactive compounds, from lipids to algal phlorotannin, various phycocolloids (alginates and carrageenans), to the sulfated polysaccharides such as Ulvan and Fucoidan.

Among the species of macroalgae studied in the articles of this special issue, we find, for example, the species harvested in Indonesia and Japan, where lipid content variations were evaluated. Susanto et al. (2019) [10] concluded that the total lipid (TL) content and fatty acid composition were strongly affected by sampling location. The TL and n-3 PUFA levels tended to be higher in temperate seaweeds compared to those in tropical seaweeds. In this study were analyzed four species of edible green algae (Chlorophyta; edible seaweeds of the world): *Caulerpa lentillifera* and *Ulva reticulata*, harvested in Indonesia; *Ulva intestinalis*, *U. australis*, and *U. reticulata*, harvested in Japan; five species of red algae (Rhodophyta): *Gracilariopsis longissima*, harvested in Indonesia, *Gloiopeltis furcata*, *Chondrus yendoi*, *Mazzaella japonica* and *Chondria crassicaulis*, harvested in Japan; and six species of brown algae (Ochrophyta, Phaeophyceae): *Sargassum aquifolium*, harvested in Indonesia, *Costaria costata*, *Undaria pinnatifida*, *Saccharina japonica*, *Sargassum fusiforme* and *S. horneri*, harvested in Japan [5,10].

Brown (Phaeophyceae) and red (Rhodophyta) algal sulfated polysaccharides have been widely described as anticoagulant agents. However, data on green (Chlorophyta) seaweed, especially on the *Ulva* genus, are limited. In the work conducted by Adrien et al. (2019) [11], the anticoagulant activity of the ulvan extracted from *U. rigida* is evaluated. The authors of the paper conclude that the chemically-sulfated ulvan fraction could be a very interesting alternative to heparins, with different targets and high anticoagulant activity [11].

Ecklonia cava, an edible brown alga growing abundantly on the shores of Japan and Korea where it is consumed as part of the daily diet, is studied by the team of Oh et al. (2019) [12], which evaluated phlorotannin, phlorofuroeckol A (PFF-A), produced by this alga, and the osteoblastogenesis enhancing effects, which can be utilized against bone-remodeling imbalances and osteoporosis-related complications [12].

Natural compounds can be effective candidates for various skin diseases, especially phlorotannins extracted from seaweeds, which have interesting properties that make them useful for cosmeceutical applications. This is the theme explored by Zhen et al. (2019) [13] in an article on the effect of Eckol, extracted from edible brown seaweed *Ecklonia cava* harvested in Jeju, Korea, on the protection of dermal cells from apoptosis by inhibiting the MAPK signaling pathway. This was further reinforced by detailed investigations using MAPK inhibitors [13].

Another edible brown seaweed, *Cystoseira barbata* [5], harvested from the Romanian Black Sea coastal zone, is used in an experimental work by Trica et al. (2019) [14] for alginate extraction and a subsequent test to determine the adsorption capacity of Cu^{2+} and Pb^{2+} heavy metals [14].

The chemical, structural, and cytotoxic characterization of the invasive brown alga *Sargassum muticum* and the red alga *Osmundea pinnatifida*, collected on the Portuguese coast (Atlantic), was determined from enzymatic extractions. The team of Rodrigues et al. (2019) [15] determined from FTIR-ATR and ^{1}H-NMR spectra the presence of important polysaccharide structures in the extracts, namely, fucoidans from *S. muticum* or agarans as sulfated polysaccharides from *O. pinnatifida*. No cytotoxicity against normal mammalian cells was observed, making these seaweed extracts very interesting functional ingredients, which could be explored as a food ingredient (salt replacer, nutrient vector) or nutraceutical supplement [15].

In the experimental work of Hwang et al. (2019) [16], the immune-modulatory effects of orally administered crude fucoidan extracted from the edible brown algae *Saccharina japonica* (formerly *Laminaria japonica*) [5], collected in Taiwan, on the innate immune response, adaptive immune response, and MP antigen-stimulated immune response, was investigated. The hope of the researchers is that fucoidan, a natural food supplement, can enhance the immune responses needed for immunopotentiation and attenuate the *Mycoplasma pneumoniae* (MP) infectious disease [16].

In the work of Hofer et al. [17], seven bromophenols are isolated from a methanolic extract of the epiphytic red alga *Vertebrata lanosa*, an edible seaweed [5] collected in Brittany, France. Bioactivity of seven isolated bromophenols was tested in agar diffusion tests against *Staphylococcus aureus* and *Escherichia coli* bacteria. Three compounds showed a small zone of inhibition against both tested organisms [17].

As described by Valado and collaborators [18], changes in lipid profile constitute the main risk factor for cardiovascular diseases. The daily intake of a vegetable jelly (with carrageenan E-407) for 60 days showed a reduction in serum total cholesterol (TC) and low-density lipoprotein cholesterol (LDL-C) levels in women, allowing them to conclude that carrageenan has bioactive potential in reducing TC concentrations [18].

In the study by Cotas et al. (2019) [19], *Gigartina pistillata* (Rhodophyta) carrageenans, from specimens harvested from the western coast of Portugal, are evaluated against colorectal cancer stem cell (CSC)-enriched tumor-spheres. Carrageenans extracted from two *G. pistillata* life cycle phases have antitumor potential against colorectal cancer stem-like cells, especially the Lambda-family carrageenans extracted from the tetrasporophyte (T) phase [19].

Cho and Rhee's review [20] focuses on research on the health benefits of consuming substances present in high concentrations in the laver, such as porphyran, vitamin B_{12}, and taurine, with an evaluation of the expected effects of the consumption of these red algae. Mitigation of chemical and microbiological hazards and the adoption of new technologies to preserve and exploit the biochemical characteristics present in the *Porphyra/Pyropia* are reviewed as key strategies to further improve the quality of products based on these species (Laver/Nori).

Funding: This work was supported by Foundation for Science and Technology (FCT) through the strategic projects granted to MARE—Marine and Environmental Sciences Centre UID/MAR/04292/2019.

References

1. Lloréns, J.L.P.; Carrero, I.H.; Oñate, J.J.V.; Murillo, F.G.B.; González, A.L. *¿Las Algas Se Comen? Un Periplo por La Biología, La Historia, Las Curiosidades y La Gastronomía*; Servicio de Publicaciones de la Universidad de Cádiz: Cádiz, Spain, 2016; 336p, ISBN 10 8498285674.

2. Pereira, L. Cytological and cytochemical aspects in selected carrageenophytes (*Gigartinales, Rhodophyta*). In *Advances in Algal Cell Biology*; Heimann, K., Katsaros, C., Eds.; DeGruyter: Berlin, Germany, 2012; Chapter 4; pp. 81–104. ISBN 978-3-11-022960-8. [CrossRef]

3. Pereira, L. A review of the nutrient composition of selected edible seaweeds. In *Seaweed: Ecology, Nutrient Composition and Medicinal Uses*; Pomin, V.H., Ed.; Nova Science Publishers Inc.: New York, NY, USA, 2011; Chapter 2; pp. 15–47. ISBN 9781614708780.

4. Mouritsen, O.G. *Seaweeds: Edible, Available, and Sustainable*; University of Chicago Press: Chicago, IL, USA, 2013; 304p, ISBN 13 978-0-226-04436-1.

5. Pereira, L. *Edible Seaweeds of the World*, 1st ed.; CRC Press, Taylor & Francis Group: Boca Raton, FL, USA, 2016; 453p, ISBN 978-149-87-3047-1.

6. Pereira, L. Nutritional composition of the main edible algae. In *Therapeutic and Nutritional Uses of Algae*, 1st ed.; CRC Press, Taylor & Francis Group: Boca Raton, FL, USA, 2018; Chapter 2; pp. 65–127. ISBN 9781498755382. [CrossRef]

7. Pereira, L.; Correia, F. *Algas Marinhas da Costa Portuguesa—Ecologia, Biodiversidade e Utilizações*; Nota de Rodapé Editores: Paris, France, 2015; 341p, ISBN 978-989-20-5754-5.

8. Garcia, I.A.F.; Castroviejo, R.A.; Neira, C.D. *Las Algas en Galicia. Alimentación y Otros Usos*; Xunta de Galicia: Galicia, Spain, 1993; 231p, ISBN 84-453-0719-3.

9. Leandro, A.; Pereira, L.; Gonçalves, A.M.M. Diverse applications of marine macroalgae. *Mar. Drugs* **2020**, *18*, 17. [CrossRef] [PubMed]

10. Susanto, E.; Fahmi, A.S.; Hosokawa, M.; Miyashita, K. Variation in lipid components from 15 species of tropical and temperate seaweeds. *Mar. Drugs* **2019**, *17*, 630. [CrossRef] [PubMed]

11. Adrien, A.; Bonnet, A.; Dufour, D.; Baudouin, S.; Maugard, T.; Bridiau, N. Anticoagulant activity of sulfated ulvan isolated from the green macroalga *Ulva rigida*. *Mar. Drugs* **2019**, *17*, 291. [CrossRef] [PubMed]

12. Oh, J.H.; Ahn, B.-N.; Karadeniz, F.; Kim, J.-A.; Lee, J.I.; Seo, Y.; Kong, C.-S. Phlorofucofuroeckol A from edible brown alga *Ecklonia cava* enhances osteoblastogenesis in bone marrow-derived human mesenchymal stem cells. *Mar. Drugs* **2019**, *17*, 543. [CrossRef] [PubMed]

13. Zhen, A.X.; Hyun, Y.J.; Piao, M.J.; Fernando, P.D.S.M.; Kang, K.A.; Ahn, M.J.; Yi, J.M.; Kang, H.K.; Koh, Y.S.; Lee, N.H.; et al. Eckol inhibits particulate matter 2.5-induced skin keratinocyte damage via MAPK signaling pathway. *Mar. Drugs* **2019**, *17*, 444. [CrossRef] [PubMed]

14. Trica, B.; Delattre, C.; Gros, F.; Ursu, A.V.; Dobre, T.; Djelveh, G.; Michaud, P.; Oancea, F. Extraction and characterization of alginate from an edible brown seaweed (*Cystoseira barbata*) harvested in the Romanian Black Sea. *Mar. Drugs* **2019**, *17*, 405. [CrossRef] [PubMed]

15. Rodrigues, D.; Costa-Pinto, A.R.; Sousa, S.; Vasconcelos, M.W.; Pintado, M.M.; Pereira, L.; Rocha-Santos, T.A.; Costa, J.P.; Silva, A.M.; Duarte, A.C.; et al. *Sargassum muticum* and *Osmundea pinnatifida* enzymatic extracts: Chemical, structural, and cytotoxic characterization. *Mar. Drugs* **2019**, *17*, 209. [CrossRef] [PubMed]

16. Hwang, P.-A.; Lin, H.-T.V.; Lin, H.-Y.; Lo, S.-K. Dietary supplementation with low-molecular-weight fucoidan enhances innate and adaptive immune responses and protects against *Mycoplasma pneumoniae* antigen stimulation. *Mar. Drugs* **2019**, *17*, 175. [CrossRef] [PubMed]

17. Hofer, S.; Hartmann, A.; Orfanoudaki, M.; Ngoc, H.N.; Nagl, M.; Karsten, U.; Heesch, S.; Ganzera, M. Development and validation of an HPLC method for the quantitative analysis of bromophenolic compounds in the red alga *Vertebrata lanosa*. *Mar. Drugs* **2019**, *17*, 675. [CrossRef] [PubMed]

18. Valado, A.; Pereira, M.; Caseiro, A.; Figueiredo, J.P.; Loureiro, H.; Almeida, C.; Cotas, J.; Pereira, L. Effect of carrageenans on vegetable jelly in humans with hypercholesterolemia. *Mar. Drugs* **2020**, *18*, 19. [CrossRef] [PubMed]

19. Cotas, J.; Marques, V.; Afonso, M.B.; Rodrigues, C.M.P.; Pereira, L. Antitumour potential of *Gigartina pistillata* carrageenans against colorectal cancer stem cell-enriched tumourspheres. *Mar. Drugs* **2020**, *18*, 50. [CrossRef] [PubMed]

20. Cho, T.J.; Rhee, M.S. Health functionality and quality control of laver (*Porphyra*, *Pyropia*): Current issues and future perspectives as an edible seaweed. *Mar. Drugs* **2020**, *18*, 14. [CrossRef] [PubMed]

 *marine drugs*

Review
Diverse Applications of Marine Macroalgae

Adriana Leandro [1], Leonel Pereira [1] and Ana M. M. Gonçalves [1,2,*]

[1]   MARE (Marine and Environmental Sciences Centre), Department of Life Sciences, Faculty of Sciences and
      Technology, University of Coimbra, 3004-517 Coimbra, Portugal; adrianaleandro94@hotmail.com (A.L.);
      leonel.pereira@uc.pt (L.P.)
[2]   Department of Biology and CESAM, University of Aveiro, 3810-193 Aveiro, Portugal
*   Correspondence: amgoncalves@uc.pt; Tel.: +351-239-240-700 (ext. 262-286)

Received: 27 November 2019; Accepted: 22 December 2019; Published: 24 December 2019

Abstract: The aim of this paper is to review the multiplicity of the current uses of marine macroalgae. Seaweeds are already used in many products and for different purposes, from food products to medicine. They are a natural resource that can provide a number of compounds with beneficial bioactivities like antioxidant, anti-inflammatory, anti-aging effects, among others. Despite studies directed in prospecting for their properties and the commodities already marketed, they could, surely, be even more researched and sustainably explored.

Keywords: macroalgae/seaweed; natural resources; health; food; feed; agriculture

---

1. Marine Macroalgae Diversity and Ecology

The marine environment is home for many diverse organisms such as algae, molluscs, sponges, corals, tunicates. Currently, oceans are already considered the "lungs of the Earth" due to Cyanobacteria and other algae that live in seawater. In fact, these beings provide up to 80% of the atmospheric oxygen, which we rely on to breathe. Cyanobacteria are blue colored, aquatic, photosynthetic, and because they are bacteria, they usually are unicellular, but they often grow in colonies large enough to be seen. Cyanobacteria are prokaryotic organisms while algae are eukaryotic organisms. Algae are almost ubiquitous, between microscopic and macroscopic species, they can be found in every wet environment in land, in fresh water or in oceans [1].

In this review, the focus is the marine macroalgae or seaweeds, which are multicellular, macroscopic, eukaryotic, and autotrophic organisms. They are taxonomically organized in three large and distinct groups, based on the color of the thallus: Chlorophyta (green algae), Rhodophyta (red algae), and Ochrophyta—Phaeophyceae (brown algae). All of them accumulate starch in the interior of their cells as energy store, and other different polysaccharides of large molecular chain. The green algae produce ulvan and contain carotene and xanthophylls and chlorophylls *a* and *b* (what sustains the idea that they are the ancestors of the plants) as pigments. The red algae (most common in hot seas) have chlorophylls *a* and *d* and carotenoids and their staining is due to the presence of phycoerythrin (pigment) in their cells. In the brown algae are found the pigments fucoxanthin, chlorophylls *a* and *c* and carotenoids and, as reserve substances, oils, and polysaccharides (such as laminarin) [2,3].

Like plants in terrestrial land, seaweeds have similar ecological roles but in aquatic territory. Some macroalgae species may serve as bioindicators of the quality of water and some can do bioremediation by bioabsorption and bioaccumulation [4–6].

As other vegetables, seaweeds are primary producers, the base of the marine food chain, sustaining several benthic animal communities [7]. They also compete for light, nutrients, and space, in addition to the need of carbon dioxide and water to develop. Inclusively algae and plants produce the same storage compounds and use similar defence strategies against predators and parasites [2]. They have also developed effective mechanisms to survive many biotic threats, like bacteria, virus, or fungal infections.

Because they are sessile organisms, seaweeds have evolved to live in variable, extreme, and hostile abiotic environmental and stress conditions, like temperature changes, salinity, environmental pollutants, or UV radiation exposure. That caused these beings to be able to produce a wide range of compounds called 'secondary metabolites', like pigments, vitamins, phenolic compounds, sterols, and other bioactive agents. Besides these, they also produce amino acids and proteins, saturated/unsaturated fatty acids and all kinds of polysaccharides which are directly implicated in the development, growth, or reproduction conditions to perform physiological functions. So, based on the production of these molecules, in addition to its ecological importance, marine macroalgae also have great importance at commercial level. That's why, a few years ago, the interest in the cultivation and exploitation of macroalgae in the most varied forms increased. Seaweeds are already used in many countries for very different purposes, like industrial phycocolloids extraction or extraction of compounds with antiviral, antibacterial, or antitumor activity [8]. They can also be, directly or indirectly, used for human and animal nutrition (livestock) or farming (biofertilizers) [9].

Although there is still much to investigate and find out about these living beings, it is known that several of the substances they synthesize have great potential to be used in areas such as pharmaceutical, cosmetics and the food industry. As their interest, cultivation and applications increase, their value in the market rises too. It is estimated that in 2024 this value will exceed twice the achieved in 2017 (see Figure 1) [10].

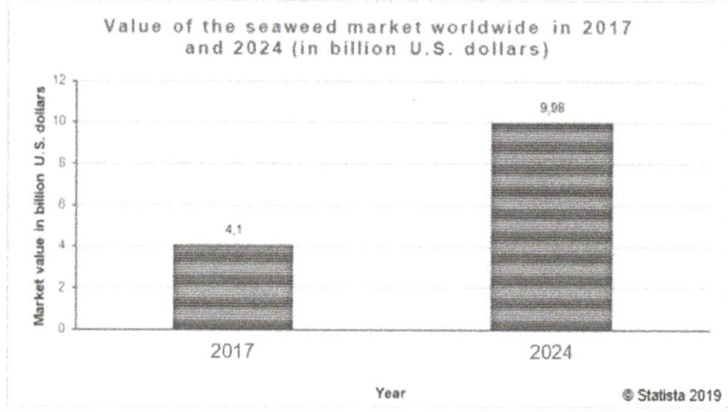

Figure 1. Value of the seaweed market worldwide in 2017 and 2024 (in billion U.S. dollars) [10].

2. Marine Macroalgae Applications

2.1. Human Food

Asian countries, especially China and Japan, are known for being large seaweed consumers for many centuries. The first records show that the harvesting of macroalgae, such as *Laminaria* spp., *Undaria pinnatifida*, *Sargassum fusiforme* (formerly *Hizikia fusiforme*) (commonly known as kombu, wakame, and hiziki, respectively), for human consumption was already carried out by people in China, at least from 500 B.C. [7]. While in Europe it occurred a thousand years later [9].

More than 600 species of edible seaweeds are categorized. Now it is recognized that edibles seaweeds have great nutritional content as they are a low caloric food, but rich in vitamins, minerals, and dietary fibers [7,9]. Their nutritive value may vary depending on the geographic location, season of the year, growth stage, part of the seaweed harvested, etc [11]. Thus, to assure the nutritional value of seaweeds, they need to be evaluated before being used as supplements.

Seaweeds draw from the sea an incomparable wealth of mineral elements, macro elements, and trace elements. They are known as an excellent source of vitamins and minerals, especially potassium and iodine (i.e., *Palmaria palmata*, *Fucus vesiculosus*, *Laminaria* sp.), and potentially good sources of proteins (i.e., red algae such as *Pyropia tenera* (formerly *Porphyra tenera*), *Grateloupia filicina*), long-chain polysaccharides (i.e., *P. tenera*) and soluble and insoluble dietary fibers (i.e., *G. filicina*, *Chondrus crispus*, *Ulva lactuca*) [12]. It was found that the ashes of edible seaweeds contained higher amounts of macrominerals (8.083–17.875 mg/100 g; Na, K, Ca, Mg) and trace elements (5.1–15.2 mg/100 g; Fe, Zn, Mn, Cu), than those reported for edible land plants. So edible brown and red seaweeds could be used as a food supplement to reach the recommended daily intake of some essential minerals and trace elements [13]. For example, the consumption of 10 g of the green alga *Ulva lactuca* provides 70% of the body's daily magnesium requirements and over half of its iron requirements [12,13].

Macroalgae can be used like other vegetables, being equally or even more versatile than them. Sea vegetables, as they are known, could be commercialized and/or eaten in many forms, such as fresh, dried, in flakes, flour or powder, or incorporated in other food products (added-value products) (see Table 1).

Table 1. Examples of seaweed-derived food products.

Seaweed species	Extract	Product(s)	Reference(s)
Chlorophyta (green seaweed)			
Cladophora sp.	Fresh or dry	Sea vegetable	[14]
Ulva (formerly *Enteromorpha*) sp., *Monostroma* sp.	Fresh or dry	Sea vegetable	[14]
Ulva pertusa	Fresh or dry	Sea vegetable (known as green nori)	[15]
Ulva rigida	Fresh or dry	Sea vegetable Seasoning in ready to eat canned fish Added to marine salt	[16]
Ochrophyta, Phaeophyceae (brown seaweed)			
Fucus vesiculosus	Extract [1] Fresh or dry	Incorporated in honey Seasoning in ready to eat canned fish Added to marine salt	[16,17]
Himanthalia elongata	Dry	Sea vegetable Pasta	[18,19]
Himanthalia elongata, *Undaria pinnatifida*	Fresh or dry (whole or in flakes) Sodium alginate extract	Wrap Tartar with olives Kelp noodles	[14,18–20]
Sargassum fusiforme	Fresh or dry	Sea vegetable	[15]
Undaria pinnatifida	Fresh or dry	Sea vegetable Pasta	[14,19]
Rhodophyta (Red seaweed)			
Chondrus crispus	Fresh or dry	Sea vegetable	[14,16]
Meristotheca papulosa	Fresh or dry	Sea vegetable	[14]
Palmaria palmata	Fresh or dry (whole or in powder)	Sea vegetable Bacon substitute	[16,18]
Pyropia spp. (*P. dioica, P. tenera, P. yezoensis*), *Porphyra umbilicalis*	Fresh or dry	Sea vegetable Nori sheets Laverbread Crispy nougat Crispy thins/snacks Added to marine salt	[15,16,19,21]

[1] The extract/compound used in the product(s) is not specified in the reference.

The entire organism can be eaten freshly harvested or after dried and then re-hydrated and cooked [7,16]. They are already commercialized under multiple brands, and labeled with "fat-free", "gluten-free", "mineral rich", "low carbohydrates", and "low calories" [14,15,18,21]. There are natural and healthier substitutes of pasta or bacon (i.e., *Himanthalia elongata*, as spaghetti, and *Palmaria palmata*, as sea bacon, both from Seamore food company), the well-known nori sheets (genus *Porphyra/Pyropia*) to prepare sushi rolls or crispy thin snacks, and many other recipes such as wraps with *Undaria pinnatifida* (wakame) and *H. elongata*, or the laverbread, a paste prepared with boiled nori (also recognized as laver) [15,18] and in desserts like in innovative Spanish nougats with crushed nori algae [21]). Above all this nutritional value, macroalgae are donors of a number of great 'side-effects' acting as nutraceuticals. A study with Japanese children revealed that seaweeds intake in the diet was significantly negatively related to systolic blood pressure in girls and with diastolic blood pressure in boys. This study suggests that seaweeds have beneficial effects on blood pressure [22].

As we can see, macroalgae can be incorporated in food products after processed into flakes, flour, powder or even in more specific extracts. Their pigments, like carotenoids, are, in fact, one of the products of interest for the food industry [23]. Traditionally, carotenoids have been used in that industry due to their properties as natural color enhancers. However, those that are synthetically obtained are now suspected of being promoters of carcinogenesis and liver and renal toxicity. So, there is a strong market demand to replace them with natural pigments. Seaweeds are a great source of many pigments, especially β-carotene which besides its anticancer activity, has been reported that it is absorbed 10 times more easily by the body than the synthetic one [24].

Pigments are important in this industrial range, but there are some constitutes of algae that are even more, their hydrocolloids, such as carrageenan, alginic acids, and agar. These are the main constituents of red and brown algal cell walls and are widely used in several food industries (see Table 2).

Table 2. Phycocolloids utilized in food industries and its properties.

Seaweed species	Compound	Product(s)	Properties	Reference(s)
		Rhodophyta (red seaweed)		
Gigartina skottsbergii	Carrageenan: [A] Iota [B] Kappa [C] Lambda	yoghurts, flans, jellies, ice creams, meat products (ham)	[A] and [B]—gelling [C]—thickening/viscosifier	[9,25,26]
Gelidium sp., *Gracilaria* sp., *Pterocladiella* sp.	Agar	vegetal jelly	Gelling	[7,9]
		Ochrophyta, Phaeophyceae (brown seaweed)		
Lessonia spp., *Macrocystis* sp.	Alginate	yoghurts, ice creams	Emulsifying, Gelling, Stabilizer	[27]

Carrageenan is a natural phycocolloid and is one of the main additives used by the food industry, in many dairy products (e.g. yoghurts, flavoured milkshakes, flans, jellies, ice creams, and beers) and meat products (e.g. hams), as thickening, emulsifier or stabilizing agent [3,25,26]. Extracted from several families of the order Gigartinales. These polygalactans are sulfated and have a linear structure formed by galactose residues with alternating α (1–3) and β (1–4) bonds. Regulatory authorities (FDA) have established a minimum value for the molecular weight of the carrageenan to be used in food preparations. The commercial carrageenans usually range from 400 to 600 kDa, having the minimum of 100 kDa. This minimum value was established by the response to reports of highly degraded carrageenan-induced ulceration of the colon. There are three main varieties of carrageenan, differing in their sulfation degree. Kappa (κ)-carrageenan has one sulfate group per disaccharide, Iota (ι)-carrageenan has two sulfates and Lambda (λ)-carrageenan has three sulfates per disaccharide. The type of carrageenan selected is dependent on the desired finished product characteristics. Iota and kappa carrageenans are gelling carrageenans, while lambda is a thickening/viscosifier carrageenan [28].

Agar is other phycocolloid, composed of a variable combination of agarose and agaropectin, depending of the species and seasonal factors. Agarose, which is the primary component of agar, is a linear polymer of agarobiose, a disaccharide composed of D-galactose and 3,6-anhydro-Lgalactopyranose. Agaropectin, which occurs usually in minor amounts, is a heterogeneous mixture of β-1,3-linked D-galactose which contains substituted sulfate and pyruvate moieties. Like carrageenan, agar has a similar application, so it also has gelling properties, but while carrageenan gels by both ionic and hydrogen bonds, agar gels only by hydrogen bonds. Extracted from several species of red algae, mainly the *Gelidium* sp., *Gracilaria* sp. and *Pterocladiella* sp., it is frequently used as thickener in food products and a vegetarian substitute for gelatine [7,9]. Currently, agar is also being used to develop a new biomaterial for packaging. Made from agar and other natural raw materials, these new wrappers are sustainable, biodegradable, and constitute an alternative to plastics [29].

Alginate is also a gelling agent found and extracted from brown seaweed (eg. *Ascophyllum* sp., *Laminaria* sp., *Lessonia* sp., *Macrocystis* sp.). In fact, it is the most abundant marine biopolymer and, next to cellulose, the most abundant biopolymer in the world. Alginate is a linear acidic polysaccharide that can be a homopolymer or a heteropolymer of β-d-mannuronate and/or α-l-guluronate [27]. It is used as a stabilizer in many food products like ice cream, yogurt, cream, and cheese. It is also used in the food industry as a thickener, emulsifier for sauces, dressings, and jam, and it needs no heat to gel. It is most commonly used with calcium lactate or calcium chloride in the spherification process, a technique performed in molecular cooking.

2.2. Livestock and Agriculture

The consumption of macroalgae is not only for humankind, but also for other animal species. European usage of seaweeds in animal husbandry has come since the time of the Romans. Countries such as Iceland, France, and Norway usually use them in domestic animal nutrition [30]. In fact, the first seaweed meal for animal feed was produced in Norway. It was made from brown seaweeds that were collected, dried, and milled [7].

Besides its direct uses as feed, macroalgae are already introduced in other type of feed as a nutritive additive, and as a nutraceutical compound. Currently the feeding of the animals is supplemented by algae to fill the deficiency in mineral pastures in the U.S.A., Australia and New Zealand. Seaweed meal, used principally as a vitamin and mineral supplement, is produced mainly from the kelps *Ascophyllum nodosum*, *Fucus* spp., *Laminaria* spp., *Macrocystis* spp. [30].

Extracts like macroalgae-derived sulfated polysaccharides are added to animal feed. It was proven that these meals can improve animal intestinal integrity and efficient immune response [31].

On the other hand, feeding seaweeds and macroalgal products has been shown to reduce enteric methane emission from rumen fermentation, [32] which makes this type of feeding a promising candidate as a biotic methane mitigation strategy in the largest milk or beef producing [33].

Moreover, seaweed and seaweed-derived products have been widely used in agriculture to improve crop production systems due to the existence of a number of plant growth-stimulating compounds [34–36]. Inclusively, since ancient times, they were traditionally used to fertilize the fields, they have long been used to augment plant productivity and food production in various regions of the world [30].

Seaweeds and their compounds can promote early seed germination, root and plant growth, confer tolerance to freezing, resistance to biotic stresses, and increase the plant nutrient absorption capacity [37,38]. For example, auxins, a plant hormone responsible for the vegetative growth, and auxin-like compounds were detected in some seaweeds [39].

However, the biostimulator potential of many of these compounds has not been fully exploited due to the lack of scientific data on growth factors present in seaweeds and their mode of action in affecting plant growth [36]. The effects are complex and dependent on the crop, the local environmental conditions and on the interactions of the algae species with the soil community [30]. Seaweeds' extracts, like laminarin, have been shown to stimulate natural defence responses in plants and are involved in the induction of genes encoding various pathogenesis-related proteins with antimicrobial properties. Also, it has been demonstrated that alginate oligomers show growth-promoting effects on certain higher plant species [35].

Studies suggest that adding strongly polar degraded fucoidan, alginate, etc., to soils improves crumb structure and aeration, thus stimulating microorganisms and root systems which improves plant growth [40].

The current commercial extracts are manufactured mainly from the brown seaweeds *Ascophyllum nodosum, Laminaria* spp., *Saccorhiza* spp., *Ecklonia maxima, Fucus* spp., *Sargassum* spp., and *Durvillaea* spp., although other species such as *Ulva intestinalis, Ulva lactuca, Codium* sp. (Chlorophyta), *Gelidium* sp., and *Chondrus crispus* (Rhodophyta) are also used [30].

2.3. Cosmetics

The definition of cosmetic product, according to the European Commission, is: "Any substance or mixture intended to be placed in contact with the external parts of the human body (epidermis, hair system, nails, lips, and external genital organs) or with the teeth and the mucous membranes of the oral cavity with a view exclusively or mainly to cleaning them, perfuming them, changing their appearance, protecting them, keeping them in good condition, or correcting body odours" [41].

More recently, there is another category—the 'cosmeceuticals'—which is attracting the industry's attention and is of interest to the most attentive consumers. Despite still being without legal meaning nowadays, the industry continues to use this designation referring to a product that lies between the benefits of cosmetics and pharmaceuticals [42,43].

There is a growing demand for more natural cosmetics, those made with natural/organic ingredients, due to the benefits they offer, and the absence of many harmful chemicals which are present in conventional cosmetics products. Consequently, the cosmetic industry is rapidly expanding to meet these increased demands. Some of the key active-based natural ingredients used in cosmetics are extracted from marine organisms, like seaweeds (see Table 3).

Marine macroalgae are one of the most abundant sources of vitamins, minerals, amino acids, antioxidants, and essential fatty acids. Seaweeds are unique in containing bioavailable ingredients, meaning that its active, nutrient-rich compounds are more readily absorbed by the skin and the body. Because of its bioavailable nature, seaweeds provide a multitude of benefits including reducing the appearance of redness and blemishes, brightening, hydrating, re-mineralizing, reducing the appearance of sun damage, and firming skin [44–46].

Algae can be incorporated into these products as algal extracts of selected elements or, alternatively, pieces of dried seaweeds can be crushed and ground and incorporated into skin care products such as exfoliating lotions, face masks, face washes and soaps. Cosmetic products, such as creams and lotions, sometimes show on their labels that the contents include "marine extract", "extract of alga", "seaweed extract", or similar [7,44]. Usually this means that one of the hydrocolloids extracted from seaweeds was added to the product. Alginate or carrageenan are water-binding agents, which means they help hold water onto the skin and hair, increasing the moisture balance [47]. Both can be found in multiple products like lotions, creams, shampoos, conditioners, and toothpastes [48].

Seaweeds can be used in two ways in cosmetics: they can either be a vehicle, serving as a stabilizing, emulsifying, or other type of agent necessary for product preparation; or as the active therapeutic ingredient in the product, for example in anti-aging skin treatments or after-sun skin care products [13,48,49].

Algae are rich in saturated and unsaturated fatty acids that are bioactive compounds. For example, palmitic acid and other fatty acids, that are present in large quantities in marine seaweeds, are used in cosmetics as emulsifiers, and its derivated ascorbyl palmitate is an antioxidant that is effective for anti-aging and anti-wrinkle effects [48,50].

Purified phlorotannins extracted from brown seaweeds are included in cosmetics, since these molecules have the role of preventing and slowing down the skin aging process, which is mainly associated with free radical damage and with the reduction of hyaluronic acid concentration [51].

Wang et al. [52] compared the moisture-absorption and retention properties of polysaccharides extracts from five different seaweed species [52]. Marine algae are reported to produce different polysaccharides, including alginates, ulvans, laminarans, and fucoidans [53]. These molecules usually contain large proportions of L-fucose and sulfate, together with minor amounts of other sugars such as xylose, galactose, mannose, and glucuronic acid [45]. In their study, Wang et al. [52] reported that the polysaccharides extracted from brown seaweed (more precisely the fucoidan obtained from the *Saccharina japonica*) exhibited the best moisture-absorption and retention capacity, while the green ones were the worst. This ability of polysaccharides is influenced by its sulfated content, molecular weight (length of chain), and by the type of algae that they are extracted from [52]. An example of it is a cosmetic, CODIAVELANE®, composed of propylene glycol, water, and *Codium tomentosum* extract. It is proven that it normalizes and balances skin's moisture content by adding oligo-elements and increasing surface hydration [49].

A group of small water-soluble compounds, mycosporine-like amino acids (MAA), found in marine algae, is biologically relevant because of its photo-protective potential. In addition, its antioxidant and skin protective strategies raise the interest for possible pharmaceutical and cosmetic applications [54,55]. An extract of *Asparagopsis armata* (ASPAR'AGE™) containing this MAA molecules is already incorporated in some lotions with anti-aging properties [56].

Besides the numerous existing and marketed cosmetics and cosmeceuticals, there are many other seaweed extracts that are under investigation. Kamei et al. [57] discovered a compound from *Sargassum macrocarpum*, Sargafuran, that was bactericidal and completely killed Propionibacterium acnes by lysing bacterial cells [57]. The results suggest that this substance could be applied in new skin care cosmetics to prevent or improve acne.

Table 3. Cosmetical products containing seaweed parts or extracts.

Seaweed species.	Extract	Product(s)	Properties	Reference(s)
Chlorophyta (green seaweed)				
Caulerpa lentillifera	Extract (Rich in unsaturated fatty acids and vitamin A and C)	Hair and skin care products (shampoo, shower gel, soaps, lotions)	Moisturizing; anti-aging; whitening/lightening agent	[58]
Cladophora glomerata	Extract (Rich in unsaturated fatty acids and polyphenols)	Skin care products (emulsion, cream, lotion)	Moisturizing; anti-aging	[48,50]
Codium tomentosum	Extract Codiavelane®	Skin care products (creams, lotions)	Anti-aging; moisturizing	[49,52,59]
Monostroma sp.	Extract (rich in water-soluble polysaccharides) Extract [1]	Skin care products (e.g. slimming and anti-cellulitis formulations) Hair and nails care products (hair and nails growth) Facial Mask	Moisturizing; anti-inflammatory agent; anti-aging	[49]
Ulva compressa (as *Enteromorpha compressa*)	Extract [1]	Skin care products (creams, lotions)	Moisturizing	[59,60]
Ulva lactuca	Hydrolysed extract Aosaine® (three-quarters of aosaine consists of amino acids very similar those responsible for the skin's elasticity)	Skin care products (creams, lotions)	Anti-aging (anti-wrinkle and collagen stimulation)	[49,56,58]
Ulva spp.	Aqueous extract (rich in ulvans) Extracts [1]	Skin care product (creams, lotions) Bath salts (thalassotherapy kit)	Moisturizing; whitening/lightening; antioxidative; chelating; anti-inflammatory; calming	[61–63]
Ochrophyta, Phaeophyceae (brown seaweed)				
Alaria esculenta	Extract (rich in fatty acids and trace elements)	Skin care products (creams, lotions)	Moisturizing; anti-aging	[59,64]

Table 3. *Cont.*

Seaweed species.	Extract	Product(s)	Properties	Reference(s)
Ascophyllum nodosum	Extract [1]	Skin care product (cream)	Anti-ageing; skin softness and elasticity restoring	[59]
Bifurcaria bifurcata	Extract [1]	Bath salts, gel and facial mask (thalassotherapy kit)	Exfoliant; detoxifying; nourishing	[61]
Fucus serratus	Extract [1]	Oral-care product	Protecting agent (reduces gingivorrhagia)	[58]
Fucus spiralis	Extract [1]	Facial mask and (imperfection corrector) gel	Skin purification; oiliness and pore dilatation reduction	[65]
Fucus vesiculosus	Extract (rich in muco-polysaccharides)	Slimming and anti-cellulitis cosmetic formulations Facial Mask	Skin softness and elasticity properties; exfoliant; brightening; detoxifying	[49,58]
Halopteris scoparia	Extract (rich in anti-oxidative polyphenols, cytokines and betaines)	Skin care products (cream, lotion)	Skin softness and elasticity restoring	[59]
Sargassum fusiforme (as *Hizikia fusiforme*)	Extract [1]	Skin care products (creams)	Whitening/lightening;	[58]
Laminaria digitata	Extract (rich in trace elements, like iodine)	Skin care products (lotions, anti-cellulitis formulations)	Anti-aging (prevent lines and wrinkles. Collagen and elastin stimulation); anti-cellulitis; moisturizing	[58–60]
Laminaria hyperborea	Extracts [1]	Skin care product (cream) Facial masks	Anti-aging; moisturizing; anti-acne	[59,65]
Laminaria ochroleuca	Extract ANTILEUKINE 6™ Extracts [1]	Hair and skin care products (body lotion, shampoo and conditioner)	Anti-aging; sun-protector; anti-acne; moisturizing	[56,59,65]
Saccharina latissima (as *Laminaria saccharina*)	Extract (w/ hyaluronic acid and polysaccharides; sodium and potassium ions; phlorotannins (polyphenols))	Skin care product (cream)	Antioxidant; anti-aging; anti-blemishes	[59]
Macrocystis pyrifera	Extract (rich in polysaccharides) Extract [1]	Skin care product (anti-aging balm)	Moisturizing; antioxidant; anti-aging; anti-blemishes	[44,59]
Pelvetia canaliculata	Extracts [1]	Hair and skin care products (creams, lotions, shampoo)	Moisturizing; anti-aging (anti-wrinkle and collagen stimulation)	[58,59]
Saccharina japonica	Polysaccharide extract (rich in fucoidan)	Skin care products (anti-cellulitis formulations)	Moisturizing; anti-aging; anticellulite	[52,53,66]
Kjellmaniella crassifolia	Fucoidan extract	Hair and skin care products (creams, lotions, shampoo)	Moisturizing; anti-aging; nourishing; preventing hair loss	[67]
Sargassum muticum	Extract (rich in proteins)	Skin care products (creams, lotions)	Anti-aging (anti-wrinkle, antioxidant, and collagen stimulation. Reduce skin damage caused by UVB and chemical stress)	[59]
Undaria pinnatifida	Extract [1] Powder, whole leaf and extract forms (rich in fucoidan)	Skin care products (aromatherapy oil; face and body oil; body scrub)	Anti-aging (anti-wrinkle); whitening/lightening; moisturizing; nourishing	[44,58,59]
Rhodophyta (red seaweed)				
Asparagopsis armata	Extract ASPAR'AGE™	Skin care products (creams)	Moisturizing; anti-aging	[56]
Chondrus crispus	Extracts [1] Powder	Hair and skin care products (lotions; creams; make-up removers; body scrub; shampoo; conditioner) Lipsticks and deodorants Algae and sea salt soap	Moisturizing; cleaning; exfoliant; Emulsifier and thickener; cleaning; exfoliant	[58,59,65,68]
Corallina officinalis	Extract [1]	Skin care product (cream)	Anti-redness	[59]
Gelidium corneum (as *Gelidium sesquipedale*)	Extract (rich in minerals, trace elements and amino acids)	Skin care product (lotion)	Skin softness and elasticity restoring	[59]

Table 3. *Cont.*

Seaweed species.	Extract	Product(s)	Properties	Reference(s)
Gigartina skottsbergii	Powder, whole leaf and extract [1] Extract (rich in polysaccharides, vitamins and minerals)	Bath and skin care products (mineral-rich seaweed bath soak)	Moisturizing; whitening/lightening	[44]
Gracilaria conferta	Extract [1]	Skin care products (creams)	Moisturizing; nourishing	[58]
Palmaria palmata	Extract [1]	Skin care products (Facial clarifier gels and emulsions)	Skin clarification (reduction of pigmentation imperfections), and uniformization (skin grain homogenization)	[65]
Pyropia tenera (as *Porphyra tenera*)	Extract [1]	Skin care products (creams)	Sun protector	[58]
Porphyra umbilicalis	Extract [1]	Skin care products (cream; facial scrub masks)	Moisturizing; exfoliant; brightening; detoxifying	[59,69]
Vertebrata lanosa (as *Polysiphonia lanosa*)	Extract [1]	Skin care products (creams)	Moisturizing; nourishing	[52,65]

[1] The extract/compound used in the product(s) is not specified in the reference.

2.4. Pharmaceutics

The overuse of antibiotics can lead to the development of resistant pathogenic bacteria. New antibiotics that are effective against new and resistant bacterial strains are needed. As previously mentioned, seaweeds have evolved to survive many environmental stresses and threats. Besides the predators/herbivores, they have to continuously face high concentrations of infectious and surface-fouling bacteria that are indigenous to ocean waters [8]. So, the macroalgae have evolved and developed certain mechanisms of defence like the production of bioactive compounds. Substances such as phlorotannins, polysaccharides, and peptides allow seaweeds to avoid bacterial invasion [8], and some have been investigated about other potential pharmacological effects (antiviral, antitumoral, immunogenic effects). One example is the peptide kahalalide F and its isomer, iso-kahalalide F, extracted from a green macroalga, *Bryopsis pennata*, which present cytotoxic effects and were used in anticancer clinical trials. Despite its great potential, this molecule is under modification tests to improve its water solubility, stability, and effectiveness [70].

Sometimes the extract used can be obtained from a mix of various algae species, and even of different seaweed groups. For example, there is a patent of green and/or brown seaweed extract for the treatment of type 2 diabetes and its complications. This has brown seaweeds such as *Fucus vesiculosus* or *Ascophylum nodosum* and green algae, selected from the group consisting of *Cladophora* sp., *Monostroma* sp., *Ulva compressa* (as *Entoromorpha compressa*), *Codium* sp., among others [71].

According to another study, methanolic extracts of some brown, red and green algae are effective at inhibiting the growth of pathogenic Gram-positive (*Staphylococcus aureus*, *Micrococcus luteus*, *Enterococcus faecalis*) and Gram-negative bacteria (*Enterobacter aerogenes*, *Escherichia coli*) [72]. The species were *Corallina officinalis* (Rhodophyta), *Cystoseira barbata*, *Dictyota dichotoma*, *Halopteris filicina*, *Cladostephus spongiosus* (Ochrophyta, Phaeophyceae), and *Ulva rigida* (Chlorophyta).

The seaweed-derived substances that received most attention from pharmaceutical companies are the sulfated polysaccharides (negatively charged sugar polymers due to the presence of sulfate groups). Sulfated polysaccharides are extracted from red algae (carrageenans and agarans), brown algae (e.g. fucoidans) and green algae (e.g. ulvans). Their value lies on their bioactivities, namely their antibacterial, antiviral activity, antitumoral, and immunomodulatory potential [8,70,72,73].

On the other hand, other polysaccharides, like alginate, are also used in pharmaceutical formulations as excipients. Alginate polymers have a wide potential in drug formulation due to their lack of toxicity and they can be tailor-made to suit the demands of applicants in both the pharmaceutical and biomedical areas. This brown seaweed—derived group of polymers owns a few characteristics that makes it useful as a formulation aid, both as a conventional excipient and more specifically as a tool in polymeric-controlled drug delivery [8,74], and it is commonly used as bio-adhesive in pharmaceutical applications [75]. Other application of alginate is in wound healing dressing due to the excellent swelling properties and biocompatibility [76]. In fact, not only the alginate, but seaweed extracts—like the *Laminaria* spp.—are being studied and used for the development of biodegradable wound care products, since they contain healing accelerator substances: alpha keto isovalerate, alpha keto glutarate, and alpha keto oxaloacate [77].

It has been demonstrated that alginate has therapeutic effects in mammalian systems such as anticoagulants and antitumor activities. Also containing alginate, there is some gastrointestinal formulations and protectors (i.e., Gaviscon), that neutralize the acids, prevents the contact of stomach contents with the oesophagus (reflux), and relieve symptoms of heartburn and indigestion [78].

Agar, which was initially used as a laxative agent in the preparation of medicines, in western countries [26], is now used as an ingredient in tablets and capsules, as well as in different types of emulsions. Like alginate, the main role of agar in the pharmaceutical industry is as an excipient.

The three main types (ι, κ, λ) of carrageenan form thermo-reversible gels in aqueous solutions and in the presence of cations. Therefore, they are used in pharmaceutical formulations for stabilization of disperse systems and viscosity modification [75]. In addition to its hydrating properties, it has also been found in some studies to block the growth of viruses like human papillomavirus, making it potentially even more protective in sexual lubricants (in which it is already included). Studies in vitro demonstrated that carrageenan, even when diluted a million-fold, presents activity against a range of common sexually transmitted HPV types that can cause cervical cancer and genital warts [47]. So, due to these properties, carrageenans might have a great interest in the composition of sexual lubricant. Also the polymer galactofuran (extracted from *Undaria pinnatifida*) was proven as an effective Herpes virus inhibitor [79].

Among polysaccharides, fucoidans were particularly studied as they showed interesting biological activities (anti-thrombotic, anti-coagulant, anticancer, anti-proliferative, and anti-inflammatory) [66,80–82].

Other group of small molecules (previously indicated in this article), the MAAs, have skin protective and wound healing effects. Like the porphyra-334 was able to suppress ROS (reactive oxygen species) production in human skin fibroblast cells [83]. Pigments isolated from seaweeds also have bioactivities. Like the fucoxanthin, obtained from *Saccharina japonica* (as *Laminaria japonica*), that has been reported to suppress tyrosinase activity in UVB-irradiated guinea pig and melanogenesis in UVB-irradiated mice. Oral treatment of fucoxanthin significantly suppressed skin mRNA expression related to melanogenesis, suggesting that fucoxanthin negatively regulated melanogenesis factor at transcriptional level [45].

Seaweed phlorotannin extracts from *Ascophylum nodosum* are reported to have potential in the treatment of diabetes [84] while those from *Ecklonia cava* are now marketed for potential health benefits due to their antioxidant activities [85]. These phlorotannins are phenols structurally different from those obtained from plants, since these are oligomers and polymers of phloroglucinol (1,3,5-tri-hhydroxybenzene) and the terrestrial ones are based on gallic acids or flavones [86]. The brown algal polyphenols were investigated in an SKH-1 hairless mouse skin model with UVB-induced skin carcinogenesis. This in vivo report demonstrated that both dietary feeding and topical treatment of brown algal polyphenols has suppressed cyclooxygenase-2 (COX-2) expression and cell proliferation [87]. These results suggest the role of brown algae polyphenols, phlorotannins, as potential cancer chemo-preventive agents against photo-carcinogenesis and other adverse effects of UVB exposure. That reveals these compounds may be used as active ingredients in drugs or cosmetic/cosmeceutical formulations, like in sunscreen or anti-aging creams [87].

Marine brown algae-derived phlorotannins have also been investigated for their human beneficial aspects that include hypoallergenic, anti-inflammatory, and hyaluronidase inhibitory activities. In vitro studies with the methanol extracts from marine brown algae *Eisenia arborea* have shown inhibition of histamine release from rat basophile leukaemia cells (RBL-2H3) sensitized with anti-dinitrophenyl (DNP) IgE and stimulated with DNP-BSA [88]. Shibata et al. [89] also studied some length-varied phlorotannins obtained from *Ecklonia bicyclis* (as *Eisenia bicyclis*) and *Ecklonia kurome* in their ability to inhibit hyaluronidase activity in vitro. In fact, they proved that those molecules have a stronger inhibitory effect on hyaluronidase than the well-known inhibitors catechins and sodium cromoglycate [89].

3. Conclusions

This review intended to demonstrate the versatility and the multiple applications of marine macroalgae.

Many products we consume or use daily contain seaweed extracts in their composition, such as ham, ice cream, bottled chocolate drinks, and toothpaste or deodorizers, although most people probably do not even imagine such thing. Nowadays, there is a growing interest in seaweeds due to the recognition of numerous new bioactive compounds. Antioxidants, antimicrobials, anti-inflammatory, anti-aging, anticancer, are just some of its amazing properties to use as pharmaceuticals, cosmeceuticals, nutraceuticals, or even in agriculture or feeding.

There is more and more awareness of sustainable use of natural resources, rather than synthetic and processed products with eventual harmful side effects to the consumer. All the growing interest in these potentialities led to the fostering of macroalgae production, as well as to do research on them. Seaweeds are a resource to maintain and preserve with unique properties.

Author Contributions: A.M.M.G. conceived and designed the idea; A.M.M.G. and L.P. contributed to the idea; A.L. wrote the paper; A.M.M.G. and L.P. contributed to the writing of the paper. All authors have read and agreed to the published version of the manuscript.

Funding: This work had the support of Foundation for Science and Technology (FCT), through the strategic projects UID/MAR/04292/2019 granted to MARE and UID/AMB/50017/2019 granted to CESAM. This research was also co-financed by the project MENU - Marine Macroalgae: Alternative recipes for a daily nutritional diet (FA_05_2017_011), funded by the Blue Fund under Public Notice No. 5—Blue Biotechnology.

Acknowledgments: This work was supported by Foundation for Science and Technology (FCT) through the strategic projects granted to MARE—Marine and Environmental Sciences Centre UID/MAR/04292/2019 and granted to CESAM Centro de Estudos do Ambiente e do Mar UID/AMB/50017/2019. This research was also co-financed by the project MENU—Marine Macroalgae: Alternative recipes for a daily nutritional diet (FA_05_2017_011), funded by the Blue Fund under Public Notice No. 5—Blue Biotechnology. Adriana Leandro thanks FCT for the financial support provided through the doctoral grant SFRH/BD/143649/2019 funded by National Funds and Community Funds through FSE. Ana M.M. Gonçalves acknowledges University of Coimbra for the contract IT057-18-7253.

Conflicts of Interest: The authors declare no conflict of interest.

References

1. Lewin, R.A.; Andersen, R.A. Algae. Encyclopedia Britannica. Algae. Available online: https://www.britannica.com/science/algae (accessed on 14 November 2019).
2. Barsanti, L.; Gualtieri, P. *Algae Anatomy, Biochemistry and Biotechnology*; CRC Press: Boca Raton, FL, USA, 2014.
3. Vidotti, E.C.; Rollemberg, M.; Do, C.E. Algas: Da economia nos ambientes aquáticos à biorremediação e à química analítica. *Quim. Nova* **2014**, *27*, 139–145. [CrossRef]
4. Neveux, N.; Bolton, J.J.; Bruhn, A.; Roberts, D.A.; Ras, M. The Bioremediation Potential of Seaweeds: Recycling Nitrogen, Phosphorus, and Other Waste Products. *Blue Biotechnol.* **2018**, *1*, 217–239. [CrossRef]
5. Yu, Z.; Robinson, S.M.C.; Xia, J.; Sun, H.; Hu, C. Growth, bioaccumulation and fodder potentials of the seaweed *Sargassum hemiphyllum* grown in oyster and fish farms of South China. *Aquaculture* **2016**, *464*, 459–468. [CrossRef]
6. Henriques, B.; Lopes, C.; Figueira, P.; Rocha, L.; Duarte, A.; Vale, C.; Pardal, M.; Pereira, E. Bioaccumulation of Hg, Cd and Pb by *Fucus vesiculosus* in single and multi-metal contamination scenarios and its effect on growth rate. *Chemosphere* **2017**, *171*, 208–222. [CrossRef] [PubMed]
7. Klnc, B.; Cirik, S.; Turan, G.; Tekogul, H.; Koru, E. Seaweeds for Food and Industrial Applications. *Food Ind.* **2013**. [CrossRef]
8. Shannon, E.; Abu-Ghannam, N. Antibacterial derivatives of marine algae: An overview of pharmacological mechanisms and applications. *Mar. Drugs* **2016**, *14*, 81. [CrossRef]
9. Pereira, L. *Edible Seaweeds of the World*; CRC Press: Boca Raton, FL, USA, 2016. [CrossRef]

10. Statista-The Statistic Portal. The Statistic Portal Value of the Seaweed Market Worldwide in 2017 and 2024 (in Billion U.S. Dollars). Available online: https://www.statista.com/ (accessed on 6 January 2019).

11. Kim, S.K. *Handbook of Marine Macroalgae: Biotechnology and Applied Phycology*; Wiley-blackwell: Hoboken, NJ, USA, 2011. [CrossRef]

12. Yuan, Y.V.; Westcott, N.D.; Hu, C.; Kitts, D.D. Mycosporine-like amino acid composition of the edible red alga, *Palmaria palmata* (dulse) harvested from the west and east coasts of Grand Manan Island, New Brunswick. *Food Chem.* **2009**, *112*, 321–328. [CrossRef]

13. Rupérez, P. Mineral content of edible marine seaweeds. *Food Chem.* **2002**, *79*, 23–26. [CrossRef]

14. Tangles, S. Kelp Noodles. Available online: https://kelpnoodles.com/ (accessed on 10 January 2019).

15. Clearspring Clearspring. Authentic Japanese Specialities and Organic Fine Foods. Available online: www.clearspring.co.uk (accessed on 22 December 2018).

16. ALGAplus. Tok de Mar. Available online: www.algaplus.pt (accessed on 20 January 2019).

17. Beesweet. Available online: https://beesweet.pt/ (accessed on 7 January 2019).

18. Food, S. Seamore Food. Available online: https://seamorefood.com (accessed on 16 December 2018).

19. PortoMuiños. Seaweed. Available online: http://www.portomuinos.com (accessed on 21 January 2019).

20. The Whole Foodies. Available online: https://thewholefoodies.com.au/ (accessed on 10 January 2019).

21. Vicens, T. Torrons Vicens. Turrones Artesanales. Available online: https://www.vicens.com/ (accessed on 20 December 2018).

22. Wada, K.; Nakamura, K.; Tamai, Y.; Tsuji, M.; Sahashi, Y.; Watanabe, K.; Ohtsuchi, S.; Yamamoto, K.; Ando, K.; Nagata, C. Seaweed intake and blood pressure levels in healthy pre-school Japanese children. *Nutr. J.* **2011**, *10*, 83. [CrossRef]

23. Kristinsson, G.; Jónsdóttir, R. *Novel Bioactive Seaweed Based Ingredients and Products*; Norden: Heerup, Denmark; Nordic Innovation: Oslo, Norway, 2015.

24. Christaki, E.; Bonos, E.; Giannenasa, I.; Florou-Paneria, P. Functional properties of carotenoids originating from algae. *J. Sci. Food Agric.* **2013**, *93*, 5–11. [CrossRef]

25. Pereira, L.; Van De Velde, F. Portuguese carrageenophytes: Carrageenan composition and geographic distribution of eight species (Gigartinales, Rhodophyta). *Carbohydr. Polym.* **2011**, *84*, 614–623. [CrossRef]

26. Armisen, R. World-wide use and importance of *Gracilaria*. *J. Appl. Phycol.* **1995**, *7*, 231–243. [CrossRef]

27. Stiger-Pouvreau, V.; Bourgougnon, N.; Deslandes, E. Carbohydrates from Seaweeds. In *Seaweed in Health and Disease Prevention*; Elsevier Inc.: Amsterdam, The Netherlands, 2016. [CrossRef]

28. Pereira, L.; Gheda, S.F.; Ribeiro-Claro, P.J.A. Analysis by Vibrational Spectroscopy of Seaweed Polysaccharides with Potential Use in Food, Pharmaceutical, and Cosmetic Industries. *Int. J. Carbohydr. Chem.* **2013**, *7*. [CrossRef]

29. Talep, M. Desintegra Me. Available online: https://margaritatalep.com/ (accessed on 24 January 2019).

30. Craigie, J.S. Seaweed extract stimuli in plant science and agriculture. *J. Appl. Phycol.* **2015**, *23*, 371–393. [CrossRef]

31. OlmixGroup. Algimun. Available online: https://www.olmix.com/animal-care/algimun (accessed on 4 January 2019).

32. Li, X.; Norman, H.; Kinley, R.; Laurence, M.; Wilmot, M.; Bender, H.; Nys, R.; Tomkins, N. *Asparagopsis taxiformis* decreases enteric methane production from sheep. *Anim. Prod. Sci.* **2018**, *58*, 681–688. [CrossRef]

33. Pereira, L.; Bahcevandziev, K.; Joshi, N.H. *Seaweeds as Plant Fertilizer. Agricultural Biostimulants and Animal Fodder*; CRC Press: Boca Raton, FL, USA, 2019. [CrossRef]

34. Nabti, E.; Jha, B.; Hartmann, A. Impact of seaweeds on agricultural crop production as biofertilizer. *Int. J. Environ. Sci. Technol.* **2017**, *14*, 1119–1134. [CrossRef]

35. Khan, W.; Menon, U.; Subramanian, S.; Jithesh, M.; Rayorath, P.; Hodges, D.M.; Critchley, A.T.; Craigie, J.; Norrie, J.; Prithiviraj, B. Seaweed extracts as biostimulants of plant growth and development. *J. Plant Growth Regul.* **2009**, *28*, 386–399. [CrossRef]

36. Tuchy, L.; Chowańska, J.; Chojnacka, K. Seaweed extracts as biostimulants of plant growth: Review. *Chemik* **2013**, *67*, 636–641.

37. Fernandes, A.L.T.; Oliveira Silva, R. Avaliação do extrato de algas (*Ascophyllum nodosum*) no desenvolvimento vegetativo e produtivo do cafeeiro irrigado por gotejamento e cultivado em condições de cerrado. *Enciclopédia Biosf. Cent. Científico Conhecer Goiânia* **2011**, *7*, 147–157.

38. Akila, N.; Jeyadoss, T. The potential of seaweed liquid fertilizer on the growth and antioxidant enhancement of *Helianthus annuus* L. *Orient. J. Chem.* **2010**, *2*, 19–23.

39. Crouch, I.; van Staden, J. Evidence for the presence of plant growth regulators in commercial seaweed products. *Plant Growth Regul.* **1993**, *13*, 21–29. [CrossRef]

40. Milton, R. Liquid seaweed as a fertilizer. *Proc. Int. Seaweed Symp.* **1964**, *4*, 428–431.

41. *Regulation (EC) No 1223/2009 of the European Parliament and of the Council on Cosmetic Products*; European Union: Brussels, Belgium, 2009.

42. Brandt, F.S.; Cazzaniga, A.; Hann, M. Cosmeceuticals: Current trends and market analysis. *Semin. Cutan. Med. Surg.* **2011**, *30*, 141–143. [CrossRef] [PubMed]

43. Vermeer, B.J.; Gilchrest, B.A.; Friedel, S.L. A proposal for rational definition, evaluation, and regulation. *Arch. Dermatol.* **1996**, *132*, 337–340. [CrossRef] [PubMed]

44. Osea Malibu. Non-Toxic Seaweed Skin Care. Available online: https://oseamalibu.com/ (accessed on 14 December 2018).

45. Thomas, N.V.; Kim, S.K. Beneficial effects of marine algal compounds in cosmeceuticals. *Mar. Drugs* **2013**, *11*, 146–164. [CrossRef] [PubMed]

46. Pereira, L. Seaweeds as Source of Bioactive Substances and Skin Care Therapy—Cosmeceuticals, Algotheraphy and Thalassotherapy. *Cosmetics* **2018**, *5*, 68. [CrossRef]

47. Buck, C.B.; Thompson, C.D.; Roberts, J.N.; Müller, M.; Lowy, D.R.; Schiller, J.T. Carrageenan is a potent inhibitor of papillomavirus infection. *PLoS Pathog.* **2006**, *2*, 671–680. [CrossRef]

48. Fabrowska, J.; Łęska, B.; Schroeder, G. Freshwater *Cladophora glomerata* as a new potential cosmetic raw material. *Chemik* **2015**, *69*, 491–497.

49. Majmudar, G. Compositions of Marine Botanicals to Provide Nutrition to Aging and Environmentally Damaged Skin. U.S. Patent 8318178, 27 November 2012.

50. Yarnpakdee, S.; Benjakul, S.; Senphan, T. Antioxidant Activity of the Extracts from Freshwater Macroalgae (*Cladophora glomerata*) Grown in Northern Thailand and Its Preventive Effect against Lipid Oxidation of Refrigerated Eastern Little Tuna Slice. *Turk. J. Fish. Aquat. Sci.* **2018**, *19*, 209–219.

51. Ferreres, F.; Lopes, G.; Gil-Izquierdo, A.; Andrade, P.B.; Sousa, C.; Mouga, T.; Valentão, P. Phlorotannin Extracts from Fucales Characterized by HPLC-DAD-ESI-MSn: Approaches to Hyaluronidase Inhibitory Capacity and Antioxidant Properties. *Mar. Drugs* **2012**, *10*, 2766–2781. [CrossRef]

52. Wang, J.; Jin, W.; Hou, Y.; Niu, X.; Zhang, H.; Zhang, Q. Chemical composition and moisture-absorption/retention ability of polysaccharides extracted from five algae. *Int. J. Biol. Macromol.* **2013**, *57*, 26–29. [CrossRef] [PubMed]

53. Wijesinghe, W.A.J.P.; Jeon, Y.-J. Biological activities and potential industrial applications of fucose rich sulfated polysaccharides and fucoidans isolated from brown seaweeds: A review. *Carbohydr. Polym.* **2012**, *88*, 13–20. [CrossRef]

54. Hartmann, A.; Gostner, J.; Fuchs, J.E.; Chaita, E.; Aligiannis, N.; Skaltsounis, L.; Ganzera, M. Inhibition of collagenase by mycosporine-like amino acids from marine sources. *Planta Med.* **2015**, *81*, 813–820. [CrossRef] [PubMed]

55. Chrapusta, E.; Kaminski, A.; Duchnik, K.; Bober, B.; Adamski, M.; Bialczyk, J. Mycosporine-Like Amino Acids: Potential Health and Beauty Ingredients. *Mar. Drugs* **2017**, *15*, 326. [CrossRef] [PubMed]

56. Drouart, C. SEPPIC. Ingredients and Formulas. Available online: https://www.seppic.com/ (accessed on 7 December 2018).

57. Kamei, Y.; Sueyoshi, M.; Hayashi, K.; Terada, R.; Nozaki, H. The novel anti-Propionibacterium acnes compound, Sargafuran, found in the marine brown alga *Sargassum macrocarpum*. *J. Antibiot.* **2009**, *62*, 259–263. [CrossRef]

58. Cabarry, C. SpecialChem-Connect, Innovate, accelerate. The Universal Selection Source: Cosmetics Ingredients 2018. Available online: https://cosmetics.specialchem.com/ (accessed on 5 December 2018).

59. Bommers, M. La-Mer. My Skin—And What It Needs. Available online: https://www.la-mer.com/en/ (accessed on 6 December 2018).

60. Ziaja. Focus on Skin. Available online: http://ziaja.co.uk/ (accessed on 12 December 2018).

61. Lusalgae. We Innovate in Marine Biotechnology. Available online: http://www.lusalgae.pt/lusalgae_en.html (accessed on 12 December 2018).

62. Demais, H.; Brendle, J.; Herve, D.; Anca, L.L.; Lurton, L.; Brault, D. Argiles Intercalées. FR2874912B1. Available online: https://patents.google.com/patent/FR2874912B1/fr (accessed on 13 December 2019).

63. Algabase. Available online: http://www.algabase.com (accessed on 12 December 2018).

64. Verdy, C.; Branka, J.E.; Mekideche, N. Quantitative assessment of lactate and progerin production in normal human cutaneous cells during normal ageing: Effect of an *Alaria esculenta* extract. *Int. J. Cosmet. Sci.* **2011**, *33*, 462–466. [CrossRef]

65. Thalgo. La Beaute Marine. Available online: http://www.thalgo.com/ (accessed on 6 December 2018).

66. Chizhov, A.O.; Dell, A.; Morris, H.R.; Haslam, S.M.; McDowell, R.A.; Shashkov, A.S.; Nifant'ev, N.E.; Khatuntseva, E.A.; Usov, A.I. A study of fucoidan from the brown seaweed *Chorda filum*. *Carbohydr. Res.* **1999**, *320*, 108–119. [CrossRef]

67. Chizhov, A.O.; Dell, A.; Morris, H.R.; Haslam, S.M.; McDowell, R.A.; Shashkov, A.S.; Nifant'ev, N.E.; Khatuntseva, E.A.; Usov, A.I. Fucoidan-Containing Cosmetics. U.S. Patent 20060093566A1, 13 August 2000.

68. Ach Brito-SPA Collection. Available online: https://www.achbrito.com/pt/ (accessed on 27 January 2019).

69. Jeunesse, M. 7th Heaven. Available online: https://www.my7thheaven.com/ (accessed on 6 January 2019).

70. Wang, B.; Waters, A.L.; Valeriote, F.A.; Hamann, M.T. An efficient and cost-effective approach to kahalalide F N-terminal modifications using a nuisance algal bloom of *Bryopsis pennata*. *Biochim. Biophys. Acta Gen. Subj.* **2015**, *1850*, 1849–1854. [CrossRef]

71. Daniels, B.A. Seaweed Extract Composition for Treatment of Diabetes and Diabetic Complications. Available online: https://patents.google.com/patent/US20070082868A1/en (accessed on 13 December 2019).

72. Taskin, E.; Ozturk, M.; Taskin, E.; Kurt, O. Antibacterial activities of some marine algae from the Aegean Sea (Turkey). African. *J. Biotechnol.* **2007**, *6*, 2746–2751.

73. Vasconcelos, A.G.; Araújo, K.V. POLISSACARÍDEOS EXTRAÍDOS DE ALGAS MARINHAS E SUAS APLICAÇÕES BIOTECNOLÓGICAS: UMA REVISÃO. *Revista Brasileira de Inovação Tecnológica em Saúde* **2015**, *5*, 27–51. [CrossRef]

74. Tønnesen, H.H.; Karlsen, J. Alginate in drug delivery systems. *Drug Dev. Ind. Pharm.* **2002**, *28*, 621–630. [CrossRef] [PubMed]

75. Guo, J.H.; Skinner, G.W.; Harcum, W.W.; Barnum, P.E. Pharmaceutical applications of naturally occurring water-soluble polymers. *Pharm. Sci. Technol. Today* **1998**, *1*, 254–261. [CrossRef]

76. Yanagibayashi, S.; Kishimoto, S.; Ishihara, M.; Murakami, K.; Aoki, H.; Takikawa, M.; Fujita, M.; Sekido, M.; Kiyosawa, T. Novel hydrocolloid-sheet as wound dressing to stimulate healing-impaired wound healing in diabetic db/db mice. *Biomed. Mater. Eng.* **2012**, *22*, 301–310. [CrossRef] [PubMed]

77. Glynn, K.P.; Martin, A. Biodegradable Wound Care Products with Biocompatible Artificial Skin Treatment and Healing Accelerator. U.S. Patent Application No. 13/135646, 14 February 2013.

78. Mandel, K.G.; Daggy, B.P.; Brodie, D.A.; Jacoby, H.I. Review article: Alginate-raft formulations in the treatment of heartburn and acid reflux. *Aliment. Pharmacol. Ther.* **2000**, *14*, 669–690. [CrossRef] [PubMed]

79. Hemmingson, J.A.; Falshaw, R.; Furneaux, R.H.; Thompson, K. Structure and antiviral activity of the galactofucan sulfates extracted from *Undaria pinnatifida* (Phaeophyta). *J. Appl. Phycol.* **2006**, *18*, 185–193. [CrossRef]

80. Church, F.C.; Meade, J.B.; Treanor, R.E.; Whinna, H.C. Antithrombin activity of fucoidan. The interaction of fucoidan with heparin cofactor II, antithrombin III, and thrombin. *J. Biol. Chem.* **1989**, *264*, 3618–3623. [PubMed]

81. Kim, E.J.; Park, S.Y.; Lee, J.-Y.; Yoon, J.H. Fucoidan present in brown algae induces apoptosis of human colon cancer cells. *BMC Gastroenterol.* **2010**. [CrossRef]

82. Hsu, H.Y.; Takada, H.; Iha, M.; Nagamine, T. Attenuation of N-nitrosodiethylamine-induced liver fibrosis by high-molecular-weight fucoidan derived from *Cladosiphon okamuranus*. *Oncotarget* **2014**. [CrossRef]

83. Choi, Y.-H.; Yang, D.J.; Kulkarni, A.; Moh, S.H.; Kim, K.W. Mycosporine-Like Amino Acids Promote Wound Healing through Focal Adhesion Kinase (FAK) and Mitogen-Activated Protein Kinases (MAP Kinases) Signaling Pathway in Keratinocytes. *Mar. Drugs* **2015**, *13*, 7055–7066. [CrossRef]

84. Zhang, J.; Ewart, H.S.; Barrow, J.K.S.; James, C. Ascophyllum Compositions and Methods. U.S. Patent Application No. 11/660275, 13 November 2008.

85. Lee, B.-H.; Choi, B.-W.; Ryu, G.-S.; Kim, S.-K.; Shin, H.-C. Material Separated from Ecklonia cava, Method for Extracting and Purifying the Same and Use Thereof as Antioxidants. U.S. Patent 6384085, 12 December 2002.

86. Shibata, T.; Kawaguchi, S.; Hama, Y.; Inagaki, M.; Yamaguchi, K.; Nakamura, T. Local and chemical distribution of phlorotannins in brown algae. *J. Appl. Phycol.* **2004**, *16*, 291–296. [CrossRef]

87. Hwang, H.; Chen, T.; Nines, R.G.; Shin, H.; Stoner, G.D. Photochemoprevention of UVB-induced skin carcinogenesis in SKH-1 mice by brown algae polyphenols. *Int. J. Cancer* **2006**, *119*, 2742–2749. [CrossRef] [PubMed]

88. Sugiura, Y.; Takeuchi, Y.; Kakinuma, M.; Amano, H. Inhibitory effects of seaweeds on histamine release from rat basophile leukemia cells (RBL-2H3). *Fish. Sci.* **2006**, *72*, 1286–1291. [CrossRef]

89. Shibata, T.; Fujimoto, K.; Nagayama, K.; Yamaguchi, K.; Nakamura, T. Inhibitory activity of brown algal phlorotannins against hyaluronidase. *Int. J. Food Sci. Technol.* **2002**, *37*, 703–709. [CrossRef]

 *marine drugs* 

Review

Health Functionality and Quality Control of Laver (*Porphyra*, *Pyropia*): Current Issues and Future Perspectives as an Edible Seaweed

Tae Jin Cho and Min Suk Rhee *

Department of Biotechnology, College of Life Sciences and Biotechnology, Korea University, 145, Anam-ro, Seongbuk-gu, Seoul 02841, Korea; chshatria@korea.ac.kr
* Correspondence: rheems@korea.ac.kr; Tel.: +82-2-3290-3058

Received: 30 November 2019; Accepted: 20 December 2019; Published: 23 December 2019

Abstract: The growing interest in laver as a food product and as a source of substances beneficial to health has led to global consumer demand for laver produced in a limited area of northeastern Asia. Here we review research into the benefits of laver consumption and discuss future perspectives on the improvement of laver product quality. Variation in nutritional/functional values among product types (raw and processed (dried, roasted, or seasoned) laver) makes product-specific nutritional analysis a prerequisite for accurate prediction of health benefits. The effects of drying, roasting, and seasoning on the contents of both beneficial and harmful substances highlight the importance of managing laver processing conditions. Most research into health benefits has focused on substances present at high concentrations in laver (porphyran, Vitamin B_{12}, taurine), with assessment of the expected effects of laver consumption. Mitigation of chemical/microbiological risks and the adoption of novel technologies to exploit under-reported biochemical characteristics of lavers are suggested as key strategies for the further improvement of laver product quality. Comprehensive analysis of the literature regarding laver as a food product and as a source of biomedical compounds highlights the possibilities and challenges for application of laver products.

Keywords: raw laver; processed laver product; edible seaweed; nutritional value; functional substance; health functionality; processing technology; microbial risk; chemical risk; omics-based technology

---

1. Introduction

Lavers are red seaweed species mainly consumed as processed food products or used as a source of health-promoting substances. They belong to the genera *Porphyra* and *Pyropia* (which contains many species formerly included in *Porphyra*) (Phylum: Rhodophyta; Class: Bangiophyceae; Order: Bangiales; Family: Bangiaceae) [1]. Traditionally, lavers were staple foods in limited regions of Asia, but increased awareness of their health benefits and the globalization of processed food products has led to dramatic increases in consumption across the world [2]. The growth of global seaweed aquaculture as a source of pharmaceuticals and biomaterials (e.g., Alga Technologies, Cyanotech, etc.) is expected to contribute to the expansion of the laver industry [3,4]. Global laver production has increased from 517,739 t/US $945.1 billion in 1987 to 841,131 t/US $1285.0 billion in 1997; 1,510,911 t/US $1403.9 billion in 2007; and 2,563,048 t/US $2319.7 billion in 2017 (Figure 1) [5]. Commercial production of laver products (e.g., gim snack (seasoned laver), mareun-gim (dried laver), okazu nori (laver for side dish), yakinori (roasted laver), zicai tang (laver soup), etc.) are concentrated in northeastern Asia, with South Korea, China, and Japan producing 99.87% of total world production in 2017 (Figure 1) [5]. This reflects the traditional consumption of laver in these countries and also regional environmental conditions favorable for aquaculture [6].

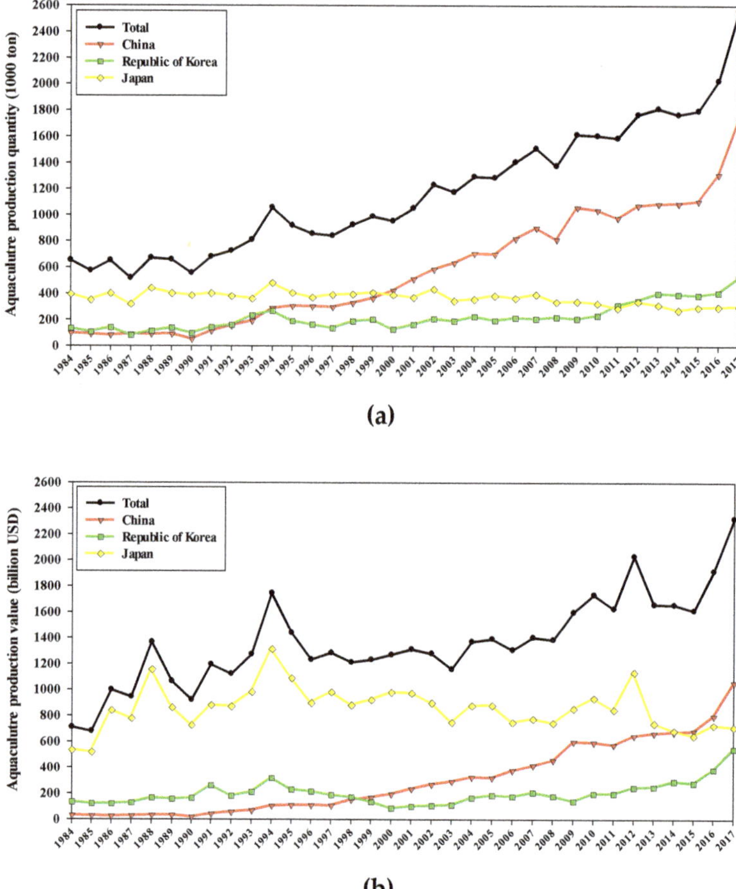

(a)

(b)

Figure 1. Global aquaculture production of the dominant production regions of the laver: (**a**) production quantity, (**b**) production value. Data (production quantity and value of 'Laver (Nori)' for Republic of Korea, 'Nori nei' for China, 'Laver (Nori)' for Japan) was obtained from FAO's Fisheries and Aquaculture statistics (FishStatJ) [5].

Currently, Algaebase [7] lists 188 and 77 species for *Porphyra* and *Pyropia*, respectively. As shown in Table 1, species found mainly in northeastern Asia and suitable for aquaculture have been used as the target organisms for research studies on the food production (raw and/or processed laver) and/or substances potentially beneficial to health (e.g., porphyran, taurine, vitamins, etc.). Previous studies on laver exploitation focused on the main species produced commercially in the Republic of Korea (*Pyropia tenera* (Kjellman) Kikuchi et al., 2011, *P. yezoensis* (Ueda) Hwang and Choi, 2011, *P. seriata* (Kjellman) Kikuchi and Miyata, 2011, *P. dentata* (Kjellman) Kikuchi and Miyata, 2011), China (*Pyropia haitanensis* (Chang and Zheng) Kikuchi and Miyata, 2011, *P. yezoensis*) and Japan (*Pyropia tenera*, *P. yezoensis*, *P. pseudolinearis* (Ueda) Kikuchi et al., 2011) [8].

Table 1. Major species of lavers.

Genus	Species [1]
Porphyra	*P. acanthophora, P. columbina, P. dentata, P. dioica, P. fucicola, P. haitanensis, P. kanakaensis, P. perforata, P. pseudolinearis, P. purpurea, P. sanjuanensis P. seriata, P. tenera, P. umbilicalis, P. vietnamensis, P. yezoensis*
Pyropia [2]	*P. acanthophora, P. columbina, P. dentata, P. fucicola, P. haitanensis, P. kanakaensis, P. nitida, P. orbicularis, P. perforata, P. pseudolinearis P. seriata, P. tenera, P. vietnamensis, P. yezoensis*

[1] Species used as the target organisms from research studies cited in this review were summarized. [2] This taxonomy was based on the generic revision of laver (*Porphyra* and *Pyropia*) [1].

Laver can be consumed as food, either raw or processed (e.g., dried, roasted, seasoned) or as a source of substances beneficial to health. In northeastern Asia, laver is consumed mainly as a side-dish and thus is generally perceived as a foodstuff rather than as a source of health functionality substances. By contrast, studies of other edible seaweeds (e.g., green or brown algae) focused mainly on their non-food roles as sources of nutraceuticals, food additives, and biomaterials. Nutritional values and bioactive components of algal species linked to major health benefits were reported, highlighting the potential for growth of the laver industry as both an edible seaweed and a source of useful compounds [4,9–18]. However, although practical studies of the utility of laver as an edible seaweed reported the distinct characteristics of various product types (i.e., material composition, effects of manufacturing processes), there is no comprehensive analysis of the literature regarding the nutritional/functional characteristics of laver and the technological basis for its quality control.

This review evaluates the results of research into the nutritional/functional characteristics of laver products (consumed as food or for health benefits) and the application of technology to those products, through the categorization of current issues (Sections 2 and 3) and discussion of future perspectives (Section 4). To analyze advances over the decade 2009 to 2019, we review literature retrieved from the following databases: PubMed, EBSCO, Research Information Sharing Service (RISS), National Digital Science Library (NDSL), SCOPUS, Web of Science, and WIPO IP Portal. The aims of this work are (1) to comprehensively review recent findings on the utility of edible lavers in both raw and processed products and (2) to identify priority areas for future research on the exploitation of lavers.

2. Food Products Containing Lavers

Food products may contain either raw or processed lavers. Studies reporting the distinct nutritional/functional values for each product type are reviewed here, the major factors determining the expected effects of consumption are identified, and their implications are discussed. Whereas it should be noted that since overall dietary habit defines total intake of nutritional substances with potential health functionality, consuming those substances is likely to have no effect on health (if total intake is already adequate) or even results in negative health consequences (in case of excessive intake).

2.1. Raw Lavers

Published analyses of the nutritional and health benefits of raw laver can be divided into studies of raw wet laver directly consumed as edible seaweed and studies of raw laver pre-treated for measurement of dry weight composition. Nutritional values vary widely among product types due to the high-water content of raw wet laver (Table 2). The water content of raw wet laver was generally reported as ca. 90% [19] (Table 2). Thus, data on compounds of nutritional and health value available in raw wet laver can only be estimated by analysis of raw wet laver itself rather than the dehydrated product used for dry weight-based analysis. However, most previous studies of nutritional values presented analytical data based on dry weight [20–27], and wet-weight values are rarely reported [19]. Since seasoned raw wet laver is often consumed without dehydration or further processing, data on nutritional value relative to wet weight are needed to accurately estimate the potential health benefits.

Table 2. Nutritional values of the raw laver.

Product Type	Raw Material (Species)	Nutritional Values from Proximate Analysis (w/w %)						Other Nutritional Substances	Reference
		Carbohydrate	Dietary Fiber	Protein	Lipid	Ash	Moisture		
Raw wet laver	*P. yezoensis*	1.2–2.7	- [1]	3.0–5.0	0.5	3.6–4.3	89.2–90.5	mineral	[19]
	Porphyra sp.	-	43.1, 38.9	25.6, 26.0	-	-	-	-	[20]
Raw laver (dry weight) [2]	*P. vietnamensis*	38.8–60.4	-	12.4–20.5	0.2–2.7	3.9–7.4	13.6–20.7	fatty acids	[21]
	P. dentata	45.7–45.9	-	36.2–37.7	0.7–1.0	7.1–8.2	8.6–8.8	mineral, amino acids	[25]
	P. purpurea	21.7 [3]	22.9	33.2	1.0	21.3	-	amino acids, fatty acids, sterol	[24]
	P. columbina	-	48.0	24.6	0.3	6.5	12.8	mineral, amino acids, fatty acids, antioxidants, phenolic compounds	[22]
	P. yezoensis	51.2–57.9	-	36.2–39.2	2.3–3.1	3.8–7.3	-	mineral, amino acids	[27]
	P. acanthophora var. robusta	35.5–61.0	-	14.1–18.4	1.7–2.6	4.2–6.8	12.5–21.5	mineral, fatty acids, pigments, vitamin	[26]
	P. purpurea	-	-	-	-	-	-	mineral, vitamin	[23]

[1] Not analyzed or not indicated. [2] Dry weight of components of raw laver. [3] Non-fibrous.

With respect to the nutritional composition of raw laver by dry weight (Table 3), most parameter values are similar to those obtained from other edible seaweeds (carbohydrate, dietary fiber, protein, lipid, and ash). Other nutritional components analyzed for raw laver are those generally used for assessing the health benefits of edible seaweeds, namely minerals, fatty acids, amino acids, sterol, antioxidants, phenolic compounds, pigments, and vitamins.

2.2. Processed Laver Products

The CODEX regional standard for laver products (CXS 323R-2017) categorizes these according to the processing methods (i.e., dried, roasted, seasoned) and raw materials (i.e., single or multiple edible seaweed species) [28]. Dried laver is the most common type and can be divided into primary dried products (i.e., washed, chopped/cut, molded, dehydrated, and dried after harvesting) and secondary dried products (made by re-drying primary dried products for long-term storage). Roasted laver is dried laver roasted without seasoning, while seasoned laver is dried material treated with a variety of ingredients using several processing methods (e.g., roasting, frying, treating with edible oil) before or after seasoning. The seasoned category also includes laver seasoned for brewing and broken and roasted/stir-fried dried laver seasoned for consumption after addition of boiling water. Maximum water contents of primary dried, secondary dried, roasted, and seasoned laver are set as 14%, 7%, 5%, and 5%, respectively. Nutritional values of processed laver products show the expected range of composition as described for raw laver in Section 2.1 (Table 3). Moreover, as noted for each product type, this variation in the nutritional/functional content implies a need for further analysis of nutritional values of specific product types to accurately estimate the potential for health benefits arising from the consumption of those products. Composition of processed laver products is standardized according to the CODEX regional standard (CXS 323R-2017) for products based on *Pyropia* spp. (and containing other optional ingredients) [28]. Edible seaweed other than *Pyropia* spp. is one of the major optional ingredients and can be intentionally or unintentionally included in processed laver products. Multiple species of edible seaweeds such as *Ulva* spp. (green laver), *Ecklonia cava* Kjellman, 1885, or *Capsosiphon fulvescens* (C. Agardh) Setchell and Gardner, 1920, can be combined in a single processed laver product to improve palatability or to create specific organoleptic characteristics. These optional ingredients have distinct nutritional and potential for health-promoting features [29]. The nutritional value of the combined product depends on raw material composition and enhancement of product quality with respect to potential health functionalities and organoleptic properties [30].

Table 3. Nutritional values of processed laver products.

Category	Product Type [1] (Species) [2]	Nutritional Values from Proximate Analysis (*w/w* %)						Other Nutritional Substances	Reference
		Carbohydrate	Dietary Fiber	Protein	Lipid	Ash	Moisture		
Processed laver products	DL (*Porphyra* spp.)	36.8	31.6	43.0	0.5	10.3	9.4	mineral, amino acids	[8]
	DL (*P. dentata, P. seriata*)	47.6	40.4	37.3	0.3	7.6	7.3	fatty acids, pigments, antioxidants	[30]
	DL	43.8–46.2	- [3]	37.8–40.0	1.5–2.3	8.0–9.0	5.7–7.4	mineral, amino acids, fatty acids, component sugar	[31]
	DL (*P. tenera*)	-	-	36.9	2.3	9.1	3.7	mineral, amino acids	[32]
	DL (*P. haitanensis*)	-	-	32.16	1.96	8.78	6.74	-	
	DL (*P. yezoensis*)	45.4–50.0	-	29.3–35.0	1.8–2.0	8.1–9.9	8.2–9.8	mineral	[19]
	DL	-	-	-	-	-	8.4	phenolic compounds	[33]
	DL	-	-	-	-	-	7.6	-	
	RL	-	-	-	-	-	8.7	-	
	DL	-	-	-	-	-	8.7	-	
	DL (*P. yezoensis*)	-	-	-	-	-	-	vitamin, organic acid, free sugar	[34]
	DL	-	-	-	-	-	-	phenolic compounds, anion, element	[35]
	DL (*P. tenera, P. yezoensis + P. dentata, P. seriata*)	41.7	33.4	38.4	0.3	8.0	11.6	-	[30]
Processed laver products with other seaweeds as optional ingredients	DL combined with green laver (*Ulva* spp.) [4]	43.7	36.6	35.0	0.8	9.1	11.4	-	[30]

[1] DL: dried laver; RL: roasted laver. [2] Species as the raw material of the product was indicated. [3] Not analyzed. [4] Nutritional values of processed laver products composed multiple species of optional ingredients.

Nutritional values obtained from processed laver products (i.e., dried, roasted, or seasoned) are broadly similar (Table 3). However, major differences in nutritional values and product quality can arise as a result of processing method, either by drying of raw laver or subsequent roasting or by seasoning. The drying process can affect various nutraceutical components including dietary fiber, phenolic compounds, pigments, and antioxidants [36]. The higher content and bioavailability of Vitamin B_{12} (VitB$_{12}$) in raw, as opposed to dried, laver may imply the conversion of VitB$_{12}$ in raw laver to its analogues (which are not bioavailable to mammals) by the air-drying process [37]. Drying of laver by lyophilization was suggested to prevent loss of bioactive VitB$_{12}$ [38]. With respect to further processing, roasting or deep-frying of dried laver results in decreased mineral content (calcium, iron, magnesium, phosphate, potassium). Deep-frying (160–180 °C for 10 sec) decreased mineral content by a factor of 2–6 compared with roasting (300 W for 2 min) [39]. Simulated domestic cooking of dried, roasted, and seasoned laver showed consistent trends in (1) decreased water content (9.69%, 3.66%, and 1.49% for dried, roasted, and seasoned laver, respectively), (2) denaturation of amino acids (mainly glycine, citrulline, valine, isoleucine, leucine, and gamma-aminobutyric acid), and (3) higher mineral contents in dried than in roasted or seasoned laver (calcium: dried 4976 mg/g, roasted 2202 mg/g, seasoned 2037 mg/g; potassium: dried 31210 mg/g, roasted 29540 mg/g, seasoned 28800 mg/g; zinc: dried 45.12 mg/g, roasted 24.33 mg/g, seasoned 18.37 mg/g; copper: dried 6.49 mg/g, roasted 6.11 mg/g, seasoned 4.71 mg/g) [40]. However, lower contents of functional substances in processed laver products than in dried laver does not always implicate the processing steps as causal factors: Although VitB$_{12}$ content was lower in seasoned (51.7 µg/100 g) than in dried laver (133.8 µg/100 g), the destruction of VitB$_{12}$ by the roasting process was not detected, and thus, the addition of optional ingredients (e.g., seasoning) is thought to be the cause of the difference [41]. With respect to other measures of product quality, roasting may cause color deterioration by its effects on pigments such as chlorophyll [42]. Since the processing of laver has acted as essential role for the quality control of products (e.g., drying greatly improves shelf-life of laver to facilitate wider accessibility for individuals as a food source), the application of optimal processing conditions ensuring both nutritional/functional values and the product quality is important.

Lavers can be a component of other food products such as seaweed chips, salads, laverbread, laver cake, onigiri (rice ball), kimbab (seasoned rice roll), and laver soup [10,13,15,16]. However, previous studies of these processed products focused on organoleptic characteristics with sensory evaluation, rather than on potential for health and/or nutritional benefits.

3. Lavers as Functional Foods: Unique Health Benefits of Laver

Seaweeds have long been regarded as a rich source of health-promoting substances. However, relevant studies mainly focused on internationally produced and consumed seaweeds, including brown [43] and green algae [44], and on red algae other than laver [45]. This section reviews the available information on the health benefits of lavers, with particular reference to active substances unique to this group of edible seaweeds. Table 4 summarizes the major components of laver linked to well-known health benefits based on the currently reported research studies investigating the putative health effects of each component, highlighting the necessity for future research regarding long-term outcomes of laver consumption on human health.

Since most of the reported health benefits of laver could also be gained by consumption of other edible seaweeds [11], this section focuses on studies that report unique health functionalities of laver derived from components not found in other edible seaweeds, namely, porphyran, VitB$_{12}$ and taurine [9,10].

Table 4. Major health functionality of laver products (raw laver and processed laver products).

Health Functionality	Major Components Linked to Health Functionalities	References
Anti-cancer	polysaccharides (dietary fiber, porphyran), phospholipids, sterol, peptide	[17,46–54]
Prevention of cardiovascular disease (e.g., hypertension, atherosclerosis, ischemia)	betaine, dietary fiber, taurine, porphyran	[17,55–61]
Antioxidant effect (e.g., Anti-ageing)	porphyran, glycoprotein, polyphenols, tocopherols, peptide	[62–67]
Anti-inflammatory effect and immunomodulation	glycoprotein, porphyran	[64,68–73]
Alcohol metabolism	glycoprotein	[74,75]
Prevention of nervous diseases (e.g., Alzheimer's diseases, methylmalonic acidemias)	taurine, porphyran	[38,76,77]
Prevention of bone disease (e.g., osteoporosis, rheumatoid arthritis)	porphyran, glycoprotein	[64,78]
Anti-diabetes mellitus	phenolic compounds (carotenoids, anthocyanins), polysaccharides (porphyran), peptide	[79–82]

Porphyran is the distinctive dietary fiber found in laver, and its health effects were intensively studied to determine the nutritional/functional quality of lavers [83]. Important bioactivities that can be attributed to porphyran include anti-cancer, antioxidant, and anti-inflammatory effects and/or immunomodulation and prevention of diseases such as cardiovascular, nervous, bone, and diabetic disorders [9,10]. Those bioactivities have been demonstrated by the examination of porphyran extracted from laver species as reported in following research studies. The anti-cancer effect of porphyran was demonstrated by using human cell lines including the hepatic carcinoma (Hep3B), cervical cancer (HeLa), and human breast carcinoma (MDA-MB-231) cell lines [53]. The anti-cancer effect of porphyran was also evaluated using human gastric cancer cells through the induction of apoptosis [46] and the inhibition of cell proliferation [47]. In dietary experiments using rats, prevention of cardiovascular disease may be achieved by the anti-hyperlipidemic effect revealed by the decrease in serum cholesterol level [57,84]. The basis of these effects was shown to be reduced secretion of the essential component of very low-density lipoprotein (VLDL) in blood (i.e., apolipoprotein B100) [58]. Properties of porphyran derivatives were also reported, especially for antioxidant effects assessed by radical scavenging and reducing power [62,63]. Immunomodulatory effects were shown as immune responses to myelosuppression by the oral administration of porphyran to rats [68]. Anti-inflammatory activity was evaluated by inhibition of secretion of inflammatory markers (nitric oxide and tumor necrosis factor alpha) by macrophages (RAW264.7 cell) [72], and the suppression of activation of immune cells [73]. Neuroprotective effects may be attributable to oligo-porphyran, with the mechanism for the protection of neurons shown to be regulatory effects linked to the inhibition of apoptosis in neuronal cells [77]. Therapeutic effects against bone diseases by suppression of osteoclast formation induced by the receptor activator of nuclear factor κB ligand were demonstrated using RAW 264.7 cells [78].

The beneficial health effects of the vitamin B complex (e.g., choline, inositol) in laver include the synthesis of major nutritional factors (i.e., carbohydrate, protein, and lipid), anti-cancer effects, and enhanced immunomodulation [18,56,85,86]. Lavers produce exceptional quantities of $VitB_{12}$ and thus can be used to counter the deficiency of $VitB_{12}$ (e.g., methylmalonic acidemias) in vegan diets by the consumption of laver [86–88]. The bioavailability of $VitB_{12}$ was also confirmed by increases in the hepatic $VitB_{12}$ level of rats by the intake of laver [38,89] and by the release of $VitB_{12}$ from laver after human consumption simulated through in vitro gastrointestinal digestion experiments [41].

Taurine is a major amino acid in laver and other red algae but rarely present in brown or green algae [90]. Decreased plasma cholesterol levels in rats after the consumption of taurine were reported [56,61]. Promotion of neuronal development by taurine in laver extracts was also experimentally demonstrated by the primary culture of hippocampal neurons [76].

4. Future Perspectives on Technical Advances in Laver Utilization

The main current issue in laver utilization is the elimination of potential risk factors in the process from farming to the manufacture of laver products. For the future, we need to consider the application of new technologies for the identification of useful constituents. This section covers the development of technologies for the control of chemical/microbiological risks and novel techniques that may promote the consumption of lavers as edible seaweeds, in particular, omics-based research linked to the health benefits of lavers.

4.1. Management of Current Issues: Control of Potential Risks from Farming to Processing of Lavers

The sequence from cultivation of laver in aquaculture farms to the final drying steps [91] is common to most processed laver products because roasted and seasoned laver are manufactured using dried laver as a basic material.

For processed laver food products, effective control of chemical and microbiological risk factors is the prerequisite for human consumption. Since major risk factors and their extent vary across the production process, it is important to establish intervention strategies based on a detailed understanding of the overall production process. Management strategies effective in the identification and control of chemical/microbiological risks should also be established for individual product types.

4.1.1. Control of Chemical Risks

The chemical risks can be defined as the consumption of excessive levels of substances that lead to side-effects and/or the exposure to toxic agents (such as heavy metals) in laver products [92,93]. Major food constituents with side-effects are iodine, fibers, and sodium (in seasoned laver). Iodine has beneficial effects on thyroid gland functioning, but excessive intake should be avoided to prevent potential adverse effects such as autoimmune thyroiditis or hypothyroidism [94–98]. Overconsumption of fibers can lead to vomiting or abdominal pain with diarrhea in people with sensitive stomachs and may cause dyspepsia even in healthy people due to generation of gases in the digestive system [99]. The much greater sodium content of seasoned (relative to dried) laver shows that salts added during processing can result in consumption of excessive levels of sodium [32]. Heavy metals (As, Cd, Cr, Cu, Hg, Ni, Pb, Zn, etc.) were detected in both raw and processed laver products (Table 5). It should be noted that the contents of heavy metals in laver have been reported as variable according to a range of factors including the cultivar, species, season, and processing conditions [100,101]. Most previous studies recorded acceptable levels of contaminants according to the hazard quotient (HQ) of heavy metals in laver products or provisional tolerable weekly intake (PTWI) set by the Food and Agriculture Organization/World Health Organization (FAO/WHO) [42,102]. Guidance values for tolerable intake [PTWI, provisional tolerable monthly intake (PTMI), provisional maximum tolerable daily intake (PMTDI)] for major heavy metals detected from laver were set as follows: Al (PTWI 2.0 mg/kg bw/week), Cd (PTMI 25 µg/kg bw/month), Cu (PMTDI 0.5 mg/kg bw/day), and Hg (PTWI 4.0 µg/kg bw/week) [103]. However, high levels of aluminum in laver (388.6–623.4 mg/kg dry weight) were reported as indicators of food pollution [104]. In addition, arsenic is the major heavy metal contaminant in laver [42,102,105,106], and the potential risk of production of toxic metabolites by the human digestive process was also stressed [107,108].

Table 5. Research studies regarding the chemical risk of laver products.

Category	Product Type [1]	Target	Results (mg/kg or µg /g of dw [2])	References
Raw laver	-	arsenic (As)	9.59–34.0	[100]
			22.9–33.8	[101]
			0.22–0.70	[105]
			12.87	[109]
		cadmium (Cd)	0.40–1.21	[100]
			2.83–3.54	[101]
		chromium (Cr)	0.32–0.86	[100]
		copper (Cu)	7.92–16.9	[100]
			1.94–6.94	[101]
		iron (Fe)	290–723	[100]
		lead (Pb)	0.78–1.30	[100]
			<LOD [3]	[101]
			0.98	[109]
		mercury (Hg)	0.005–0.006	[100]
			0.03	[109]
		nickel (Ni)	0.69–1.04	[100]
			0.74–1.51	[101]
		zinc (Zn)	18.0–57.7	[100]
			21.1–70.1	[101]

Table 5. *Cont.*

Category	Product Type [1]	Target	Results (mg/kg or µg/g of dw [2])	References
Processed laver products	DL	aluminum (Al)	388.6–623.4	[104]
			66–511	[106]
		arsenic (As)	13.5–32.8	[100]
			<LOD–29.850	[102]
			ND [4]–0.303	[106]
			30.18–39.05	[42]
		cadmium (Cd)	0.69–4.73	[100]
			0.108–3.11	[106]
			0.076–0.318	[42]
			0.501–2.421	[102]
		chromium (Cr)	0.46–0.66	[100]
		copper (Cu)	5.02–8.64	[100]
		iron (Fe)	103–214	[100]
		lead (Pb)	ND–0.86	[100]
			ND	[42]
			<LOD–2.362	[102]
			ND–0.208	[106]
		mercury (Hg)	0.004–0.008	[100]
			0.005–0.009	[42]
			0.002–0.050	[102]
		nickel (Ni)	0.17–1.49	[100]
		zinc (Zn)	27.1–57.7	[100]
	DL, RL	arsenic (As)	2.1–21.6	[107]
	_ [5]	aluminum (Al)	15.50 [6]	[110]
		arsenic (As)	2.07 [6]	[110]
		cadmium (Cd)	0.109 [6]	[110]
		lead (Pb)	0.063 [6]	[110]
		mercury (Hg)	<LOD	[110]

[1] DL: dried laver; RL: roasted laver. [2] dw: dry weight. [3] LOD: limit of detection. [4] ND: not detected. [5] Specific product type was not indicated in the cited literature. [6] Average value.

Setting recommended-intake limits to prevent overconsumption of nutritional components (i.e., iodine, dietary fiber, sodium) that pose potential chemical risks is the primary strategy for risk management, and specific control methods for these factors are generally not required. By contrast, since reducing the heavy metal content of laver can lower the risks, intervention technologies for the elimination of heavy metal contaminants were developed. Cadmium, chromium, and lead can be removed by immersion of laver in acid solution (citric, hydrochloric, or nitric) of pH 2.5–4.0 for 20 min, and this method could be applied because laver undergoes color changes only in more acidic conditions (pH 2.0) [111]. Heavy metal contents of processed (roasted or seasoned) laver products indicate a reduction in the levels of lead, mercury, and cadmium during the cooking process [40]. However, the increase in bioaccessible arsenic after human digestion may be a result of the roasting process [10,112]. A correlation between the arsenic content of laver and that of seawater in the cultivation area was also reported, suggesting environmental management as one of the risk control strategies [105].

4.1.2. Control of Microbiological Risks

Potential microbiological risks can be identified in the national standards and regulations for laver products. China, in particular, has strict regulations (GB 2733-2005) specifying maximum permitted values for aerobic plate counts (APC; 30,000 CFU/g), coliforms (30 MPN/100 g), mold (300 CFU/g), *Salmonella* spp. (not detected), *Vibrio parahaemolyticus* (Fujino et al.) Sakazaki et al., 1963 (not detected), *Staphylococcus aureus* Rosenbach, 1884 (not detected), and *Shigella* spp. (not detected) [113]. However, laver is generally contaminated with marine bacteria [114,115], and inappropriate processing conditions may allow the growth of contaminants not only from the natural habitat but also from surrounding environments [42,116]. The importance of microbiological quality was stressed by contamination reports relating to raw and processed laver [117,118] and also to food products based on laver (e.g., sushi, kimbab) [119,120].

Table 6 summarizes microbiological contamination data for laver categorized as (1) processed laver products, (2) food products containing laver, and (3) intermediate and end-products from manufacturing plants. Aerobic plate count is the main indicator of microbiological quality. An APC value of 6.5 log CFU/g recorded from commercial dried laver products [121,122] indicates a high level of contamination. Microbiological quality factors represented by viable cell counts and coliforms were also reported for dried (7.6 log CFU/g and 3.2 MPN/100 g, respectively) and roasted laver (7.5 log CFU/g and 3.7 MPN/100 g, respectively) as the end-products from manufacturing plants. The close similarity of these values suggests that the processing of dried into roasted laver has a very limited antimicrobial effect [123,124]. Microbial populations of commercial products also differ according to product type (APC: 6.9, 3.4, and 4.9 log CFU/g; coliforms: 2.1, 1.6, and 1.0 log CFU/g, for dried, roasted, and seasoned laver, respectively) [125]. Aerobic plate count, coliforms, yeast/mold, and *Bacillus cereus* Frankland and Frankland, 1887 from processed laver products were also reported as 4.3–7.2 log CFU/g, 1.9–2.2 log CFU/g, 2.1–4.9 log CFU/g, and 2.3 log CFU/g, respectively [126]. Microbiological quality factors (APC, coliforms) were used to identify laver as the main source of microbial contaminants in ready-to-eat foods (e.g., kimbab) containing this edible seaweed: dried laver APC: 8.8 log CFU/g [127–129]; dried and roasted laver APC: 6.0–7.0 log CFU/g; and coliforms: 2.0–3.0 log CFU/g [130]. *Bacillus cereus* (detection rate: 12%) and *Clostridium perfringens* (Veillon and Zuber) Hauduroy et al. 1937 (detection rate: 3%) were also reported in dried laver [131]. In the case of intermediate and end-products from the manufacture of processed laver products, changes in the microbial level for each step indicate critical control points for the management of microbiological risks. Intermediate products from the manufacture of dried laver (i.e., after primary scrubbing in salt water, primary debris elimination, secondary scrubbing in salt water, secondary debris elimination, chopping and scrubbing in fresh water, molding, drying, and packaging) from seven companies showed an increase in total viable cell count (TVC) during the drying step (final products TVC: 5.6–8.0 log CFU/g; coliforms: 54–27,600 MPN/100 g) compared with the first manufacturing step (primary scrubbing in salt water TVC: 1.5–2.8 log CFU/g; coliforms: 18–75 MPN/100 g). This indicates the drying step as a critical control point for the microbiological quality of dried laver [42]. Samples collected from six companies producing seasoned laver showed high levels of microorganisms in the dried laver raw material (APC: 4.4–7.8 log CFU/g; coliforms: 54–27,600 MPN/100 g). Changes in microbial counts during the manufacturing process (i.e., primary roasting, seasoning, secondary roasting, counting, and packaging) highlighted second roasting as the key intervention step for microbial control [116]. Sequential changes in APC at each stage in the manufacture of seasoned laver (i.e., primary roasting, secondary roasting, counting, and packaging) also indicate secondary roasting as the most effective decontamination process [132,133].

Table 6. Research studies regarding the microbiological risk of laver products.

Category	Target Microorganisms	Product Type [1]	Results	References [2]
Processed laver products	Mesophilic bacteria	Standard	4.48 log CFU/g	[113]
		DL	6.5 log CFU/g	[121]
		DL	7.6 log CFU/g	[123]
		RL	7.5 log CFU/g	
		DL	6.9 log CFU/g	[125]
		RL	3.4 log CFU/g	
		SL	4.9 log CFU/g	
		DL	5.6–7.2 log CFU/g	[126]
		RL	3.6 log CFU/g	
		SL	4.3–6.0 log CFU/g	
	Coliforms	Standard	30 MPN [3]/100 g	[113]
		DL	3.2 MPN/ 100 g	[123]
		RL	3.7 MPN/ 100 g	
		DL	2.1 log CFU/g	[125]
		RL	1.6 log CFU/g	
		SL	1.0 log CFU/g	
		DL	1.9–2.2 log CFU/g	[126]
	Yeast/mold	Standard	2.48 log CFU/g	[113]
		DL	4.3–4.9 log CFU/g	[126]
		RL	2.1 log CFU/g	
		SL	2.1–4.7 log CFU/g	
	Bacillus cereus	DL	2.3 log CFU/g	[126]
Raw materials of food products using laver	Mesophilic bacteria	Standard	4.48 log CFU/g	[113]
		DL	8.8 log CFU/g	[127]
		DL	*ca.* 7.0 log CFU/g	[130]
		RL	*ca.* 6.0 log CFU/g	
		DL	5.3 log CFU/g	[131]
	Coliforms	Standard	30 MPN/100 g	[113]
		DL	*ca.* 3.0 log CFU/g	[130]
		RL	*ca.* 2.0 log CFU/g	
		DL	detection rate 6%	[131]
	B. cereus	DL	detection rate 12%	[131]
	Clostridium perfringens	DL	detection rate 3%	[131]

Table 6. *Cont.*

Category	Target Microorganisms	Product Type [1]	Results	References [2]
Work-in-process and end-products from manufacturing plants	Mesophilic bacteria	Standard	4.48 log CFU/g	[113]
		DL	5.6–8.0 log CFU/g	[42]
		DL	4.4–7.8 log CFU/g	[116]
		SL	1.3–5.9 log CFU/g	
		DL	4.7–4.8 log CFU/g	[132]
		SL	ND [4]–1.0 log CFU/g	
		DL	3.4–3.6 log CFU/g	[133]
		SL	1.4–2.8 log CFU/g	
	Coliforms	Standard	30 MPN/100 g	[113]
		DL	54–27,600 MPN/100 g	[42]

[1] DL: dried laver; RL: roasted laver; Standard: permitted values standardized by China which has strict regulations (GB 2733-2005) were indicated as reference data [113]. [2] This table is adapted and modified from [91].[3] MPN: Most probable number.[4] ND: Not detectable.

Antimicrobial treatments to mitigate potential risks were applied to a range of product types from raw materials to processed foods (Table 7). Little research was conducted into decontamination of harvested raw laver because subsequent manufacturing steps (washing, drying, roasting) have generally been considered effective for the control of microbial contaminants. Park et al. [134] reported that exposing *Bacillus cereus* and *Escherichia coli* (Migula) Castellani and Chalmers, 1919 in raw laver to 200 ppm NaOCl with 60 min of ultrasound could achieve reductions of 2.6 and 3.2 log CFU/g, respectively. Decontamination methods can be applied to both processed laver products and to other processed foods containing lavers. Gamma irradiation (3 kGy for 24 h) used as an antibacterial treatment for kimbab [120,127,130] and dried laver [127] produced a reduction of up to 2 log in mesophilic bacteria (initial population: 6.0–8.8 log CFU/g), and an inactivation of foodborne pathogens to levels undetectable by plate-count methods (e.g., *Salmonella typhimurium*, *Staphylococcus aureus*, and *Listeria monocytogenes* (E. Murray et al.) Pirie, 1940; initial population: 6–7 log CFU/g). Corona discharge plasma jet and low-pressure air plasma achieved a 1.5 log reduction in mesophilic bacteria on dried laver without post-treatment color changes [135]. An electron beam (e-beam) can be used as a more energy-efficient alternative to gamma irradiation. A 4 kGy treatment achieved a 1.4 log CFU/g reduction of APC from dried laver [136]. Using heat-assisted low dose e-beam irradiation, coliforms could be eliminated from dried laver (> 1.5 log reduction) by 1–4 kGy without any changes in color or pigment contents [137,138]. Optimal treatment conditions ensuring product quality (i.e., thermophilic acidophilic bacterial count: ~10^3 CFU/g; water content: 5%; acceptable palatability score from sensory evaluation) were suggested by response surface analysis as an irradiation dose of 1.8–3.0 kGy and heating at 154–170 °C for 10–18 s [139]. Not only for research studies regarding the development and application of intervention methods for microorganisms, patents specified for laver processing have been reported including raw wet laver [140], dried laver [141,142], and roasted laver [143]. However, although decontamination technologies reported from both academic research cases and patents were shown to reduce microbial populations to acceptable levels, there is major limitation in burden on the adoption of additional treatment devices for manufacturers. Since most those techniques are physical treatments generally applied to final products, further research into chemical/physicochemical treatment technologies applicable from intermediate products to end-products is required to support the wider application of microbial risk management strategies by manufacturers.

Table 7. Intervention methods for microbial potential risks of processed laver products.

Target Product	Treatment Methods	Target Microorganisms	Treatment Conditions	Microbial Reduction (log CFU/g)	References
Raw harvested laver	NaOCl + ultrasound	*Escherichia coli* *Bacillus cereus*	200 ppm, 60 min	2.6 3.2	[134]
Kimbab	Gamma irradiation	Mesophilic bacteria	1–3 kGy, 24 h	1.0–2.0	[130]
Kimbab	Gamma irradiation	*Escherichia coli* *Salmonella* Typhimurium *Staphylococcus aureus* *Listeria monocytogenes*	1–3 kGy, 24 h	1.3–ND [1] 2.3–ND 3.4–ND 2.7–ND	[120]
DL [2]	UV	Mesophilic bacteria	20 W, 20 min	1.0	[123]
DL	Gamma irradiation	*Escherichia coli* *Salmonella* Typhimurium *Staphylococcus aureus* *Listeria ivanoviis*	1–3 kGy, 24 h	2.7–ND 1.7–ND 2.0–ND 1.6–ND	[127]
DL	Corona discharge plasma	Mesophilic bacteria	3312 rpm, 58Hz, 20 min	2.0	[135]
DL	Low–pressure air plasma	Mesophilic bacteria	20 min	1.5–2.0	
DL	e-beam	Mesophilic bacteria	4 kGy	1.4	[136]
DL	Heat-assisted e-beam irradiation	Mesophilic bacteria	1.8–3.0 kGy, 154–170 °C, 10–18 s	> 2.0 [3]	[139]
DL	Heat-assisted e-beam irradiation	Coliform	4 kGy, dose rate as 2.1 kGy/h	> 1.5 [4]	[137]
DL	Heat-assisted low-dosee-beam irradiation	Coliform	1 kGy, dose rate as 2.1 kGy/hg	> 1.4 [5]	[138]

[1] ND: not detected. [2] DL: dried laver. [3] Microbial reduction was calculated from the control group of this study (no treatment of e-beam irradiation) as heating 160 °C for 14 sec without e-beam irradiation. [4] Initial population level of coliform was 2.5 log CFU/g and the irradiation reduced coliform to undetectable levels with the detection limit as 1 log CFU/g. [5] Initial population level of coliform was 2.4 log CFU/g and the irradiation reduced coliform to undetectable levels with the detection limit as 1 log CFU/g.

4.2. Future Issues: Identifying the Health-Promoting Properties of Laver

Omics-technology is the most advanced research method for detailed understanding of the biological characteristics of edible seaweeds. This section covers the recent findings from the omics-technologies applied to laver as a future issue for obtaining useful information linked to the health functionalities and quality control of laver including the growth characteristics and biochemical composition. Especially since these research studies have also identified the key determinant factors on the quality of laver with the perspectives to its potential health-promoting properties, the improvement on long-term outcomes of laver consumption on human health can be expected, and thus, practical evaluation should be followed as further studies. The genome, microbiome, transcriptome, proteome, and metabolome of laver species were all analyzed (Table 8).

The plastid and mitochondrial genomes of red algae were first analyzed in *Porphyra purpurea* (Roth) C. Agardh, 1824 [144–146], with subsequent genomic research conducted using next-generation sequencing technologies [146–154]. Laver is regarded as the model genome for red algae, and detailed analysis has provided data for phylogenetic, taxonomic, and evolutionary studies [146,147,155]. Genomic data from *Porphyra* and *Pyropia* species were generated by whole genome sequencing (WGS) and comparative genomics (Table 8). WGS of laver has allowed identification of genomic features associated with nutritional/functional characteristics [147] and/or product quality (e.g., color) [154]. In *Porphyra umbilicalis* Kützing, 1843, structural characteristics are linked to major nutritional/functional constituents and are believed to confer resistance to stressful habitat conditions, such as repeated desiccation and rehydration in the intertidal zone [147]. WGS of *Pyropia yezoensis* followed by annotation of major functional genes governing photosynthesis has identified new genes involved in the control of laver color. Designation of biomarkers based on these gene sequences is expected to provide practical information to the laver industry, allowing prediction of color and also details of diseases responsible for color fading [154]. Genomic data reveal the sequences for gene sets associated with the specific metabolism of laver, and this genome-wide identification can be used to target key functional genes for further transcriptomic analysis [156]. Comparative genomic analysis is an effective method of species determination in morphologically simple organisms such as lavers. It was suggested that destructive sampling and DNA extraction from fragmentary material may be useful in the identification of type specimens [146]. Phylogenetic analysis can clarify the distinct characteristics of laver compared with other red algae [153]. Complete mitochondrial genome analysis can also allow recognition of unidentified algal species by comparison of gene sequences with those of morphologically similar organisms [148]. Phylogenetic analyses of *Pyropia* and *Porphyra* also broaden our understanding of biodiversity within these genera, with *Pyropia haitanensis* [150] and *P. yezoensis* [151] providing examples of this. Biomarkers for the differentiation of laver cultivated in different regions can be identified from gene sequences based on the divergence in genomic features [152].

From the perspective of the laver industry, transcriptomic analysis can provide a detailed understanding of the unique life cycle (linked to laver yields) and stress response (linked to laver quality) (Table 8). Advances in applied transcriptome analysis include identification of housekeeping genes involved in internal control processes [157]. With respect to the life cycle, previous studies focused on characterization of the evolutionary aspects of the distinct stages [158,159], development of applied technologies including the designation of biomarkers [160], and treatments affecting the regulation of gene expression linked to reproductive processes [161]. A major focus for genomic research has been the interpretation and/or comparative analysis of transcriptomic data relating to the responses of laver species to abiotic stress [162,163], principally environmental conditions such as high temperature [164–166], repeated desiccation-rehydration [167,168], and hypersalinity [169].

The microbiota of laver is generally host-specific and changes according to the physiological state of the alga or according to factors in the surrounding environment [170]. Because the spatial heterogeneity of microbiota on laver can make accurate characterization of microbial communities difficult, a database of the determining factors affecting microbiome analysis was established [171]. Changes in the microbiome can be regarded as indicators of environmental conditions in the farming area [172,173]. Analysis of the microbial community of laver has revealed not only seasonal variation in the microbiota but also evidence of host-microbiota interactions from the identification of bacteria affecting the morphogenesis and growth of laver [155]. In the sea water of seedling pools, the emergence of filament diseases resulted in clear changes in microbial community composition. Healthy seedling pools showed a dominance of *Sediminicola* spp. and *Roseivirga* spp. Seedling pools with diseased laver showed increased abundance of *Vibrio* spp. and *Polaribacter* spp., suggesting they may be suitable biomarker organisms for monitoring laver disease [174].

Proteomic studies of the impact of stress factors on laver have focused on stress-response mechanisms and the effects of mutation. Abiotic and biotic stresses associated with the sequential steps from farming to processing of laver include high temperature [175,176], desiccation [177,178], and pathogen infection [179]. High temperature is one of the main determining factors of nutritional/functional quality and yield for laver strains able to tolerate elevated temperature [175,176]. A major topic in proteome research is the coordinated activation of various pathways governing desiccation tolerance, which is triggered under natural conditions by exposure to low tide [177,178]. Metabolic responses to seaweed pathogens and other aspects of host-pathogen interaction can be explored by comparative proteomic analysis before and after infection [179,180]. In the case of mutation, identification of key proteins is the primary step for the production and/or isolation of stress-tolerant strains of laver. Experimental mutagenesis aims to produce cultivated laver strains capable of efficient growth with resistance to environmental stress factors and has been induced by chemical (ethyl methane sulfonate exposure) [181] or physical (gamma irradiation) treatments [182].

Lipidomic studies have mainly focused on lipid biomarkers, which play an important role in stress-response metabolism, and on the assessment of the nutritional/functional quality of lavers. Elevated habitat temperature as a result of global warming is regarded as a key stress factor for laver, and thus, lipidomic changes under high temperature were examined to determine the acclimation strategies of laver [183,184]. Clear differences in the lipidomic profiles of blade (gametophyte) and conchocelis (sporophyte) stages have identified the main health-promoting lipids characterizing the life cycle stages. This provides reference data for the selection of appropriate life cycle stages for practical uses as food and for functional ingredients and/or biotechnological applications [185].

Since metabolites can determine the nutritional quality of laver and organoleptic characteristics (such as flavor), metabolomic analysis can contribute to the identification of the major metabolites and suggest optimal strategies to produce high-quality products. Metabolomic variations reported in raw, recently harvested laver (changes in glutamine, alanine, aspartate, taurine, and isofloridoside) [186] and in processed laver products during manufacturing (changes in dominant metabolites including amino acids, carboxylic acids, choline metabolites, and sugars) [187] highlight the importance of metabolomic data with respect to the effects of these determinant factors on product quality. Moreover, since metabolite profiles of major edible seaweeds show distinct, species-specific characteristics, metabolomic data specific to laver are also required [188].

Table 8. Omics-based studies linked to the health functionality and the processing of lavers.

Omics Technology	Topic	Species	Major Findings	References
Genome	Whole genome sequencing and genomic feature	*P. umbilicalis*	- Genome governing nutritional/functional values linked to the growth and survival strategy of laver under stressful condition of natural habitat (intertidal zone)	[147]
		P. yezoensis	- First report on the genome sequence of nuclear ribosomal DNA (nrDNA) cistron	[149]
		P. yezoensis	- Genome sequence and annotated functional genes from *P. yezoensis* - Identification of photosynthesis system and key genes governing color of laver	[154]
	Genome-wide identification of functional genes	*P. yezoensis*	- Gene structure associated with mitogen-activated protein kinases from *P. yezoensis* (*PyMAPKs*)	[156]
	Comparative genomics	*P. perforata* *P. sanjuanensis* *P. fucicola* *P. kanakaensis*	- Reliable analytical method for the genomes of laver by the destructive sampling of type specimen	[146]
		P. nitida	- Recognition of new red algal species	[148]
		P. haitanensis	- Supportive data for the phylogenic differences between *Pyropia* from *Porphyra*	[150]
		P. yezoensis	- Supportive data for the phylogenic differences between *Pyropia* from *Porphyra*	[151]
		P. yezoensis	- Different genomic structure of strains according to the regions of cultivars (Korea and China)	[152]
		P. haitanensis, *P. yezoensis*	- Biodiversity and distinct phylogenies of laver compared with other red algae	[153]

Table 8. *Cont.*

Omics Technology	Topic	Species	Major Findings	References
Transcriptome	Analytical techniques	*P. haitanensis*	- Selection of housekeeping gene mostly adequate for the designation of internal control based on the stability under abiotic stresses	[157]
	Unique life cycle	*P. yezoensis*	-Transition observed in the life cycle with apospory	[158]
		P. umbilicalis, P. purpurea	- Evolutionary analysis for the growth and development of laver	[159]
		P. haitanensis	-Transcriptomic profile under different physiological conditions -Role of cSSR markers linked to the differences in the gene expressions among lifecycle stages of laver	[160]
		P. pseudolinearis	- Impact of ethylene precursor treatment to the regulation of gene expression governing reproduction	[161]
	Stress response	*P. yezoensis*	- Stress response of PyMAPK gene family	[156]
		P. yezoensis	- Identification of key response genes expressed under various abiotic stresses	[162]
		P. haitanensis	- Role of heat shock proteins against the abiotic stresses	[163]
		P. tenera	- Distinct transcriptional characteristics of gametophyte thalli by high-temperature stresses	[164]
		P. yezoensis	- Transcriptomic profiles in response to stresses associated with temperature	[165]
		P. haitanensis	- Identification of key response genes expressed under thermal stresses - Mechanisms on the adaptation of high-temperature tolerant strain	[166]
		P. columbina	- Identification of mechanisms on resistance and key response genes expressed under stresses from desiccation-hydration cycles in natural habitat	[167]
		P. tenera	- Identification of mechanisms on resistance and key response genes expressed under desiccation	[168]
		P. haitanensis	- Identification of mechanisms on resistance and maintenance of homeostasis under stresses from hypersaline conditions	[169]
	Biosynthesis	*P. yezoensis*	- Role of glycine-betaine (GB) capable of maintenance of osmotic balance in response to desiccation stresses - Identification of major enzymes involved in the biosynthesis of GB	[189]

Table 8. *Cont.*

Omics Technology	Topic	Species	Major Findings	References
Microbiome	Diversity in the microbiota	*P. umbilicalis*	- Seasonal variation to the microbial community in laver - Identification of bacterial groups which are expected to contribute to the evolution and/or function of laver	[155]
		P. yezoensis	- Seasonal variation and the effects of the yellow spot disease outbreaks to the microbial community in the seawater of laver seedling pools - Identification of disease-associated bacteria	[174]
	Analytical techniques	*P. umbilicalis*	- Microbial communities affected by the sampling position of laver and the stabilization techniques applied for the microbiome analysis	[171]
	Influencing factor (Red rot disease)	*P. yezoensis*	- Alterations of bacterial community by red dot disease - Close association between health status of algal host (uninfected or infected) and bacterial community	[190]
Proteome	Mechanism of stress-tolerance	*P. haitanensis*	- Investigation on the key metabolisms elucidating the mechanisms of resistance to high-temperature	[175,176]
		P. orbicularis *P. haitanensis*	- Investigation on the key metabolisms elucidating the mechanisms of resistance to desiccation	[177,178]
	Mechanism of infection resistance	*P. yezoensis*	- Investigation on the pathogen-responsive proteins elucidating the mechanisms of responses against the infection	[179]
	Identification of key functional protein	*P. yezoensis*	- Identification of major protein [*Pyropia yezoensis* aldehyde dehydognease (PyALDH)] which contributes to the resistance of laver against oxidative stress	[180]
	Mutation of laver strain	*P. yezoensis*	- Induction of high-growth-rate mutation by the exposure to ethyl methane sulfonate - Comparative analysis for the proteome of mutated strain with wild-type strain with the perspective to the enhanced growth	[181]
		P. yezoensis	- Induction of thermo-tolerance mutation by the exposure to gamma-irradiation - Isolation of protein from thermo-tolerant mutant which contributes to the resistance against elevated temperature	[182,191]

Table 8. *Cont.*

Omics Technology	Topic	Species	Major Findings	References
Lipidome	Lipidomic variations	P. haitanensis	- Identification of lipid biomarkers distinctly expressed under elevated temperatures	[184]
		P. dioica	- Differences in composition of major lipid molecular species according to the life cycle stages between the blade and conchocelis	[185]
Metabolome	Metabolomic variations	P. haitanensis	- Changes in the nutrient composition according to the harvest time	[186]
		P. yezoensis	- Changes in the nutrient composition which can determine the taste of laver by the food processing steps not only for seasoning but also washing, cutting, and roasting	[187]
	Metabolite profile	P. pseudolinearis	- Distinctive characteristics of metabolites among species of edible seaweeds (brown, red, and green algae) and sorbitol as the major sugar metabolite in laver	[188]

5. Conclusions

The comprehensive information regarding advances in research reporting health functionality and quality control of laver analyzed in this study demonstrates the need for consistent research studies specializing on laver. Major findings from the analysis of literatures are as follows: (1) nutritional/functional values can be variable according to the type of laver product (raw laver and processed laver products), (2) potential health functionalities linked to the unique substances in laver have been demonstrated by in vitro and in vivo research whereas long-term outcomes of laver consumption on human health should be further examined, (3) intervention methods for chemical/microbial risks should be improved for wider application to manufacturers' use, (4) omics-technologies have revealed the clues for understanding biological nature linked to product quality of laver. Although previous research studies reported the nutritional/functional values of laver products (raw laver, processed laver products) as useful edible seaweeds, most of those studies highlighted that the accumulation of laver-specific data should be accelerated for the in-depth understanding of the biological nature of laver. The current and future issues described in this study regarding the usefulness of laver are expected to contribute to the balanced progress on both the utilization of lavers as edible seaweeds and the source of the health functionality components.

Author Contributions: Contributions for each author are as follows: conceptualization, T.J.C and M.S.R.; investigation, T.J.C. and M.S.R.; writing—original draft preparation, T.J.C.; writing—review and editing, M.S.R.; supervision, M.S.R. All authors have read and agreed to the published version of the manuscript.

Funding: This research was funded by the Ministry of Food and Drug Safety, grant number 17162MFDS034. This research was also supported by a Korea University grant.

Acknowledgments: The authors thank the School of Life Science and Biotechnology of Korea University for BK 21 PLUS and the Institute of Biomedical Science and Food Safety, Korea University Food Safety Hall, for access to the equipment and facilities.

Conflicts of Interest: The authors declare no conflict of interest. The funders had no role in the design of the study; in the collection, analyses, or interpretation of data; in the writing of the manuscript; or in the decision to publish the results.

References

1. Sutherland, J.E.; Lindstrom, S.C.; Nelson, W.A.; Brodie, J.; Lynch, M.D.; Hwang, M.S.; Choi, H.G.; Miyata, M.; Kikuchi, N.; Oliveira, M.C. A new look at an ancient order: Generic revision of the *Bangiales* (*Rhodophyta*). *J. Phycol.* **2011**, *47*, 1131–1151. [CrossRef]

2. Stoyneva-Gärtner, M.P.; Uzunov, B.A. An ethnobiological glance on globalization impact on the traditional use of algae and fungi as food in Bulgaria. *J. Nutr. Food Sci.* **2015**, *5*, 1.

3. Pereira, L. A review of the nutrient composition of selected edible seaweeds. *Seaweed Ecol. Nutr. Compos. Med. Uses* **2011**, *7*, 15–47.

4. Wells, M.L.; Potin, P.; Craigie, J.S.; Raven, J.A.; Merchant, S.S.; Helliwell, K.E.; Smith, A.G.; Camire, M.E.; Brawley, S.H. Algae as nutritional and functional food sources: Revisiting our understanding. *J. Appl. Phycol.* **2017**, *29*, 949–982. [CrossRef]

5. FAO. Fishery and Aquaculture Statistics. Global Aquaculture Production. In *FAO Fisheries and Aquaculture Department*; FAO: Rome, Italy, 2019; Available online: http://www.fao.org/fishery/ (accessed on 18 September 2019).

6. Levine, I.A.; Sahoo, D. *Porphyra: Harvesting Gold from the Sea*; IK International Pvt Ltd.: New Delhi, India, 2009.

7. Guiry, M.D.; Guiry, G.M. *Algaebase, World-Wide Electronic Publication*; National University of Ireland: Galway, Ireland, 2019.

8. Admassu, H.; Abera, T.; Abraha, B.; Yang, R.; Zhao, W. Proximate, mineral and amino acid composition of dried laver (*Porphyra* spp.) seaweed. *J. Acad. Ind. Res.* **2018**, *6*, 149.

9. Cao, J.; Wang, J.; Wang, S.; Xu, X. *Porphyra* species: A mini-review of its pharmacological and nutritional properties. *J. Med. Food* **2016**, *19*, 111–119. [CrossRef]

10. Bito, T.; Teng, F.; Watanabe, F. Bioactive compounds of edible purple laver *Porphyra* sp.(Nori). *J. Agric. Food Chem.* **2017**, *65*, 10685–10692. [CrossRef] [PubMed]

11. Holdt, S.L.; Kraan, S. Bioactive compounds in seaweed: Functional food applications and legislation. *J. Appl. Phycol.* **2011**, *23*, 543–597. [CrossRef]

12. Cornish, M.L.; Garbary, D.J. Antioxidants from macroalgae: Potential applications in human health and nutrition. *Algae* **2010**, *25*, 155–171. [CrossRef]

13. Kasimala, M.B.; Mebrahtu, L.; Magoha, P.P.; Asgedom, G.; Kasimala, M.B. A review on biochemical composition and nutritional aspects of seaweeds. *Carib J. Scitech* **2015**, *3*, 789–797.

14. MacArtain, P.; Gill, C.I.; Brooks, M.; Campbell, R.; Rowland, I.R. Nutritional value of edible seaweeds. *Nutr. Rev.* **2007**, *65*, 535–543. [CrossRef] [PubMed]

15. Abowei, J.F.N.; Ezekiel, E.N. The potentials and utilization of Seaweeds. *Scientia* **2013**, *4*, 58–66.

16. Jesmi, D.; Viji, P.; Rao, B.M. Seaweeds: A promising functional food ingredient. In *Training Manual on Value Addition of Seafood*; Visakhapatnam Research Centre of ICAR-Central Institute of Fisheries Technology: Visakhapatnam, India, 2018.

17. Rajapakse, N.; Kim, S.K. Nutritional and digestive health benefits of seaweed. *Adv. Food Nutr. Res.* **2011**, *64*, 17–28. [PubMed]

18. Madhusudan, C.; Manoj, S.; Rahul, K.; Rishi, C.M. Seaweeds: A diet with nutritional, medicinal and industrial value. *Res. J. Med. Plant* **2011**, *5*, 153–157. [CrossRef]

19. Mok, J.S.; Lee, T.S.; Son, K.T.; Song, K.C.; Kwon, J.Y.; Lee, K.J.; Kim, J.H. Proximate composition and mineral content of laver *Porphyra yezoensis* from the Korean coast. *Korean J. Fish Aquat. Sci.* **2011**, *44*, 554–559. [CrossRef]

20. Patarra, R.F.; Paiva, L.; Neto, A.I.; Lima, E.; Baptista, J. Nutritional value of selected macroalgae. *J. Appl. Phycol.* **2011**, *23*, 205–208. [CrossRef]

21. Kavale, M.G.; Kazi, M.A.; Bagal, P.U.; Singh, V.; Behera, D.P. Food value of *Pyropia vietnamensis* (*Bangiales, Rhodophyta*) from India. *Indian J. Geo Mar. Sci.* **2018**, *47*, 402–408.

22. Cian, R.E.; Fajardo, M.A.; Alaiz, M.; Vioque, J.; González, R.J.; Drago, S.R. Chemical composition, nutritional and antioxidant properties of the red edible seaweed *Porphyra columbina*. *Int. J. Food Sci. Nutr.* **2014**, *65*, 299–305. [CrossRef]

23. Taboada, M.C.; Millán, R.; Miguez, M.I. Nutritional value of the marine algae wakame (*Undaria pinnatifida*) and nori (*Porphyra purpurea*) as food supplements. *J. Appl. Phycol.* **2013**, *25*, 1271–1276. [CrossRef]

24. Taboada, C.; Millan, R.; Miguez, I. Evaluation of marine algae *Undaria pinnatifida* and *Porphyra purpurea* as a food supplement: Composition, nutritional value and effect of intake on intestinal, hepatic and renal enzyme activities in rats. *J. Sci. Food Agric.* **2013**, *93*, 1863–1868. [CrossRef]

25. Jung, S.M.; Kang, S.G.; Kim, K.T.; Lee, H.J.; Kim, A.; Shin, H.W. The analysis of proximate composition, minerals and amino acid content of red alga *Pyropia dentata* by cultivation sites. *Korean J. Environ. Ecol.* **2015**, *29*, 1–6. [CrossRef]

26. Kavale, M.G.; Kazi, M.A.; Sreenadhan, N.; Murgan, P. Nutritional profiling of *Pyropia acanthophora* var. robusta (*Bangiales, Rhodophyta*) from Indian waters. *J. Appl. Phycol.* **2017**, *29*, 2013–2020. [CrossRef]

27. Jung, S.M.; Kang, S.G.; Son, J.S.; Jeon, J.H.; Lee, H.J.; Shin, H.W. Temporal and spatial variations in the proximate composition, amino acid, and mineral content of *Pyropia yezoensis*. *J. Appl. Phycol.* **2016**, *28*, 3459–3467. [CrossRef]

28. CODEX. Regional standard for laver products (CXS 323R-2017). In *Codex Alimentarius International Food Standards*; CODEX: Montreal, QC, Canada, 2017.

29. Sanjeewa, K.A.; Lee, W.; Jeon, Y.J. Nutrients and bioactive potentials of edible green and red seaweed in Korea. *Fish Aquat. Sci.* **2018**, *21*, 19. [CrossRef]

30. Oh, S.; Kim, J.; Kim, H.; Son, S.; Choe, E. Composition and antioxidant activity of dried laver, Dolgim. *Korean J. Food Sci. Technol.* **2013**, *45*, 403–408. [CrossRef]

31. Kim, K.W.; Hwang, J.H.; Oh, M.J.; Kim, M.Y.; Choi, M.R.; Park, W.M. Studies on the major nutritional components of commercial dried lavers (*Porphyra yezoensis*) cultivated in Korea. *Korean J. Food Preserv.* **2014**, *21*, 702–709. [CrossRef]

32. Hwang, E.S.; Ki, K.N.; Chung, H.Y. Proximate composition, amino acid, mineral, and heavy metal content of dried laver. *Prev. Nutr. Food Sci.* **2013**, *18*, 139. [CrossRef]

33. Lee, J.A. Comparative study on the physicochemical character of commercial dried laver in Korea. *Culi. Sci. Hos. Res.* **2018**, *24*, 92–99.

34. Park, W.M.; Kang, D.S.; Bae, T.J. Studies on organic acid, vitamin and free sugar contents of commercial dried lavers (*Porphyra yezoensis*) cultivated in Korea. *J. Korean Soc. Food Sci. Nutr.* **2014**, *43*, 172–177. [CrossRef]

35. Park, W.M.; Kim, K.W.; Kang, D.S.; Bae, T.J. Studies on anion, element, chromaticity and antioxidant activities of commercial dried lavers (*Porphyra yezoensis*) cultivated in Korea. *J. Korean Soc. Food Sci. Nutr.* **2014**, *43*, 323–327. [CrossRef]

36. Uribe, E.; Vega-Gálvez, A.; Heredia, V.; Pastén, A.; Di Scala, K. An edible red seaweed (*Pyropia orbicularis*): Influence of vacuum drying on physicochemical composition, bioactive compounds, antioxidant capacity, and pigments. *J. Appl. Phycol.* **2018**, *30*, 673–683. [CrossRef]

37. Yamada; Yamada; Fukuda; Yamada. Bioavailability of dried asakusanori (*Porphyra tenera*) as a source of cobalamin (vitamin B_{12}). *Int. J. Vitam. Nutr. Res.* **1999**, *69*, 412–418. [CrossRef] [PubMed]

38. Takenaka, S.; Sugiyama, S.; Ebara, S.; Miyamoto, E.; Abe, K.; Tamura, Y.; Watanabe, F.; Tsuyama, S.; Nakano, Y. Feeding dried purple laver (nori) to vitamin B_{12}-deficient rats significantly improves vitamin B_{12} status. *Br. J. Nutr.* **2001**, *85*, 699–703. [CrossRef] [PubMed]

39. Han, J.S.; Lee, Y.J.; Yoon, M.R. Changes of chromaticity and mineral contents of laver dishes using various cooking methods. *J. East Asian Soc. Diet. Life* **2003**, *13*, 326.

40. Hwang, E.S. Composition of amino acids, minerals, and heavy metals in differently cooked laver (*Porphyra tenera*). *J. Korean Soc. Food Sci. Nutr.* **2013**, *42*, 1270–1276. [CrossRef]

41. Miyamoto, E.; Yabuta, Y.; Kwak, C.S.; Enomoto, T.; Watanabe, F. Characterization of vitamin B_{12} compounds from Korean purple laver (*Porphyra* sp.) products. *J. Agric. Food Chem.* **2009**, *57*, 2793–2796. [CrossRef]

42. Son, K.T.; Lach, T.; Jung, Y.; Kang, S.K.; Eom, S.H.; Lee, D.S.; Lee, M.S.; Kim, Y.M. Food hazard analysis during dried-laver processing. *Fish Aquat. Sci.* **2014**, *17*, 197–201. [CrossRef]

43. Gupta, S.; Abu-Ghannam, N. Bioactive potential and possible health effects of edible brown seaweeds. *Trends Food Sci. Technol.* **2011**, *22*, 315–326. [CrossRef]

44. Rasala, B.A.; Mayfield, S.P. Photosynthetic biomanufacturing in green algae; production of recombinant proteins for industrial, nutritional, and medical uses. *Photosyn. Res.* **2015**, *123*, 227–239. [CrossRef]

45. Camacho, F.; Macedo, A.; Malcata, F. Potential industrial applications and commercialization of microalgae in the functional food and feed industries: A short review. *Mar. Drugs* **2019**, *17*, 312. [CrossRef]

46. Kwon, M.J.; Nam, T.J. Porphyran induces apoptosis related signal pathway in AGS gastric cancer cell lines. *Life Sci.* **2006**, *79*, 1956–1962. [CrossRef] [PubMed]

47. Yu, X.; Zhou, C.; Yang, H.; Huang, X.; Ma, H.; Qin, X.; Hu, J. Effect of ultrasonic treatment on the degradation and inhibition cancer cell lines of polysaccharides from *Porphyra yezoensis*. *Carbohydr. Polym.* **2015**, *117*, 650–656. [CrossRef] [PubMed]

48. Noda, H.; Amano, H.; Arashima, K.; Nisizawa, K. Antitumor activity of marine algae. *Hydrobiologia* **1990**, *204*, 577–584. [CrossRef]

49. Yamamoto, I.; Maruyama, H. Effect of dietary seaweed preparations on 1, 2-dimethylhydrazine-induced intestinal carcinogenesis in rats. *Cancer Lett.* **1985**, *26*, 241–251. [CrossRef]

50. Kazłowska, K.; Lin, H.T.V.; Chang, S.; Tsai, G. In vitro and in vivo anticancer effects of sterol fraction from red algae *Porphyra dentata*. *Evid. Based Complement. Alternat. Med.* **2013**, *2013*. [CrossRef]

51. Ichihara, T.; Wanibuchi, H.; Taniyama, T.; Okai, Y.; Yano, Y.; Otani, S.; Imaoka, S.; Funae, Y.; Fukushima, S. Inhibition of liver glutathione S-transferase placental form-positive foci development in the rat hepatocarcinogenesis by *Porphyra tenera* (Asakusa-nori). *Cancer Lett.* **1999**, *141*, 211–218. [CrossRef]

52. Park, S.J.; Ryu, J.; Kim, I.H.; Choi, Y.H.; Nam, T.J. Activation of the mTOR signaling pathway in breast cancer MCF-7 cells by a peptide derived from *Porphyra yezoensis*. *Oncol. Rep.* **2015**, *33*, 19–24. [CrossRef]

53. He, D.; Wu, S.; Yan, L.; Zuo, J.; Cheng, Y.; Wang, H.; Liu, J.; Zhang, X.; Wu, M.; Choi, J.I.; et al. Antitumor bioactivity of porphyran extracted from *Pyropia yezoensis* Chonsoo2 on human cancer cell lines. *J. Sci. Food Agric.* **2019**, *99*, 6722–6730. [CrossRef]

54. Chen, Y.-Y.; Xue, Y.-T. Optimization of microwave assisted extraction, chemical characterization and antitumor activities of polysaccharides from *porphyra haitanensis*. *Carbohydr. Polym.* **2019**, *206*, 179–186. [CrossRef]

55. Abe, S.; Kaneda, T. Occurrence of gamma-butyrobetaine in a red alga, *Porphyra yezoensis*. *Bull. Jpn. Soc. Sci. Fish.* **1973**, *39*, 239. [CrossRef]

56. Noda, H. Health benefits and nutritional properties of nori. *J. Appl. Phycol.* **1993**, *5*, 255–258. [CrossRef]

57. Tsuge, K.; Okabe, M.; Yoshimura, T.; Sumi, T.; Tachibana, H.; Yamada, K. Dietary effects of porphyran from *Porphyra yezoensis* on growth and lipid metabolism of Sprague-Dawley rats. *Food Sci. Technol. Res.* **2007**, *10*, 147–151. [CrossRef]

58. Inoue, N.; Yamano, N.; Sakata, K.; Nagao, K.; Hama, Y.; Yanagita, T. The sulfated polysaccharide porphyran reduces apolipoprotein B100 secretion and lipid synthesis in HepG2 cells. *Biosci. Biotechnol. Biochem.* **2009**, *73*, 447–449. [CrossRef] [PubMed]

59. Jiménez-Escrig, A.; Sánchez-Muniz, F. Dietary fibre from edible seaweeds: Chemical structure, physicochemical properties and effects on cholesterol metabolism. *Nutr. Res.* **2000**, *20*, 585–598. [CrossRef]

60. Wijesekara, I.; Pangestuti, R.; Kim, S.K. Biological activities and potential health benefits of sulfated polysaccharides derived from marine algae. *Carbohydr. Polym.* **2011**, *84*, 14–21. [CrossRef]

61. Yamanaka, Y.; TsujiI, K.; IchikahaE, T.; Nakaguwa, A.; Kawamura, M. Effect of dietary taurine on cholesterol gallstone formation and tissue cholesterol contents in mice. *J. Nutr. Sci. Vitaminol.* **1985**, *31*, 225–232. [CrossRef]

62. Zhang, Z.; Zhang, Q.; Wang, J.; Zhang, H.; Niu, X.; Li, P. Preparation of the different derivatives of the low-molecular-weight porphyran from *Porphyra haitanensis* and their antioxidant activities *in vitro*. *Int. J. Biol. Macromol.* **2009**, *45*, 22–26. [CrossRef]

63. Zhao, T.; Zhang, Q.; Qi, H.; Zhang, H.; Niu, X.; Xu, Z.; Li, Z. Degradation of porphyran from *Porphyra haitanensis* and the antioxidant activities of the degraded porphyrans with different molecular weight. *Int. J. Biol. Macromol.* **2006**, *38*, 45–50. [CrossRef]

64. Shin, E.S.; Hwang, H.J.; Kim, I.H.; Nam, T.J. A glycoprotein from *Porphyra yezoensis* produces anti-inflammatory effects in liposaccharide-stimulated macrophages via the TLR4 signaling pathway. *Int. J. Mol. Med.* **2011**, *28*, 809–815.

65. Choe, E.; Oh, S. Effects of water activity on the lipid oxidation and antioxidants of dried laver (*Porphyra*) during storage in the dark. *J. Food Sci.* **2013**, *78*, C1144–C1151. [CrossRef]

66. Gong, G.; Zhao, J.; Wang, C.; Wei, M.; Dang, T.; Deng, Y.; Sun, J.; Song, S.; Huang, L.; Wang, Z. Structural characterization and antioxidant activities of the degradation products from *Porphyra haitanensis* polysaccharides. *Process Biochem.* **2018**, *74*, 185–193. [CrossRef]

67. Cermeno, M.; Stack, J.; Tobin, P.; O'Keeffe, M.B.; Harnedy, P.A.; Stengel, D.; FitzGerald, R.J. Peptide identification from a *Porphyra dioica* protein hydrolysate with antioxidant, angiotensin converting enzyme and dipeptidyl peptidase IV inhibitory activities. *Food Funct.* **2019**, *10*, 3421–3429. [CrossRef]

68. Bhatia, S.; Rathee, P.; Sharma, K.; Chaugule, B.; Kar, N.; Bera, T. Immuno-modulation effect of sulphated polysaccharide (porphyran) from *Porphyra vietnamensis*. *Int. J. Biol. Macromol.* **2013**, *57*, 50–56. [CrossRef] [PubMed]

69. Liu, Q.-M.; Xu, S.-S.; Li, L.; Pan, T.-M.; Shi, C.-L.; Liu, H.; Cao, M.-J.; Su, W.-J.; Liu, G.-M. In vitro and in vivo immunomodulatory activity of sulfated polysaccharide from *Porphyra haitanensis*. *Carbohydr. Polym.* **2017**, *165*, 189–196. [CrossRef] [PubMed]

70. Lee, H.A.; Kim, I.H.; Nam, T.J. Bioactive peptide from *Pyropia yezoensis* and its anti-inflammatory activities. *Int. J. Mol. Med.* **2015**, *36*, 1701–1706. [CrossRef] [PubMed]

71. Choi, J.W.; Kwon, M.J.; Kim, I.H.; Kim, Y.M.; Lee, M.K.; Nam, T.J. *Pyropia yezoensis* glycoprotein promotes the M1 to M2 macrophage phenotypic switch via the STAT3 and STAT6 transcription factors. *Int. J. Mol. Med.* **2016**, *38*, 666–674. [CrossRef] [PubMed]

72. Yanagido, A.; Ueno, M.; Jiang, Z.; Cho, K.; Yamaguchi, K.; Kim, D.; Oda, T. Increase in anti-inflammatory activities of radical-degraded porphyrans isolated from discolored nori (*Pyropia yezoensis*). *Int. J. Biol. Macromol.* **2018**, *117*, 78–86. [CrossRef]

73. Wang, Y.; Hwang, J.Y.; Park, H.B.; Yadav, D.; Oda, T.; Jin, J.O. Porphyran isolated from *Pyropia yezoensis* inhibits lipopolysaccharide-induced activation of dendritic cells in mice. *Carbohydr. Polym.* **2020**, *229*, 115457. [CrossRef]

74. Choi, J.W.; Kim, Y.M.; Park, S.J.; Kim, I.H.; Nam, T.J. Protective effect of *Porphyra yezoensis* glycoprotein on D-galactosamine-induced cytotoxicity in Hepa 1c1c7 cells. *Mol. Med. Rep.* **2015**, *11*, 3914–3919. [CrossRef]

75. Choi, J.W.; Kim, I.H.; Kim, Y.M.; Lee, M.K.; Choi, Y.H.; Nam, T.J. Protective effect of *Pyropia yezoensis* glycoprotein on chronic ethanol consumption-induced hepatotoxicity in rats. *Mol. Med. Rep.* **2016**, *14*, 4881–4886. [CrossRef]

76. Mohibbullah, M.; Bhuiyan, M.M.H.; Hannan, M.A.; Getachew, P.; Hong, Y.K.; Choi, J.S.; Choi, I.S.; Moon, I.S. The edible red alga *Porphyra yezoensis* promotes neuronal survival and cytoarchitecture in primary hippocampal neurons. *Cell. Mol. Neurobiol.* **2016**, *36*, 669–682. [CrossRef] [PubMed]

77. Liu, Y.; Deng, Z.; Geng, L.; Wang, J.; Zhang, Q. In vitro evaluation of the neuroprotective effect of oligo-porphyran from *Porphyra yezoensis* in PC12 cells. *J. Appl. Phycol.* **2019**, *31*, 2559–2571. [CrossRef]

78. Ueno, M.; Cho, K.; Isaka, S.; Nishiguchi, T.; Yamaguchi, K.; Kim, D.; Oda, T. Inhibitory effect of sulphated polysaccharide porphyran (isolated from *Porphyra yezoensis*) on RANKL-induced differentiation of RAW264. 7 cells into osteoclasts. *Phytother. Res.* **2018**, *32*, 452–458. [CrossRef]

79. Kitano, Y.; Murazumi, K.; Duan, J.; Kurose, K.; Kobayashi, S.; Sugawara, T.; Hirata, T. Effect of dietary porphyran from the red alga, *Porphyra yezoensis*, on glucose metabolism in diabetic KK-Ay mice. *J. Nutr. Sci. Vitaminol.* **2012**, *58*, 14–19. [CrossRef]

80. Chao, P.C.; Hsu, C.C.; Liu, W.H. Renal protective effects of *Porphyra dentate* aqueous extract in diabetic mice. *Biomedicine (Taipei)* **2014**, *4*, 18. [CrossRef]

81. Cao, C.; Chen, M.; Liang, B.; Xu, J.; Ye, T.; Xia, Z. Hypoglycemic effect of abandoned *Porphyra haitanensis* polysaccharides in alloxan-induced diabetic mice. *Bioact. Carbohydr. Diet. Fibre* **2016**, *8*, 1–6. [CrossRef]

82. Admassu, H.; Gasmalla, M.A.; Yang, R.; Zhao, W. Identification of bioactive peptides with α-amylase inhibitory potential from enzymatic protein hydrolysates of red seaweed (*Porphyra* spp). *J. Agric. Food Chem.* **2018**, *66*, 4872–4882. [CrossRef]

83. Usov, A.I. Polysaccharides of the red algae. In *Advances in Carbohydrate Chemistry and Biochemistry*; Elsevier: Amsterdam, The Netherlands, 2011; Volume 65, pp. 115–217.

84. Ren, D.; Noda, H.; Amano, H.; Nishino, T.; Nishizawa, K. Study on antihypertensive and antihyperlipidemic effects of marine algae. *Fish Sci.* **1994**, *60*, 83–88. [CrossRef]

85. Dawczynski, C.; Schubert, R.; Jahreis, G. Amino acids, fatty acids, and dietary fibre in edible seaweed products. *Food Chem.* **2007**, *103*, 891–899. [CrossRef]

86. Watanabe, F.; Yabuta, Y.; Bito, T.; Teng, F. Vitamin B_{12}-containing plant food sources for vegetarians. *Nutrients* **2014**, *6*, 1861–1873. [CrossRef]

87. Škrovánková, S. Seaweed vitamins as nutraceuticals. *Adv. Food Nutr. Res.* **2011**, *64*, 357–369. [PubMed]

88. Rauma, A.L.; Törrönen, R.; Hänninen, O.; Mykkänen, H. Vitamin B-$_{12}$ status of long-term adherents of a strict uncooked vegan diet ("living food diet") is compromised. *J. Nutr.* **1995**, *125*, 2511–2515. [PubMed]

89. van den Berg, H.; Brandsen, L.; Sinkeldam, B.J. Vitamin B-$_{12}$ content and bioavailability of spirulina and nori in rats. *J. Nutr. Biochem.* **1991**, *2*, 314–318. [CrossRef]

90. Kawasaki, A.; Ono, A.; Mizuta, S.; Kamiya, M.; Takenaga, T.; Murakami, S. The taurine content of Japanese seaweed. In *Taurine 10*; Springer: New York, NY, USA, 2017; pp. 1105–1112.

91. Kwak, E.S. Combined Effect of Citric Acid and Drying for the Improvement of Microbiological Quality in Dried Laver. Master's Thesis, Korea university, Seoul, Korea, 2016.

92. Kang, S.J.; Yang, S.Y.; Lee, J.W.; Lee, K.W. Polycyclic aromatic hydrocarbons in seasoned-roasted laver and their reduction according to the mixing ratio of seasoning oil and heat treatment in a model system. *Food Sci. Biotechnol.* **2019**, *28*, 1247–1255. [CrossRef] [PubMed]

93. Hwang, Y.I.; Kim, J.G.; Kwon, S.C. A study on physical risk and chemical risk analaysis of seasoned laver. *J. Korea Acad. Industr. Coop. Soc.* **2017**, *18*, 620–626.

94. Bouga, M.; Combet, E. Emergence of seaweed and seaweed-containing foods in the UK: Focus on labeling, iodine content, toxicity and nutrition. *Foods* **2015**, *4*, 240–253. [CrossRef]

95. Yeh, T.S.; Hung, N.H.; Lin, T.C. Analysis of iodine content in seaweed by GC-ECD and estimation of iodine intake. *J. Food Drug Anal.* **2014**, *22*, 189–196. [CrossRef]

96. Jeon, M.J.; Kim, W.G.; Kwon, H.; Kim, M.; Park, S.; Oh, H.-S.; Han, M.; Kim, T.Y.; Shong, Y.K.; Kim, W.B. Excessive iodine intake and thyrotropin reference interval: Data from the Korean National Health and Nutrition Examination Survey. *Thyroid* **2017**, *27*, 967–972. [CrossRef]

97. Bulow Pedersen, I.; Knudsen, N.; Jørgensen, T.; Perrild, H.; Ovesen, L.; Laurberg, P. Large differences in incidences of overt hyper-and hypothyroidism associated with a small difference in iodine intake: A prospective comparative register-based population survey. *J. Clin. Endocrinol.* **2002**, *87*, 4462–4469. [CrossRef]

98. Li, Y.; Teng, D.; Shan, Z.; Teng, X.; Guan, H.; Yu, X.; Fan, C.; Chong, W.; Yang, F.; Dai, H. Antithyroperoxidase and antithyroglobulin antibodies in a five-year follow-up survey of populations with different iodine intakes. *J. Clin. Endocrinol.* **2008**, *93*, 1751–1757. [CrossRef]

99. Gao, Y.; Yue, J. Dietary fiber and human health. In *Cereals and Pulses: Nutraceutical Properties and Health Benefits*; Wiley: Hoboken, NJ, USA, 2012; pp. 261–271.

100. Yang, W.H.; Lee, H.J.; Lee, S.Y.; Kim, S.G.; Kim, G.B. Heavy metal contents and food safety assessment of processed seaweeds and cultured lavers. *J. Korean Soc. Mar.* **2016**, *19*, 203–210.

101. Pérez, A.A.; Farías, S.S.; Strobl, A.M.; Pérez, L.B.; López, C.M.; Piñeiro, A.; Roses, O.; Fajardo, M.A. Levels of essential and toxic elements in *Porphyra columbina* and *Ulva* sp. from San Jorge Gulf, Patagonia Argentina. *Sci. Total Environ.* **2007**, *376*, 51–59.

102. Hwang, Y.O.; Park, S.G.; Park, G.Y.; Choi, S.M.; Kim, M.Y. Total arsenic, mercury, lead, and cadmium contents in edible dried seaweed in Korea. *Food Addit. Contam. B Surveill.* **2010**, *3*, 7–13. [CrossRef] [PubMed]

103. JECFA Database for the Evaluations of Flavours, Food Additives, Contaminants, Toxicants and Veterinary Drugs. Available online: https://apps.who.int/food-additives-contaminants-jecfa-database/search.aspx (accessed on 16 December 2019).

104. Zhan, Z.; Wang, W.; Zhang, G. Risk analysis of aluminium contamination in kelp and laver. *J. Food Saf. Food Qual.* **2016**, *7*, 1330–1334.

105. Zhang, Q.; Ma, J.; Ju, M.; Lu, Q.; Shen, H. Analysis of inorganic arsenic in *Pyropia yezoensis* and determination of total arsenic in aquaculture areas. *Shanghai Hai Yang Da Xue Xue Bao* **2014**, *23*, 518–522.

106. Zhu, Y.; Shao, B.; Ding, H.; Chen, D.; Shi, Y.; Ji, G.; Fan, B. Contaminations of arsenic, aluminum, lead and cadmium in dried streak laver in Nantong area. *J. Food Saf. Food Qual.* **2018**, *9*, 1950–1954.

107. Wei, C.; Li, W.; Zhang, C.; Van Hulle, M.; Cornelis, R.; Zhang, X. Safety evaluation of organoarsenical species in edible *Porphyra* from the China Sea. *J. Agric. Food Chem.* **2003**, *51*, 5176–5182. [CrossRef]

108. Taylor, V.F.; Li, Z.; Sayarath, V.; Palys, T.J.; Morse, K.R.; Scholz-Bright, R.A.; Karagas, M.R. Distinct arsenic metabolites following seaweed consumption in humans. *Sci. Rep.* **2017**, *7*, 3920. [CrossRef]

109. Smith, J.L.; Summers, G.; Wong, R. Nutrient and heavy metal content of edible seaweeds in New Zealand. *N. Z. J. Crop Hortic. Sci.* **2010**, *38*, 19–28. [CrossRef]

110. Khan, N.; Ryu, K.Y.; Choi, J.Y.; Nho, E.Y.; Habte, G.; Choi, H.; Kim, M.H.; Park, K.S.; Kim, K.S. Determination of toxic heavy metals and speciation of arsenic in seaweeds from South Korea. *Food Chem.* **2015**, *169*, 464–470. [CrossRef]

111. Mok, J.S.; Son, K.T.; Lee, T.S.; Lee, K.J.; Jung, Y.J.; Kim, J.H. Removal of hazardous heavy metals (Cd, Cr, and Pb) from laver *Pyropia* sp. with acid treatment. *Korean J. Fish Aquat. Sci.* **2016**, *49*, 556–563. [CrossRef]

112. Oya-Ohta, Y.; Kaise, T.; Ochi, T. Induction of chromosomal aberrations in cultured human fibroblasts by inorganic and organic arsenic compounds and the different roles of glutathione in such induction. *Mutat. Res.* **1996**, *357*, 123–129. [CrossRef]

113. SAC (The Standardization Administration of China), GB 2733-2005. In *Hygienic Standard for Fresh and Frozen Marine Products of Animal Origin*; SAC: Beijing, China, 2005.

114. Harakeh, S.; Yassine, H.; El-Fadel, M. Antimicrobial-resistant patterns of *Escherichia coli* and *Salmonella* strains in the aquatic Lebanese environments. *Environ. Pollut.* **2006**, *143*, 269–277. [CrossRef] [PubMed]

115. Iwamoto, M.; Ayers, T.; Mahon, B.E.; Swerdlow, D.L. Epidemiology of seafood-associated infections in the United States. *Clin. Microbiol. Rev.* **2010**, *23*, 399–411. [CrossRef] [PubMed]

116. Choi, E.S.; Kim, N.H.; Kim, H.W.; Kim, S.A.; Jo, J.I.; Kim, S.H.; Lee, S.H.; Ha, S.D.; Rhee, M.S. Microbiological quality of seasoned roasted laver and potential hazard control in a real processing line. *J. Food Prot.* **2014**, *77*, 2069–2075. [CrossRef] [PubMed]

117. Röhr, A.; Lüddecke, K.; Drusch, S.; Müller, M.J.; Alvensleben, R.V. Food quality and safety—consumer perception and public health concern. *Food Control* **2005**, *16*, 649–655. [CrossRef]

118. Wang, F.; Zhang, J.; Mu, W.; Fu, Z.; Zhang, X. Consumers' perception toward quality and safety of fishery products, Beijing, China. *Food Control* **2009**, *20*, 918–922. [CrossRef]

119. Kim, N.H.; Yun, A.R.; Rhee, M.S. Prevalence and classification of toxigenic *Staphylococcus aureus* isolated from refrigerated ready-to-eat foods (sushi, kimbab and California rolls) in Korea. *J. Appl. Microbiol.* **2011**, *111*, 1456–1464. [CrossRef]

120. Chung, H.J.; Lee, N.Y.; Jo, C.; Shin, D.H.; Byun, M.W. Use of gamma irradiation for inactivation of pathogens inoculated into Kimbab, steamed rice rolled by dried laver. *Food Control* **2007**, *18*, 108–112. [CrossRef]

121. Ahn, H.J.; Yook, H.S.; Kim, D.H.; Kim, S.; Byun, M.W. Identification of radiation-resistant bacterium isolated from dried laver (*Porphyra tenera*). *J. Korean Soc. Food Sci. Nutr.* **2001**, *30*, 193–195.

122. Costello, M.; Rhee, M.S.; Bates, M.P.; Clark, S.; Luedecke, L.O.; Kang, D.H. Eleven-year trends of microbiological quality in bulk tank milk. *Food Prot. Trends* **2003**, *23*, 393–400.

123. Lee, T.S.; Lee, H.J.; Byun, H.S.; Kim, J.H.; Park, M.J.; Park, H.Y.; Jung, K.J. Effect of heat treatment in dried lavers and modified processing. *Korean J. Fish Aquat. Sci.* **2000**, *33*, 529–532.

124. Choi, Y.M.; Park, H.J.; Jang, H.I.; Kim, S.A.; Imm, J.Y.; Hwang, I.G.; Rhee, M.S. Changes in microbial contamination levels of porcine carcasses and fresh pork in slaughterhouses, processing lines, retail outlets, and local markets by commercial distribution. *Res. Vet. Sci.* **2013**, *94*, 413–418. [CrossRef] [PubMed]

125. Noh, B.Y.; Hwang, S.H.; Cho, Y.S. Microbial contamination levels in *Porphyra* sp. distributed in Korea. *Korean J. Fish Aquat. Sci.* **2019**, *52*, 180–184.

126. Lee, E.J.; Kim, G.R.; Lee, H.J.; Kwon, J.H. Monitoring microbiological contamination, pre-decontamination, and irradiation status of commercial dried laver (*Porphyra* sp.) products. *Korean J. Food Sci. Technol.* **2017**, *49*, 20–27. [CrossRef]

127. Lee, N.Y.; Jo, C.H.; Chung, H.J.; Kang, H.J.; Kim, J.K.; Kim, H.J.; Byun, M.W. The prediction of the origin of microbial contamination in Kimbab and improvement of microbiological safety by gamma irradiation. *Korean J. Food Sci. Technol.* **2005**, *37*, 279–286.

128. Jang, H.G.; Kim, N.H.; Choi, Y.M.; Rhee, M.S. Microbiological quality and risk factors related to sandwiches served in bakeries, cafés, and sandwich bars in South Korea. *J. Food Prot.* **2013**, *76*, 231–238. [CrossRef]

129. Rhee, M.S.; Kang, D.H. Rapid and simple estimation of microbiological quality of raw milk using chromogenic Limulus amoebocyte lysate endpoint assay. *J. Food Prot.* **2002**, *65*, 1447–1451. [CrossRef]

130. Kim, D.H.; Song, H.P.; Kim, J.K.; Byun, M.W.; Kim, J.O.; Lee, H.J. Determination of microbial contamination in the process of rice roled in dried laver and improvement of shelf-life by gamma irradiation. *J. Korean Soc. Food Sci. Nutr.* **2003**, *32*, 991–996.

131. Kim, J.S. Microflora Analysis and Pathogen Control of Agricultural Products and Ready-to-Eat Foods. Master's Thesis, Kyungwon University, Busan, Korea, 2008.

132. Kang, M.J.; Lee, H.T.; Kim, J.Y. Hazard analysis, determination of critical control points, and establishment of critical limits for seasoned laver. *Culi. Sci. Hos. Res.* **2015**, *21*, 1–10.

133. Kim, K.Y.; Yoon, S.Y. A study on microbiological risk assessment for the HACCP system construction of seasoned laver. *Korean J. Environ. Health Sci.* **2013**, *39*, 268–278. [CrossRef]

134. Park, S.Y.; Song, H.H.; Ha, S.D. Synergistic effects of NaOCl and ultrasound combination on the reduction of *Escherichia coli* and *Bacillus cereus* in raw laver. *Foodborne Pathog. Dis.* **2014**, *11*, 373–378. [CrossRef] [PubMed]

135. Puligundla, P.; Kim, J.W.; Mok, C. Effect of low-pressure air plasma on the microbial load and physicochemical characteristics of dried laver. *LWT Food Sci. Technol.* **2015**, *63*, 966–971. [CrossRef]

136. Kim, Y.J.; Oh, H.S.; Kim, M.J.; Kim, J.H.; Goh, J.B.; Choi, I.Y.; Park, M.K. Identification of electron beam-resistant bacteria in the microbial reduction of dried laver (*Porphyra tenera*) subjected to electron beam treatment. *Korean J. Food Preserv.* **2016**, *23*, 139–143. [CrossRef]

137. Lee, E.J.; Ameer, K.; Kim, G.R.; Chung, M.S.; Kwon, J.H. Effects of approved dose of e-beam irradiation on microbiological and physicochemical qualities of dried laver products and detection of their irradiation status. *Food Sci. Biotechnol.* **2018**, *27*, 233–240. [CrossRef]

138. Lee, E.J.; Kim, G.R.; Ameer, K.; Kyung, H.K.; Kwon, J.H. Application of electron beam irradiation for improving the microbial quality of processed laver products and luminescence detection of irradiated lavers. *Appl. Biol. Chem.* **2018**, *61*, 79–89. [CrossRef]

139. Lee, E.J.; Ameer, K.; Jo, Y.; Kim, G.R.; Kyung, H.K.; Kwon, J.H. Effects of heat-assisted irradiation treatment on microbial and physicochemical qualities of dried laver (*Porphyra* spp.) and optimization by response surface methodology. *Aquac. Res.* **2019**, *50*, 464–473. [CrossRef]

140. Zichightech. Apparatus for Sterilization of Watery. Laver. Patent 1,020,180,018,019, 21 February 2018.

141. Shirako, K.K. Apparatus for Producing Dried Laver and Dried. Laver. Patent 2,012,034,664, 23 February 2012.

142. Gachon University. Food Surface Sterilization Method Using Non-Thermal. Plasma. Patent 1,020,140,002,357, 8 January 2014.

143. Koasa Corporation. Method for Producing Roasted. Nori. Patent WO/2017/077738 (PCT/JP2016/070306), 11 May 2017.

144. Reith, M. Complete uncleotide sequence of the *Porphyra purpurea* chloroplast genome. *Plant Mol. Biol. Rep.* **1995**, *13*, 327–332. [CrossRef]

145. Burger, G.; Saint-Louis, D.; Gray, M.W.; Lang, B.F. Complete sequence of the mitochondrial DNA of the red alga *Porphyra purpurea*: Cyanobacterial introns and shared ancestry of red and green algae. *Plant Cell* **1999**, *11*, 1675–1694. [CrossRef]
146. Hughey, J.R.; Gabrielson, P.W.; Rohmer, L.; Tortolani, J.; Silva, M.; Miller, K.A.; Young, J.D.; Martell, C.; Ruediger, E. Minimally destructive sampling of type specimens of *Pyropia* (*Bangiales, Rhodophyta*) recovers complete plastid and mitochondrial genomes. *Sci. Rep.* **2014**, *4*, 5113. [CrossRef]
147. Brawley, S.H.; Blouin, N.A.; Ficko-Blean, E.; Wheeler, G.L.; Lohr, M.; Goodson, H.V.; Jenkins, J.W.; Blaby-Haas, C.E.; Helliwell, K.E.; Chan, C.X. Insights into the red algae and eukaryotic evolution from the genome of *Porphyra umbilicalis* (*Bangiophyceae, Rhodophyta*). *Proc. Natl. Acad. Sci. USA* **2017**, *114*, E6361–E6370. [CrossRef]
148. Harden, L.K.; Morales, K.M.; Hughey, J.R. Identification of a new marine algal species *Pyropia nitida* sp. nov.(*Bangiales: Rhodophyta*) from Monterey, California. *Mitochondrial DNA A DNA Mapp. Seq. Anal.* **2016**, *27*, 3058–3062. [PubMed]
149. Li, X.; Xu, J.; He, Y.; Shen, S.; Zhu, J.; Shen, Z. The complete nuclear ribosomal DNA (nrDNA) cistron sequence of *Pyropia yezoensis* (*Bangiales, Rhodophyta*). *J. Appl. Phycol.* **2016**, *28*, 663–669. [CrossRef]
150. Mao, Y.; Zhang, B.; Kong, F.; Wang, L. The complete mitochondrial genome of *Pyropia haitanensis* Chang et Zheng. *Mitochondrial DNA A DNA Mapp. Seq. Anal.* **2012**, *23*, 344–346. [CrossRef]
151. Kong, F.; Sun, P.; Cao, M.; Wang, L.; Mao, Y. Complete mitochondrial genome of *Pyropia yezoensis*: Reasserting the revision of genus *Porphyra*. *Mitochondrial DNA A DNA Mapp. Seq. Anal.* **2014**, *25*, 335–336. [CrossRef]
152. Hwang, M.S.; Kim, S.O.; Ha, D.S.; Lee, J.E.; Lee, S.R. Complete mitochondrial genome sequence of *Pyropia yezoensis* (*Bangiales, Rhodophyta*) from Korea. *Plant Biotechnol. Rep.* **2014**, *8*, 221–227. [CrossRef]
153. Wang, L.; Mao, Y.; Kong, F.; Li, G.; Ma, F.; Zhang, B.; Sun, P.; Bi, G.; Zhang, F.; Xue, H. Complete sequence and analysis of plastid genomes of two economically important red algae: *Pyropia haitanensis* and *Pyropia yezoensis*. *PLoS ONE* **2013**, *8*, e65902. [CrossRef]
154. Nakamura, Y.; Sasaki, N.; Kobayashi, M.; Ojima, N.; Yasuike, M.; Shigenobu, Y.; Satomi, M.; Fukuma, Y.; Shiwaku, K.; Tsujimoto, A. The first symbiont-free genome sequence of marine red alga, Susabi-nori (*Pyropia yezoensis*). *PLoS ONE* **2013**, *8*, e57122. [CrossRef]
155. Miranda, L.N.; Hutchison, K.; Grossman, A.R.; Brawley, S.H. Diversity and abundance of the bacterial community of the red macroalga *Porphyra umbilicalis*: Did bacterial farmers produce macroalgae? *PLoS ONE* **2013**, *8*, e58269. [CrossRef]
156. Li, C.; Kong, F.; Sun, P.; Bi, G.; Li, N.; Mao, Y.; Sun, M. Genome-wide identification and expression pattern analysis under abiotic stress of mitogen-activated protein kinase genes in *Pyropia yezoensis*. *J. Appl. Phycol.* **2018**, *30*, 2561–2572. [CrossRef]
157. Li, B.; Chen, C.; Xu, Y.; Ji, D.; Xie, C. Validation of housekeeping genes as internal controls for studying the gene expression in *Pyropia haitanensis* (*Bangiales, Rhodophyta*) by quantitative real-time PCR. *Acta Oceanol. Sin.* **2014**, *33*, 152–159. [CrossRef]
158. Mikami, K.; Li, C.; Irie, R.; Hama, Y. A unique life cycle transition in the red seaweed *Pyropia yezoensis* depends on apospory. *Commun. Biol.* **2019**, *2*, 1–10. [CrossRef] [PubMed]
159. Chan, C.X.; Blouin, N.A.; Zhuang, Y.; Zäuner, S.; Prochnik, S.E.; Lindquist, E.; Lin, S.; Benning, C.; Lohr, M.; Yarish, C. *Porphyra* (*Bangiophyceae*) transcriptomes provide insights into red algal development and metabolism. *J. Phycol.* **2012**, *48*, 1328–1342. [CrossRef] [PubMed]
160. Xie, C.; Li, B.; Xu, Y.; Ji, D.; Chen, C. Characterization of the global transcriptome for *Pyropia haitanensis* (*Bangiales, Rhodophyta*) and development of cSSR markers. *BMC Genom.* **2013**, *14*, 107. [CrossRef] [PubMed]
161. Yanagisawa, R.; Sekine, N.; Mizuta, H.; Uji, T. Transcriptomic analysis under ethylene precursor treatment uncovers the regulation of gene expression linked to sexual reproduction in the dioecious red alga *Pyropia pseudolinearis*. *J. Appl. Phycol.* **2019**, *31*, 3317–3329. [CrossRef]
162. Gao, D.; Kong, F.; Sun, P.; Bi, G.; Mao, Y. Transcriptome-wide identification of optimal reference genes for expression analysis of *Pyropia yezoensis* responses to abiotic stress. *BMC Genom.* **2018**, *19*, 251. [CrossRef] [PubMed]
163. Wang, W.; Chang, J.; Zheng, H.; Ji, D.; Xu, Y.; Chen, C.; Xie, C. Full-length transcriptome sequences obtained by a combination of sequencing platforms applied to heat shock proteins and polyunsaturated fatty acids biosynthesis in *Pyropia haitanensis*. *J. Appl. Phycol.* **2019**, *31*, 1483–1492. [CrossRef]

164. Choi, S.; Hwang, M.S.; Im, S.; Kim, N.; Jeong, W.J.; Park, E.J.; Gong, Y.G.; Choi, D.W. Transcriptome sequencing and comparative analysis of the gametophyte thalli of *Pyropia tenera* under normal and high temperature conditions. *J. Appl. Phycol.* **2013**, *25*, 1237–1246. [CrossRef]

165. Sun, P.; Mao, Y.; Li, G.; Cao, M.; Kong, F.; Wang, L.; Bi, G. Comparative transcriptome profiling of *Pyropia yezoensis* (Ueda) MS Hwang & HG Choi in response to temperature stresses. *BMC Genom.* **2015**, *16*, 463.

166. Wang, W.; Teng, F.; Lin, Y.; Ji, D.; Xu, Y.; Chen, C.; Xie, C. Transcriptomic study to understand thermal adaptation in a high temperature-tolerant strain of *Pyropia haitanensis*. *PLoS ONE* **2018**, *13*, e0195842. [CrossRef]

167. Contreras-Porcia, L.; López-Cristoffanini, C.; Lovazzano, C.; Flores-Molina, M.R.; Thomas, D.; Núñez, A.; Fierro, C.; Guajardo, E.; Correa, J.A.; Kube, M. Expresión diferencial de genes en *Pyropia columbina* (*Bangiales, Rhodophyta*) bajo hidratación y desecación natural. *Lat. Am. J. Aquat. Res.* **2013**, *41*, 933–958. [CrossRef]

168. Im, S.; Jung, H.S.; Lee, H.N.; Park, E.J.; Hwang, M.S.; Jeong, W.J.; Choi, D.W. Transcriptome-based identification of the desiccation response genes in the marine red alga *Pyropia tenera* (*Rhodophyta*) and enhancement of abiotic stress tolerance by PtDRG2 in *chlamydomonas*. *Phycologia* **2017**, *56*, 83.

169. Wang, W.; Xu, Y.; Chen, T.; Xing, L.; Xu, K.; Ji, D.; Chen, C.; Xie, C. Regulatory mechanisms underlying the maintenance of homeostasis in *Pyropia haitanensis* under hypersaline stress conditions. *Sci. Total Environ.* **2019**, *662*, 168–179. [CrossRef] [PubMed]

170. Egan, S.; Kumar, V.; Nappi, J.; Gardiner, M. Microbial diversity and symbiotic interactions with macroalgae. In *Algal and Cyanobacteria Symbioses*; World Scientific: Singapore, 2017; pp. 493–546.

171. Quigley, C.T.; Morrison, H.G.; Mendonça, I.R.; Brawley, S.H. A common garden experiment with *Porphyra umbilicalis* (*Rhodophyta*) evaluates methods to study spatial differences in the macroalgal microbiome. *J. Phycol.* **2018**, *54*, 653–664. [CrossRef] [PubMed]

172. Goecke, F.; Labes, A.; Wiese, J.; Imhoff, J.F. Chemical interactions between marine macroalgae and bacteria. *Mar. Ecol. Prog. Ser.* **2010**, *409*, 267–299. [CrossRef]

173. Harley, C.D.; Anderson, K.M.; Demes, K.W.; Jorve, J.P.; Kordas, R.L.; Coyle, T.A.; Graham, M.H. Effects of climate change on global seaweed communities. *J. Phycol.* **2012**, *48*, 1064–1078. [CrossRef]

174. Guan, X.; Zhou, W.; Hu, C.; Zhu, M.; Ding, Y.; Gai, S.; Zheng, X.; Zhu, J.; Lu, Q. Bacterial community temporal dynamics and disease-related variations in the seawater of *Pyropia* (laver) seedling pools. *J. Appl. Phycol.* **2018**, *30*, 1217–1224. [CrossRef]

175. Xu, Y.; Chen, C.; Ji, D.; Hang, N.; Xie, C. Proteomic profile analysis of *Pyropia haitanensis* in response to high-temperature stress. *J. Appl. Phycol.* **2014**, *26*, 607–618. [CrossRef]

176. Shi, J.; Chen, Y.; Xu, Y.; Ji, D.; Chen, C.; Xie, C. Differential proteomic analysis by iTRAQ reveals the mechanism of *Pyropia haitanensis* responding to high temperature stress. *Sci. Rep.* **2017**, *7*, 44734. [CrossRef]

177. López-Cristoffanini, C.; Zapata, J.; Gaillard, F.; Potin, P.; Correa, J.A.; Contreras-Porcia, L. Identification of proteins involved in desiccation tolerance in the red seaweed *Pyropia orbicularis* (*Rhodophyta, Bangiales*). *Proteomics* **2015**, *15*, 3954–3968. [CrossRef]

178. Xu, K.; Xu, Y.; Ji, D.; Xie, J.; Chen, C.; Xie, C. Proteomic analysis of the economic seaweed *Pyropia haitanensis* in response to desiccation. *Algal Res.* **2016**, *19*, 198–206. [CrossRef]

179. Khan, S.; Mao, Y.; Gao, D.; Riaz, S.; Niaz, Z.; Tang, L.; Khan, S.; Wang, D. Identification of proteins responding to pathogen-infection in the red alga *Pyropia yezoensis* using iTRAQ quantitative proteomics. *BMC Genom.* **2018**, *19*, 842. [CrossRef] [PubMed]

180. Lee, H.J.; Choi, J.I. Identification, characterization, and proteomic studies of an aldehyde dehydognease (ALDH) from *Pyropia yezoensis* (*Bangiales, Rhodophyta*). *J. Appl. Phycol.* **2018**, *30*, 2117–2127. [CrossRef]

181. Lee, H.J.; Choi, J.I. Isolation and characterization of a high-growth-rate strain in *Pyropia yezoensis* induced by ethyl methane sulfonate. *J. Appl. Phycol.* **2018**, *30*, 2513–2522. [CrossRef]

182. Lee, H.J.; Choi, J. Proteomic analysis of *Pyropia yezoensis* mutant induced by gamma irradiation. *Phycologia* **2017**, *56*, 113.

183. Kumari, P. Seaweed lipidomics in the era of 'omics' biology: A contemporary perspective. In *Systems Biology of Marine Ecosystems*; Springer: New York, NY, USA, 2017; pp. 49–97.

184. Chen, J.; Li, M.; Yang, R.; Luo, Q.; Xu, J.; Ye, Y.; Yan, X. Profiling lipidome changes of *Pyropia haitanensis* in short-term response to high-temperature stress. *J. Appl. Phycol.* **2016**, *28*, 1903–1913. [CrossRef]

185. da Costa, E.; Azevedo, V.; Melo, T.; Rego, A.; V Evtuguin, D.; Domingues, P.; Calado, R.; Pereira, R.; Abreu, M.; Domingues, M. High-resolution lipidomics of the early life stages of the red seaweed *Porphyra dioica*. *Molecules* **2018**, *23*, 187. [CrossRef] [PubMed]

186. Wei, D.; Chen, D.; Lou, Y.; Ye, Y.; Yang, R. Metabolomic profile characteristics of *Pyropia haitanensis* as affected by harvest time. *Food Sci. Technol. Res.* **2016**, *22*, 529–536. [CrossRef]

187. Ye, Y.; Yang, R.; Lou, Y.; Chen, J.; Yan, X.; Tang, H. Effects of food processing on the nutrient composition of *Pyropia yezoensis* products revealed by NMR-based metabolomic analysis. *J. Food Nutr. Reslov.Iissn* **2014**, *2*, 749–756. [CrossRef]

188. Hamid, S.S.; Wakayama, M.; Ichihara, K.; Sakurai, K.; Ashino, Y.; Kadowaki, R.; Soga, T.; Tomita, M. Metabolome profiling of various seaweed species discriminates between brown, red, and green algae. *Planta* **2019**, *249*, 1921–1947. [CrossRef]

189. Mao, Y.; Chen, N.; Cao, M.; Chen, R.; Guan, X.; Wang, D. Functional characterization and evolutionary analysis of glycine-betaine biosynthesis pathway in red seaweed *Pyropia yezoensis*. *Mar. Drugs* **2019**, *17*, 70. [CrossRef]

190. Yan, Y.; Yang, H.; Tang, L.; Li, J.; Mao, Y.; Mo, Z. Compositional shifts of bacterial communities associated with *Pyropia yezoensis* and surrounding seawater co-occurring with red rot disease. *Front. Microbiol.* **2019**, *10*, 1666. [CrossRef] [PubMed]

191. Lee, H.J.; Park, E.J.; Choi, J.I. Isolation, morphological characteristics and proteomic profile analysis of thermo-tolerant *Pyropia yezoensis* mutant in response to high-temperature stress. *Ocean Sci. J.* **2019**, *54*, 65–78. [CrossRef]

Article

Development and Validation of an HPLC Method for the Quantitative Analysis of Bromophenolic Compounds in the Red Alga *Vertebrata lanosa*

Stefanie Hofer [1], Anja Hartmann [1,*], Maria Orfanoudaki [1], Hieu Nguyen Ngoc [1,†], Markus Nagl [2], Ulf Karsten [3], Svenja Heesch [3] and Markus Ganzera [1]

[1] Department of Pharmacognosy, University of Innsbruck, Innrain 80-82, 6020 Innsbruck, Austria; stefanie.hofer@uibk.ac.at (S.H.); maria.orfanoudaki@uibk.ac.at (M.O.); hieu.nguyenngoc@phenikaa-uni.edu.vn (H.N.N.); markus.ganzera@uibk.ac.at (M.G.)

[2] Institute of Hygiene and Medical Microbiology, Medical University of Innsbruck, Schöpfstraße 41, 6020 Innsbruck, Austria; m.nagl@i-med.ac.at

[3] Institute of Biological Sciences, Applied Ecology & Phycology, University of Rostock, Albert-Einstein-Str. 3, 18059 Rostock, Germany; ulf.karsten@uni-rostock.de (U.K.); svenja.heesch@uni-rostock.de (S.H.)

* Correspondence: anja.hartmann@uibk.ac.at

† Current Address: (H.N.N.) Faculty of Pharmacy, PHENIKAA University, Hanoi 12116, Vietnam and PHENIKAA Research and Technology Institute (PRATI), A&A Green Phoenix Group JSC, No. 167 Hoang Ngan, Trung Hoa, Cau Giay, Hanoi 11313, Vietnam.

Received: 25 October 2019; Accepted: 27 November 2019; Published: 29 November 2019

Abstract: Bromophenols are a class of compounds occurring in red algae that are thought to play a role in chemical protection; however, their exact function is still not fully known. In order to investigate their occurrence, pure standards of seven bromophenols were isolated from a methanolic extract of the epiphytic red alga *Vertebrata lanosa* collected in Brittany, France. The structures of all compounds were determined by nmR and MS. Among the isolated substances, one new natural product, namely, 2-amino-5-(3-(2,3-dibromo-4,5-dihydroxybenzyl)ureido)pentanoic acid was identified. An HPLC method for the separation of all isolated substances was developed using a Phenomenex C8(2) Luna column and a mobile phase comprising 0.05% trifluoroacetic acid in water and acetonitrile. Method validation showed that the applied procedure is selective, linear ($R^2 \geq 0.999$), precise (intra-day $\leq 6.28\%$, inter-day $\leq 5.21\%$), and accurate (with maximum displacement values of 4.93% for the high spikes, 4.80% for the medium spikes, and 4.30% for the low spikes). For all standards limits of detection (LOD) were lower than 0.04 µg/mL and limits of quantification (LOQ) lower than 0.12 µg/mL. Subsequently, the method was applied to determine the bromophenol content in *Vertebrata lanosa* samples from varying sampling sites and collection years showing values between 0.678 and 0.005 mg/g dry weight for different bromophenols with significant variations between the sampling years. Bioactivity of seven isolated bromophenols was tested in agar diffusion tests against *Staphylococcus aureus* and *Escherichia coli* bacteria. Three compounds showed a small zone of inhibition against both test organisms at a concentration of 100 µg/mL.

Keywords: marine algae; macro algae; bromophenols; HPLC; quantification; isolation

---

1. Introduction

Naturally occurring organobromine compounds are very unique in the plant kingdom, as bromination does not take place in terrestrial plants. Marine organisms are able to biosynthesize such substances because they have direct access to bromides naturally occurring in seawater but also by virtue of a rare enzyme called vanadium bromoperoxidase [1]. These brominated molecules are thought to be responsible for the typical sea-like taste and flavor of seafood [2]. Bromophenols as a subtype of this group can mainly be found in red algae; for example, in the genera *Odonthalia*, *Polysiphonia*, *Rytiphlaea*, *Vadalia*, and *Symphocladia* [3]. A broad range of pharmacologically relevant activities has been reported for this class of substances in vitro and in vivo, ranging from antioxidant, antimicrobial, antithrombotic, and antidiabetic to anticancer effects in humans [4]. Although some of these phytochemicals are thought to play a role in the chemical protection of marine organisms, their functional role is still not fully understood, as most studies on brominated molecules focus on their bioactivity rather than their ecological relevance [5,6].

Various bromophenols have been isolated and structurally identified from different red algae in the past, but little information is available on their actual content, their seasonal variation [7], and the possible influences of their sampling location and other environmental factors. Phillips and Towers (1980) developed an HPLC diode array detection (HPLC-DAD) method for the determination of different bromophenols in red algae. However, baseline separation could not be achieved for all of their standards, and therefore only one bromophenol (lanosol) was determined quantitatively [8,9]. More recently, HPLC-MS/MS and HPLC with fluorescence detection have been used for the quantification of bromophenols in other matrices such as seafood or water samples [10,11].

In our study, a fully validated HPLC method with simple and cost-efficient UV detection was used for the determination and quantification of seven bromophenolic compounds in the red alga *Vertebrata lanosa* (L.) T.A.Christensen (formerly *Polysiphonia lanosa* (L.) Tandy). This species grows as an obligate epiphyte on the brown alga *Ascophyllum nodosum* (L.) Le Jolis, which inhabits intertidal rocky shores of Northern Atlantic coasts [12]. Pure standards were isolated from the methanolic extract of a sample collected in Brittany (France) by using various chromatographic techniques followed by structural elucidation of all isolated compounds by nmR spectroscopy and MS. The isolated standards include one new bromophenolic substance and six that have previously been reported from other red algae [13–16]. Their content was determined in *Vertebrata lanosa* from different collection years and sampling sites. To investigate the hypothesis that bromophenols might protect against microbial infestations, antimicrobial activity of the seven standards was tested using agar diffusion tests.

2. Results

2.1. Structural Elucidation of Compound 1

Seven bromophenolic compounds (see Figure 1 for structures) could be isolated from *V. lanosa*; among them was one new natural product, compound **1**.

Compound **1** (Figure 2a) was obtained as a light purple amorphous powder and its purity was confirmed by HPLC-MS and nmR. The positive ESI-MS spectrum of the substance showed the typical isotopic pattern for dibrominated molecules at *m/z* 454/456/458 in the ratio of 1:2:1 [M + H]$^+$ as shown in Figure 2b. High-resolution ESI-MS data ([M + H]$^+$ = 455.9618) corresponded to the molecular formula $C_{13}H_{17}{}^{79}Br^{80}BrN_3O$. The substance showed optical rotation ($[\alpha]_D^{20}$ = –53.03 (*c* 0.7, MeOH)) and a UV λ_{max} at 216 and 292 nm.

The ^{1}H nmR spectrum of **1** displayed one aromatic proton at δ 6.84 (s, H-6′), protons of methylene groups as two doublets at δ_H 4.28 (1H, *J* = 16.0 Hz) and 4.23 (16.0 Hz), and multiplets at δ_H 3.23, 3.11, 1.89, and 1.60. One proton resonance was also observed at δ_H 3.57 (t, 6.0 Hz). The nmR shifts of the protons and the corresponding carbons of the substructure were unambiguously assigned by ^{1}H-^{1}H COSY, HSQC, and HMBC experiments. The main HMBC and ^{1}H-^{1}H COSY correlations are shown in Figure 2a. In the HMBC spectrum, correlations from H-6′ to C-2′ and C-4′ and C-7′, in addition to signals typical for two brominated quaternary carbons (δ_C < 110 ppm) and two oxygenated carbons on the aromatic ring (δ_C < 140 ppm) revealed the substructure of a (2,3-dibromo-4,5-hydroxybenzyl)amino group. The ^{1}H-^{1}H COSY signals between H-2 and H-3, H-3 and H-4, and H-4 and H-5, in combination with signals of a primary amino group at C-2 (δ_C 55.88) and a carboxyl group at C-1 (δ_C 174.38), which were assigned by their HMBC correlations between H-3/H-2 and C-1, confirmed the presence of an ornithyl moiety. These two units were connected via a ureido part structure which could be determined by HMBC correlations from H-5 to C-6 and H-7′ to C-6, as well as the characteristic shift value of C-6 (δ_C 160.95). Collectively, the structure of **1** can be described as 2-amino-5-(3-(2,3-dibromo-4,5-dihydroxybenzyl)ureido)pentanoic acid; nmR shift values can be found in Table 1 and original nmR spectra of this compound in the Supplementary Material (Figures S4–S5). It is a new natural product and we named this new bromophenol Vertebratol. The nmR data of the known bromophenolic compounds **2** to **7** (see Table S1, Supplementary Materials) were in good agreement with values from the literature [13–16].

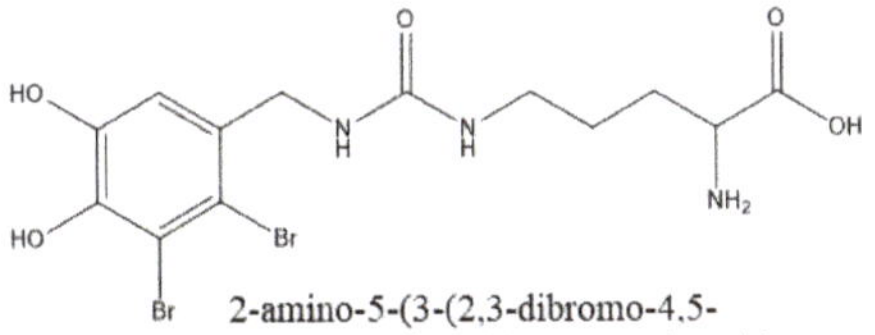

2-amino-5-(3-(2,3-dibromo-4,5-dihydroxybenzyl)ureido)pentanoic acid;
new compound (1)

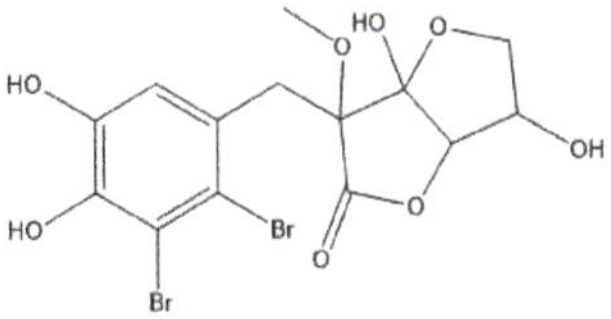

Methylrhodomelol **(2)**

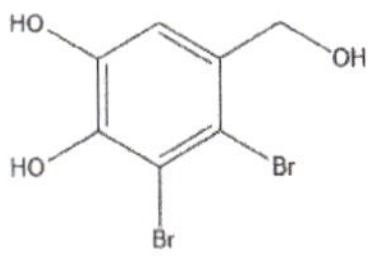

Lanosol **(3)**

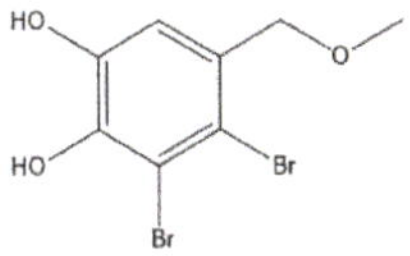

Lanosol methyl ether **(4)**

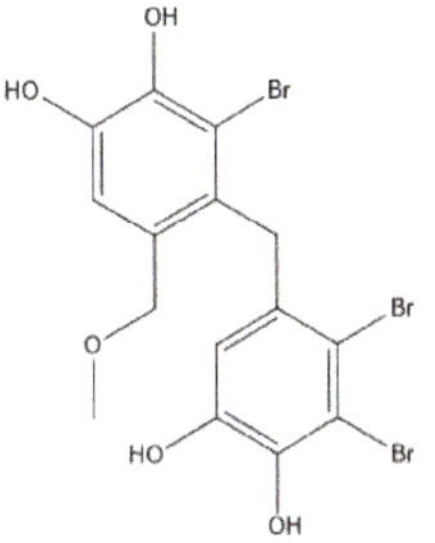

3-Bromo-4-(2,3-dibromo-4,5-dihydroxybenzyl)-5-methoxymethylpyrocatechol **(5)**

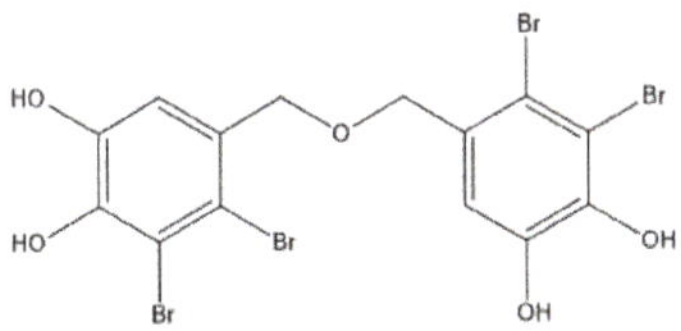

5-((2,3-Dibromo-4,5-dihydroxybenzyloxy)methyl)-3,4-dibromobenzene-1,2-diol **(6)**

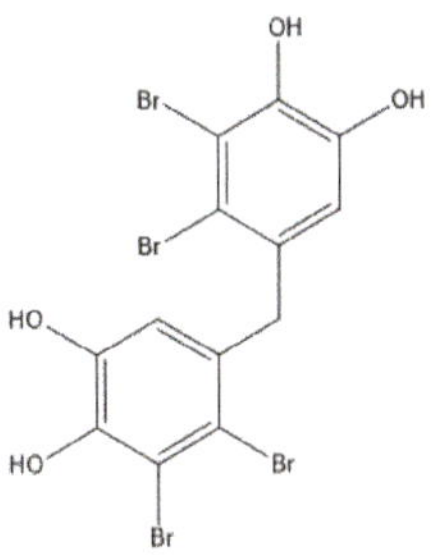

2,2′,3,3′-Tetrabromo-4,4′,5,5′-tetrahydroxydiphenylmethane **(7)**

Figure 1. Structures of the isolated standard compounds.

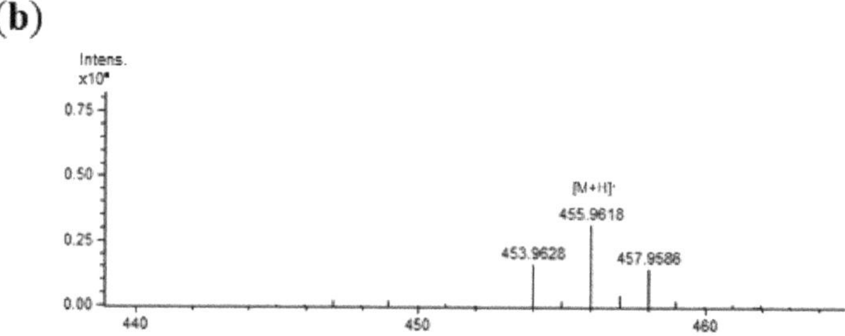

Figure 2. (a) Key HMBC correlations (H→C) and ^{1}H-^{1}H COSY (bold line) correlations and (b) high resolution ESI-MS data of compound 1 (Vertebratol).

Table 1. NMR shift values for compound 1 (Vertebratol) in MeOD; the spectra were recorded on a 600 MHz nmR instrument.

	Bromophenol m/z = 455.9618 [M + H]$^+$		
	^{13}C	^{1}H	HMBC ^{a}
1	174.38, C		
2	55.87, CH	3.57, t (6.0)	1, 3, 4
3	29.40, CH$_2$	1.89, m	1, 2, 4, 5
4	27.07, CH$_2$	1.60, m	3, 5
5	40.19, CH$_2$	3.23, m; 3.11, m	3, 4, 6
6	160.94, C		
1'	132.37, C		
2'	114.42, C		
3'	114.13, C		
4'	144.87, C		
5'	146.63, C		
6'	114.78, CH	6.84, s	2', 4', 7'
7'	46.010, CH$_2$	4.28, d (16.0); 4.23, d (16.0)	1', 2', 6

^{a} HMBC correlations are stated from proton(s) to the indicated carbon.

2.2. HPLC Method Development

The HPLC separation of seven bromophenols contained in *Vertebrata lanosa* in less than 25 min was possible with the setup shown in Figure 3. The previously described compounds (Figure 1) were used as standards for method development. During its initial phase, four stationary phases, namely, Phenomenex Synergi MAX-RP (150 mM × 2.0 mM), Phenomenex Synergi POLAR-RP (150 mM × 4.6 mM), YMC Triart (150 mM × 3.0 mM)m and Phenomenex Luna C8(2) (150 mM × 2.0 mM,), all with comparable particle sizes (3–4 μm), were tested with two different mobile phase systems, i.e., methanol/water and acetonitrile/water. It was observed that the latter column yielded the overall best results together with acetonitrile/water, not only concerning separation of the standards but also for the analysis of the crude *V. lanosa* extract to avoid interference with other naturally occurring substances. The addition of 0.05% trifluoroacetic acid to the mobile phase was advantageous, whereas using 0.1% phosphoric acid or a 10 mM ammonium acetate buffer decreased the quality of separation. The optimum temperature was found to be 30 °C, and the elution gradient was carefully optimized. Ideal elution was performed at a flow rate of 0.25 mL/min by starting at a concentration of 2% B and rapidly increasing it to 20% within only 0.1 min, followed by an increase to 50% B in 15 min and to 70% B in another 20 min. The acetonitrile content was then raised to 98% B in a further 5 min; this composition was maintained for an additional 10 min, leading to a total runtime of 55 min. Of note, changes in the starting conditions in particular (e.g., leaving the concentration at 2% B for longer than 0.1 min (see Supplementary Figure S1b) or immediately starting the gradient at 20% B (see Supplementary Figure S1c) had a negative impact on the separation of peaks **2** and **3**. During analysis the DAD was set to 210 nm and the injection volume was adjusted to 5 µL.

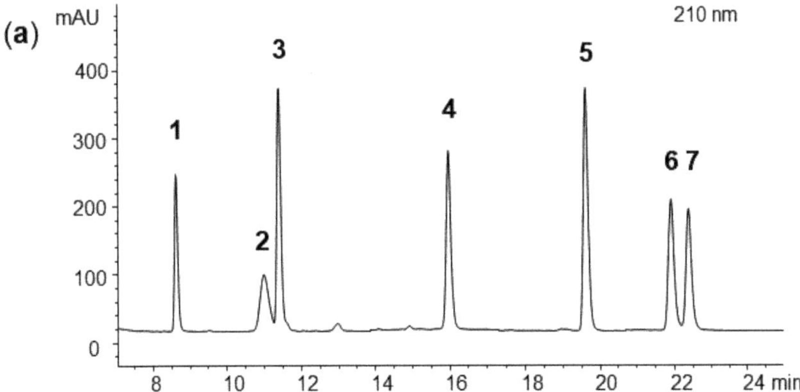

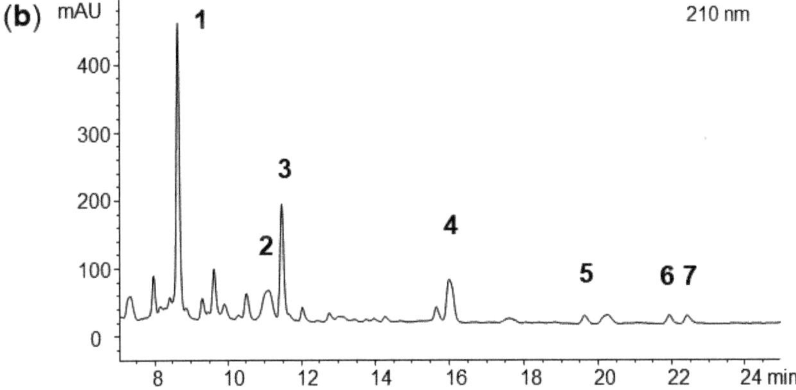

Figure 3. (a) HPLC separation of seven bromophenols under optimized conditions and (b) HPLC separation of a methanolic extract of *Vertebrata lanosa* (sample 1). The applied analytical conditions were column: Phenomenex Luna C8(2) (150 mM × 2.0 mM, 3 µm particle size); mobile phase: water (A) and acetonitrile (B), both containing 0.05% trifluoroacetic acid; gradient: 2% B at 0 min, 20% B at 0.1 min, 50% B at 15 min, and 70% B at 35 min; injection volume: 5 µL; flow rate: 0.25 mL/min; column temperature: 30 °C; and detection: 210 nm.

2.3. Method Validation

Following assay development, the method was validated. Selectivity was deduced from consistent UV spectra within the peaks of interest (by using the peak purity function available in the software) and no signs of co-eluting compounds, visible by e.g., peak shoulders, were observed. The slightly wider peak of compound **2** (Methylrhodomelol) in the extract was compared to that of the corresponding standard, which was found to be highly pure based on HPLC, HPLC-MS, and nmR experiments. Additionally, the standard of Methylrhodomelol was eluted as a broader peak, indicating that this was a substance specific effect not caused by impurity. Calibration curves were established for all of the seven standards. Coefficients of determination were always 0.999 or higher in a concentration range of 102 to 0.05 µg/mL for compound **1**, of 181 to 0.06 µg/mL for compound **2**, of 95 to 0.02 µg/mL for compound **7**, and of approximately 50 to 0.02 µg/mL for compounds **3**, **4**, **5**, and **6**. LOD values varied from 0.008 and 0.038 µg/mL and LOQ ranged from 0.024 to 0.116 µg/mL. All summarized calibration data can be found in Table 2. Assay precision was assured by repeatedly extracting and analyzing sample 1 (sample *Vertebrata lanosa*, Brittany, France 2018) under optimized conditions (see Table 2). Intra-day (relative standard deviation ≤ 6.28 %) and inter-day (relative standard deviation ≤ 5.21%) variance was shown to be in the acceptable range, especially as algal material always shows a certain degree of inhomogeneity, which is reflected in the slight variations observed in the samples collected at the same time and location. Accuracy was determined by spiking a defined amount of grounded, dried material with three different concentrations of each of the standards (low, medium, and high) prior to sample preparation. The observed recovery rates showed maximum displacement rates of 4.93% (**2**, high spike), confirming the validity of this parameter, too.

Table 2. Validation parameters of the HPLC method.

Calibration Data for the Seven Bromophenol Standards					
Substance	Regression Equation	Coefficient of Determination	Range (µg/mL)	LOD [1] (µg/mL)	LOQ [2] (µg/mL)
1	y = 83407x + 5.5502	R^2 = 1.0000	102–0.05	0.017	0.053
2	y = 98794x + 18.324	R^2 = 0.9999	181–0.06	0.038	0.116
3	y = 156713x + 34.799	R^2 = 0.9991	47–0.02	0.008	0.024
4	y = 126302x + 44.688	R^2 = 0.9990	49–0.02	0.011	0.034
5	y = 164091x + 25.645	R^2 = 0.9998	56–0.03	0.014	0.043
6	y = 126669x - 28.684	R^2 = 0.9995	49–0.05	0.032	0.097
7	y = 114941x + 42.574	R^2 = 0.9996	95–0.02	0.015	0.047

Accuracy and Precision of the Assay						
Accuracy, High Spike (n = 3)				Accuracy, Medium Spike (n = 3)		
Substance	Measured Values [3]	Theoretical Values [3]	Displace-ment [4]	Substance	Measured Values [3]	Theoretical Values [3]	Displace-ment [4]
1	89.5	97.3	2.67	1	69.0	71.3	3.29
2	135.1	142.1	4.93	2	49.5	51.3	3.51
3	41.1	41.6	1.23	3	28.1	26.8	4.68
4	48.3	49.5	2.45	4	27.9	27.3	1.96
5	49.8	51.7	3.60	5	29.8	28.5	4.80
6	43.9	45.2	2.85	6	25.3	25.1	1.12
7	87.9	87.7	0.24	7	46.0	45.0	2.15

Accuracy, Low Spike (n = 3) [3]				Precision (n = 5) [5]			
Substance	Measured Values [3]	Theoretical Values [3]	Displace-ment [4]	Day 1	Day 2	Day 3	Inter-day
1	50.7	51.4	1.33	1.83	0.71	1.21	2.33
2	29.1	30.4	4.07	1.26	0.76	1.16	2.35
3	12.5	13.0	3.57	2.15	0.62	1.06	2.93
4	9.7	10.1	4.30	2.59	1.71	3.86	2.93
5	6.9	7.0	0.54	3.83	6.28	2.56	5.21
6	5.8	5.6	3.56	1.23	0.84	2.87	1.74
7	23.6	23.5	0.51	1.23	0.43	2.28	2.11

[1] LOD: limit of detection determined with the purified standards. [2] LOQ: limit of quantification determined with the purified standards. [3] Values in µg/mL. [4] Displacement in percent (sample *Vertebrata lanosa*, Brittany, 2018). [5] Maximum relative standard deviation based on the peak area given in percent.

2.4. Quantitative Analysis of Samples

Three *V. lanosa* samples collected in Brittany, together with one commercial sample and specimens from Ireland and Norway (Table 3) were used for quantitative analysis. Two of the samples from Brittany were collected at the same sampling site and season but across two different years. The third sample originated from a different place but was also collected in June 2019. The compounds were assigned at 210 nm by matching retention times and UV spectra compared to the standards; additionally, LC-MS experiments were performed. For quantification, each of the samples was measured at least in triplicates with maximum relative standard deviations of 5.21% (see Table 4).

Table 3. Origin of analyzed *Vertebrata lanosa* samples.

Sample	Collection Site and Date
1	2018, Roscoff, Brittany, 48.727559° N; 3.987924° W, host alga *Ascophyllum nodosum*, collected and identified by U. Karsten and S. Heesch, University of Rostock
2	2019, Roscoff, Brittany, 48.727559° N; 3.987924° W, host alga *Ascophyllum nodosum*, collected and identified by S. Hofer, University of Innsbruck
3	2019, Saint Pol de Léon, 48.676896° N; 3.966533° W, host alga *Ascophyllum nodosum*, Brittany, collected and identified by S. Hofer, University of Innsbruck
4	Commercially available; Arctic Algae AS, Bodø, Norway
5	2019, Lettermore, Ireland, 53.298468° N; 9.712704° W, host alga *Ascophyllum nodosum*, collected and identified by R. Bermejo, National University of Ireland, Galway
6	2019, Bukken, Raunefjorde, Norway, 60.241183° N, 05.20225° W, host alga *Ascophyllum nodosum*, collected and identified by K. Sjøtun, University of Bergen

Table 4. Determination of bromophenol content in six different *Vertebrata lanosa* samples.

	Content of Bromophenols					
	Sample 1		Sample 2		Sample 3	
Substance	Content (µg/mg Dried Algal Material)	Relative Standard Deviation (%)	Content (µg/mg Dried Algal Material)	Relative Standard Deviation (%)	Content (µg/mg Dried Algal Material)	Relative Standard Deviation (%)
1	**0.678** (*)	2.33	**0.035**	0.55	**0.029**	4.52
2	0.175 (*)	2.35	0.027	0.64	0.013	4.08
3	0.158	2.93	0.032	0.66	0.008	4.18
4	0.094	2.93	0.015	0.68	T [1]	-
5	0.028 (*)	5.21	-	-	-	-
6	0.029	1.74	0.007	2.75	-	-
7	0.030	2.11	T [1]	-	-	-
	Sample 4		Sample 5		Sample 6	
Substance	Content (µg/mg Dried Algal Material)	Relative Standard Deviation (%)	Content (µg/mg Dried Algal Material)	Relative Standard Deviation (%)	Content (µg/mg Dried Algal Material)	Relative Standard Deviation (%)
1	0.045	0.41	0.025	0.67	0.006	2.85
2	**0.048**	2.66	0.140	0.36	0.105	0.30
3	0.035	0.40	0.313 (*)	0.76	**0.259**	0.64
4	0.014	1.99	0.339 (*)	0.51	0.059	0.46
5	0.014	1.37	0.006	2.05	0.005	1.24
6	0.011	1.91	**0.445** (*)	0.43	0.052	0.29
7	T [1]	1.76	0.062 (*)	1.08	0.055	0.27

[1] T = found in traces [2] Main compounds are shown in bold writing (*) = Highest measured value of the given compound, when compared to the other samples.

All samples contained compounds **1** to **4**. However, compounds **5** to **7** could not be found in all specimens. For example, the sample collected in Roscoff 2019 (sample 2) did not exhibit a peak corresponding to **5**, and compound **7** could only be detected in traces. The second sample from Brittany from 2019 (sample 3, collected in Saint Pol de Léon) did not show any signals for compounds **5–7**, and compound **4** was present only in minor amounts. On the other hand, sample 1, collected in the previous year in Brittany (Roscoff, same sampling site and season as sample 2) as well as samples 4–6 contained all of the seven isolated bromophenols. In addition, in quantitative terms the bromophenol content varied significantly. This becomes clear when comparing the quantitative results of the three samples from Brittany. Here, the main bromophenol was always compound **1**. However, its content was approximately 20 times higher in sample 1 (0.678 mg/g) compared to samples 2 (0.035 mg/g) and 3

(0.029 mg/g). When comparing the results for sample 1 with those of other origins, fluctuations were found to be even larger, e.g., reaching up to 113 times lower values for compound **1** in sample 6.

The main bromophenol in sample 4 was compound **2**, which had a concentration of 0.048 mg/g dried algal material, while in sample 5 the predominant bromophenol was compound **6** (0.445 mg/g), and in sample 6 it was compound **3** (0.259 mg/g). The overall highest concentrations of compounds **1** (0.678 mg/g), **2** (0.175 mg/g), and **5** (0.028 mg/g) were found in sample 1 (2018, Roscoff, Brittany). However, the highest contents of compounds **3** (0.313 mg/g), **4** (0.339 mg/g), **6** (0.445 mg/g), and **7** (0.062 mg/g) were observed in sample 5 (2019, Lettermore, Ireland).

In the course of our investigations 20 additional samples were analyzed with the given method (19 species of macroalgae and one lichen, see Table S2). The results showed that the seven bromophenols could not be found in any of the other samples.

2.5. Antimicrobial Activity

Compounds **5–7** (all tested at a concentration of 100 µg/mL) showed a small zone of inhibition in agar diffusion tests, while compounds **1–4** (100 µg/mL) and the controls did not inhibit bacterial growth. The diameter of the zone of inhibition was 11 mM for compound **5** and 9 mM for compounds **6** and **7** against *Staphylococcus aureus* ATCC 6538. *Escherichia coli* ATCC 11229, inhibition was minimal, with a 7 mM diameter observed for compounds **5–7**, and again, there was no inhibition for the other bromophenols and the controls.

3. Discussion

While halogenated compounds commonly occur in marine organisms, especially in macroalgae, the role of bromide-containing compounds in red algae is not well understood and data regarding their occurrence are inconsistent. To better understand their functional role in the plant kingdom but also the triggers for bromophenol formation, a fully validated HPLC method for the determination and quantification of seven bromophenolic compounds in the red alga *Vertebrata lanosa* (L.) T.A.Christensen was developed which offers the potential to monitor bromophenol concentrations of larger sample sets.

To obtain the standards seven bromophenols occurring in *V. lanosa* were isolated. Among them was one new natural product, 2-amino-5-(3-(2,3-dibromo-4,5-dihydroxybenzyl)ureido)pentanoic acid (Vertebratol). In the past, two related compounds with a 3-(2,3-dibromo-4,5-dihydroxybenzyl)ureido core structure have already been isolated from two other marine red algae of the family Rhodomelaceae, namely *Rhodomela confervoides* (Hudson) P.C.Silva and *Symphocladia latiuscula* (Harvey) Yamada [17,18]; however, in contrast to those molecules, the new molecule is linked to an ornithyl moiety. Interestingly, this compound was observed to occur in very high concentrations of up to 0.678 mg/g dried algal material in some of the *V. lanosa* samples. The concentrations of all seven bromophenols were determined in the methanolic extracts of several samples of this alga with our newly developed and fully validated HPLC method, which enabled their identification and precise quantification in less than 24 min. To further verify that peaks of interest in unknown samples correspond to the known standards, the method can easily be coupled to MS; however, 0.05% trifluoroacetic acid in the mobile phase has to be replaced with 0.1% formic acid in order to avoid ion suppression. Excellent determination coefficients ($R^2 \geq 0.999$) were achieved with low LOD values of between 0.008 and 0.038 µg/mL and limits of quantification (LOQ) ranging from 0.024 to 0.116 µg/mL. Thus, the method offers the potential to provide not only qualitative but also quantitative information about the occurrence of those bromophenols in algae even if only small amounts of material are available.

It is known that bromophenols are not randomly distributed. They usually only occur in particular genera and species as their biosynthesis requires certain enzymes and thus the corresponding genetic information. Consequently, the presence of a specific set of bromophenolic compounds can be used as a taxonomic marker for the discrimination of different species [19]. In our study, six *Vertebrata lanosa* samples were measured, and, interestingly, the determined concentrations of the different bromophenols showed great differences within the samples. For example, the values for the new natural product, substance **1** (Vertebratol), varied by up to 23 times within samples from Brittany from different years and sampling sites and even by up to 115 times when compared to samples from other countries. Additionally, the concentrations of the other six bromophenols were not uniform within the different samples. In qualitative terms, samples from Norway and Ireland showed different main bromophenols compared to the samples from Brittany. These results are in accordance with the literature because it has been reported that the content of brominated compounds like Isorhodolaureol in certain red algae such as *Laurencia dendroidea* J. Agardh vary extensively from year to year, even if the collection site is the same, showing that there have to be specific triggers that enhance the formation of bromophenols in some Rhodomelacean species [19,20]. As the number of samples was limited in the current work, we cannot deduce any environmental parameters which may determine the occurrence of bromophenolic compounds in *Vertebrata lanosa*. Further research will be necessary to identify them and their functional role in the algae. However, the method presented herein might serve as a useful tool for the analysis of samples in a larger number, as it is a cost-efficient and fast approach that was moreover fully validated according to the ICH guidelines.

In the past, bromophenols have been reported to possess radical scavenging activities [13], even though their antioxidant properties have so far been only determined in vitro [21,22]. These compounds have, moreover, been attributed with deterring herbivores and epiphytes: Shoeib et al. (2006) have noted that *V. lanosa* does not seem to be grazed upon much and appears remarkably free of epiphytes, both of which the authors suggested could be due to the presence of bromophenolic compounds [7]. However, a recent study on *V. lanosa* from Norway shows that microscopic epiphytes are present all year round, while (limited) macroscopic fouling does occur, especially during late autumn [23]. While some bromophenols have been shown to inhibit competitors or deter predators, for example in marine invertebrates such as worms [24,25], conclusive studies demonstrating the deterrence of herbivores and epiphytes in vivo are still lacking for *V. lanosa* as well as for other red algae. From published bioactivity data [4], it can be hypothesized that bromophenols might potentially play a role in defense against microbial infestations. Additionally, in our study compounds **5** (3-bromo-4-(2,3-dibromo-4,5-dihydroxybenzyl)-5-methoxymethylpyrocatechol), **6** (5-((2,3-dibromo-4,5-dihydroxybenzyloxy)methyl)-3,4-dibromobenzene-1,2-diol), and **7** (2,2′,3,3′-tetrabromo-4,4′,5,5′-tetrahydroxydiphenylmethane) showed weak antimicrobial activity against *E.coli* and *S. aureus* in the micro molar concentration range, which is similar to natural but also synthetic bromophenols that have been tested in the literature [26,27]. Interestingly, only substances that possess two 2,3-dibromo-4,5-dihydroxybenzyloxyl moieties were active. Whether this aspect is really essential for activity will be studied in further investigations.

4. Materials and Methods

4.1. Samples and Reagents

The origin of all *V. lanosa* samples is given in Table 3. The algal material used to isolate the standards was morphologically and genetically identified by two of the authors (Ulf Karsten and Svenja Heesch GenBank accession number LR738856), and voucher specimens were deposited at the Institute of Pharmacy, Pharmacognosy, at the University of Innsbruck, Austria. The solvents used for analytical work were of analytical grade, and they were obtained from Merck (Darmstadt, Germany). HPLC grade water was prepared by a Satorius arium 611 UV water purification system (Göttingen, Germany).

4.2. Isolation of Bromophenols from Vertebrata Lanosa

The methanol soluble fraction of the methanol extract of sample 1 was used for the isolation of the seven bromophenols shown in Figure 1. To achieve a targeted isolation, fractions were analyzed by HPLC-DAD-MS on an Agilent Technologies 1260 Infinity II instrument with an InfinityLab LC/MSD detector (Waldbronn, Germany) prior to further isolation steps using the optimum HPLC conditions (gradient as described in Section 2.2) on the Phenomenex C8(2) Luna column. However, for that purpose 0.05% trifluoroacetic acid had to be replaced with 0.1% formic acid in the mobile phase. A respective chromatogram of the crude extract is shown in the Supplementary Information in Figure S2. This step was useful, firstly because the brominated molecules show very characteristic isotopic patterns due to the relative abundance of the brome isotopes (see Figure 2b), and secondly because all of the isolated bromophenols exhibit two absorption maxima, one at 210 nm and a weaker one around 280–290 nm. The structures of all compounds were elucidated based on nmR and MS experiments and a comparison to literature values (if possible). MS data as well as characteristic nmR shift values can be found in the Supplementary Information. For the new compound **1** (Vertebratol), the original nmR spectra are shown there as well, while those of the other compounds can be provided upon request.

Air-dried and powdered *Vertebrata lanosa* material (600 g) was extracted with methanol at room temperature to obtain 70 g of crude extract. The methanol soluble part of this extract (40 g) was fractionated on a silica gel column using gradient elution (EtOAc to methanol to methanol:water 80:20), resulting in 14 fractions. Fractions 2 and 3 were combined and further fractionated on a Sephadex LH-20 column in methanol to afford 19 individual fractions. Fractions 10 and 12 were individually processed using a 4 g silica cartridge and a Reveleris® X2 iES flash chromatography system (both from Büchi, Flawil, Switzerland). Elution was performed using a gradient of toluol (A) and EtOAc (B): 0–5 min: 2% B, 10 min: 30% B, 29 min: 70% B, 35 min: 100 % B, and 35–45 min: 100% B. The flow rate was set to 5 mL/min and detection was carried out at 254, 280, and 320 nm.

Flash chromatography of fraction 10 (123 mg) directly resulted in the isolation of compound **4** (9 mg) and the same procedure with fraction 12 led to the isolation of compound **3** (50 mg). Another flash chromatography fraction (fraction 7) required an additional purification step using a semi-preparative HPLC with a Synergi 4 u Polar-RP (250 mM × 10 mM, 4 μm; Phenomenex, Torrance, CA, USA) column and a Dionex UltiMate 3000 HPLC (Thermo, Waltham, MA, USA). Water (A) and methanol (B) were used as mobile phases with the following gradient: 0 min: 20% B to 96.5% B in 38 min at 45 °C. After a final purification step on a small Sephadex LH-20 column in methanol the isolation of compound **2** (16 mg) could be achieved.

The fractions 16, 17, and 18 of the first Sephadex column were subjected to semi-preparative HPLC on the same Synergi Polar-RP column using water (A) and acetonitrile (B), both containing 0.1% formic acid, as the mobile phase. The temperature was set at 30 °C and the flow rate to 2 mL/min. The applied gradient was: 0 min: 30% B, 13 min: 51.5% B, 40 min: 51.5% B, and 45 min: 98% B. This step resulted in the isolation of compounds **5** (6.6 mg), **6** (9 mg), and **7** (6 mg).

Furthermore, the initial fraction 12 (6.5 g) also showed a main peak with an isotopic pattern typical for brominated molecules. The fraction was dissolved in water and used for liquid–liquid extraction in a separatory funnel with EtOAc followed by butanol (BuOH). The BuOH fraction was then used for a further separation step with flash chromatography, applying a C-18 40 g cartridge from Büchi on the same flash chromatography system as previously mentioned. The flow rate was set to 12 mL/min and the following elution gradient (solvent A: water; solvent B: methanol) was applied: 0–5 min: 2% B, 10 min: 30% B, 35 min: 100% B, and 35–45 min: 100% B. The flash sub-fraction 13 (79 mg) was finally purified by semi-preparative HPLC using the Synergi Polar-RP (250 mM × 10 mM, 4 μm) column with water (A) and methanol (B). Isocratic elution at 30% B led to the isolation of compound **1** (30 mg).

4.3. Structure Elucidation

NMR experiments were performed on a Bruker Avance II 600 spectrometer (Karlsruhe, Germany) at 600.19 (1H) and 150.91 MHz (13C). The pure isolated compounds were dissolved in deuterated methanol from Euriso-Top (Saint Aubin, France). An Agilent InfinityLab LC/MSD System (Santa Clara, CA, USA) was used to measure the low-resolution mass spectra in positive mode at a capillary energy of 4000 V and nebulizer gas of 40.0 psi (dry gas), at 10.0 L/min and at a temperature of 300 °C; the recorded scan range was 100–1500 *m/z*. The high-resolution mass spectra were recorded on a micrOTOF-Q II mass spectrometer (Bruker-Daltonics, Bremen, Germany). The experiments were conducted in the positive as well as negative ESI mode by applying the following parameters: capillary energy, 4500 V for positive mode and 3500 V for negative mode; nebulizer gas, 6.4 psi; dry gas, 4.0 L/min at a temperature of 180 °C; and recorded scan range of 100–600 *m/z*.

4.4. Sample Preparation

The dried and finely milled algal material (0.1 g) was extracted five times with 1.5 mL of MeOH for 20 min each on an ultrasonic bath to achieve an exhaustive extraction of all compounds of interest. The supernatants were combined after centrifugation ($1000 \times g$ for 5 min) and the solvent was evaporated.

The bromophenols were quantified at 210 nm, a wavelength where all the respective compounds show strong UV absorbance. However, when analyzing the samples, a broad and partially interfering peak was observed at min 7.6 (see Supplementary Figure S3a). The peak could successfully and selectively be removed by re-dissolving the crude extract in water and applying it onto a solid phase extraction cartridge (Phenomenex Strata C18-E, 55 μm, Torrance, CA, USA). The cartridge was then washed with water to elute the impurities (Supplementary Figure S3b), followed by an elution step with methanol to recover the enriched analytes as shown in Supplementary Figure S3c. After evaporation, samples were re-dissolved in 2 mL of methanol.

4.5. Analytical Conditions

All analytical experiments were performed on an Agilent 1100 HPLC (Waldbronn, Germany) using a Luna C8(2) column (150 mM × 2.0 mM, 3 μm particle size; Phenomenex, Torrance, CA, USA) as the stationary phase and a mobile phase comprising 0.05% trifluoroacetic acid in water (A) and acetonitrile (B).

Elution started at 2% B. This proportion was rapidly increased to 20% B in only 0.1 min, followed by an increase of up to 50% B in the first 15 min and to 70% in another 20 min. In the next 5 min elution was increased to 98% B and this composition was maintained for additional 10 min. The column was then re-equilibrated for 10 min prior to the next run. The DAD-wavelengths were adjusted to 210, 280, and 310 nm, whereas flow rate, sample injection volume, and column temperature were set at 0.25 mL/min, 5 μL, and 30 °C, respectively.

4.6. Method Validation

The newly developed HPLC method was validated according to ICH guidelines in order to ensure that it fulfilled regulatory standards. All validation results are summarized in Table 2. To establish calibration curves and to determine the linear range, a stock solution of all standards (approximately 2 mg, accurately weighed, per compound dissolved in 5.00 mL MeOH) was prepared. This solution was used for serial dilution in the ratio of 1:1. LOD and LOQ values were calculated according to the guidelines based on the given standard deviations of the response and the slope of calibration curves for each compound. The selectivity of the method was estimated by evaluating the photodiode array data and assessing the peak purity with the respective option in the operating software. Precision was confirmed by individual preparation of five sample solutions of sample 1 on each of three consecutive days and analyzing them. Variations in peak areas were calculated for all the relevant analytes and given for one day (intra-day precision) and for the three-day period (inter-day precision). Accuracy was assured by spiking sample 1 with three different concentrations of all the standards (high, medium, and low concentrations). Samples were then prepared as previously described and recovery rates determined as percentages of the actually present concentrations compared to the theoretically calculated ones.

4.7. Determination of Antimicrobial Activity

The test bacteria *Staphylococcus aureus* ATCC 6538 and *Escherichia coli* ATCC 11229 were grown on Mueller-Hinton (MH) agar plates. Overnight cultures from a single colony from these plates were grown at 37 °C in tryptic soy broth. Bacteria were spread separately on fresh MH plates with a swab. A punch with a diameter of 5 mM was made in each plate and filled with the respective test solution. Compounds **1–4** were dissolved in 50% methanol and compounds **5–7** dissolved in 50% dimethyl sulfoxide to a concentration of 1 mg/mL. These solutions were tenfold diluted in distilled water so that the final concentration of the compounds in the test solution was 100 μg/mL. Inoculated agar plates were grown at 37 °C overnight, and the zone of inhibition was observed. Controls containing 5% methanol and 5% DMSO were done in parallel.

Supplementary Materials: The following are available online at http://www.mdpi.com/1660-3397/17/12/675/s1, Table S1: nmR shift values for Compounds 2–7, Figure S1: HPLC separation of 7 standards with different gradients, Figure S2: HPLC-MS analysis of the methanolic Vertebrata lanosa extract. Figure S3: HPLC analysis after sample enrichment on an SPE cartridge, Figure S4: 1H-NMR and 13C-NMR spectra of the new compound, Figure S5: HMBC, COSY and HSQC spectra of the new compound.

Author Contributions: All authors substantially contributed to this work. Contributions: investigation, S.H. (Stefanie Hofer), M.O., H.N.N., and M.N.; resources, S.H. (Svenja Heesch), and U.K.; supervision of the practical work and conceptualization, A.H.; writing–original draft preparation, S.H. (Stefanie Hofer), writing–review and editing and supervision, A.H., U.K., S.H. (Svenja Heesch), and M.G.; project administration, M.G.

Funding: This research was funded by the Austrian Science Fund (FWF), project no. P296710.

Acknowledgments: The authors gratefully thank Vivien Hotter (Institute of Biological Sciences, Applied Ecology and Phycology, University of Rostock), Ricardo Bermejo (Department of Earth and Ocean Sciences, National University of Ireland, Galway) and Inga Kjersti Sjøtun (Department of Biology, University of Bergen) for their help with the collection of the algal material and the employees of the Station Biologique de Roscoff, France for their assistance during the stay in Roscoff. Open access funding was provided by the FWF.

Conflicts of Interest: The authors declare no conflict of interest.

References

1. Carter-Franklin, J.N.; Butler, A. Vanadium Bromoperoxidase-Catalyzed Biosynthesis of Halogenated Marine Natural Products. *J. Am. Chem. Soc.* **2004**, *126*, 15060–15066. [CrossRef] [PubMed]

2. Fuller, S.C.; Frank, D.C.; Fitzhenry, M.J.; Smyth, H.E.; Poole, S.E. Improved approach for analyzing bromophenols in seafood using stable isotope dilution analysis in combination with SPME. *J. Agric. Food Chem.* **2008**, *56*, 8248–8254. [CrossRef] [PubMed]

3. Amsler, C.D. Natural Products Chemistry of Rhodophyta. In *Algal Chemical Ecology*; Amsler, C.D., Ed.; Springer: Berlin/Heidelberg, Germany, 2009; Volume 1, p. 14.

4. Liu, M.; Hansen, P.E.; Lin, X. Bromophenols in marine algae and their bioactivities. *Mar. Drugs* **2011**, *9*, 1273–1292. [CrossRef] [PubMed]

5. Gribble, G.W. The diversity of naturally occurring organobromine compounds. *Chem. Soc. Rev.* **1999**, *28*, 335–346. [CrossRef]

6. Amsler, C.D. Tropical Chemically Defended Macroalgae. In *Algal Chemical Ecology*; Amsler, C.D., Ed.; Springer: Berlin/Heidelberg, Germany, 2009; Volume 1, p. 29.

7. Shoeiba, N.A.; Bibbyb, M.C.; Blundenc, G.; Linleya, P.A.; Swaineb, D.A.; Wrighta, C.A. Seasonal Variation in Bromophenol Content of *Polysiphonia lanosa*. *Nat. Prod. Commun.* **2005**, *1*, 47–49. [CrossRef]

8. Phillips, D.W.; Towers, G.H.N. Reversed-phase high-performance liquid chromatography of red algal bromophenols. *J. Chrom. A* **1981**, *206*, 573–580. [CrossRef]

9. Phillips, D.W.; Towers, G.H.N. Chemical ecology of red algal bromophenols. 1. Temporal, interpopulational and within-thallus measurements of lanosol levels in *Rhodomela larix* (Turner). *J. Exp. Mar. Biol. Ecol.* **1982**, *58*, 285–293. [CrossRef]

10. Chi, X.; Liu, J.; Yu, M.; Xie, Z.; Jiang, G. Analysis of bromophenols in various aqueous samples using solid phase extraction followed by HPLC-MS/MS. *Talanta* **2017**, *164*, 57–63. [CrossRef]

11. Zhang, S.; Li, Y.; You, J.; Wang, H.; Zheng, Y.; Suo, Y. Improved Method for the Extraction and Determination of Bromophenols in Seafood by High-Performance Liquid Chromatography with Fluorescence Detection. *J. Agric. Food Chem.* **2012**, *60*, 10985–10990. [CrossRef]

12. Penot, M.; Hourmant, A.; Penot, M. Comparative study of metabolism and forms of transport of phosphate between *Ascophyllum nodosum* and *Polysiphonia lanosa*. *Physiol. Plant.* **1993**, *87*, 291–296. [CrossRef]

13. Glombitza, K.W.; Sukopp, I.; Wiedenfeld, H. Antibiotics from Algae XXXVII. Rhodomelol and Methylrhodomelol from *Polysiphonia lanosa*. *Planta Med.* **1985**, *51*, 437–440. [CrossRef]

14. Lijun, H.; Nianjun, X.; Jiangong, S.; Xiaojun, Y.; Chengkui, Z. Isolation and pharmacological activities of bromophenols from *Rhodomela confervoides*. *Chin. J. Oceanol. Limnol.* **2005**, *23*, 226–229. [CrossRef]

15. Kurihara, H.; Mitani, T.; Kawabata, J.; Takahashi, K.V. Two new bromophenols from the red alga *Odonthalia corymbifera*. *J. Nat. Prod.* **1999**, *62*, 882–884. [CrossRef] [PubMed]

16. Suzuki, M.; Kowata, N.; Kurosawa, E. Bromophenols from the Red Alga *Rhodomela larix*. *Bull. Chem. Soc. Jpn.* **1980**, *53*, 2099–2100. [CrossRef]

17. Xu, X.; Yin, L.; Gao, J.; Chen, J.; Li, J.; Song, F. Two New Bromophenols with Radical Scavenging Activity from Marine Red Alga *Symphyocladia latiuscula*. *Mar. Drugs* **2013**, *11*, 842–847. [CrossRef] [PubMed]

18. Ma, M.; Zhao, J.; Wang, S.; Li, S.; Yang, Y.; Shi, J.; Fan, X.; He, L. Bromophenols Coupled with Methyl γ-Ureidobutyrate and Bromophenol Sulfates from the Red Alga *Rhodomela confervoides*. *J. Nat. Prod.* **2006**, *69*, 206–210. [CrossRef]

19. Wang, B.G.; Gloer, J.B.; Ji, N.Y.; Zhao, J.C. Halogenated Organic Molecules of Rhodomelaceae Origin: Chemistry and Biology. *Chem. Rev.* **2013**, *113*, 3632–3685. [CrossRef]

20. Coll, J.C.; Wright, A.D. Tropical Marine Algae. III New Sesquiterpenes from *Laurencia majuscula* (Rhodophyta, Rhodophyceae, Ceramiales, Rhodomelaceae). *Aust. J. Chem.* **1989**, *42*, 1591–1603. [CrossRef]

21. Choi, J.S.; Park, H.J.; Jung, H.A.; Chung, H.Y.; Jung, J.H.; Choi, W.C. Cyclohexanonyl bromophenol from the red alga *Symphyocladia latiuscula*. *J. Nat. Prod.* **2000**, *63*, 1705–1706. [CrossRef]

22. Duan, X.J.; Li, X.M.; Wang, B.G. Highly brominated mono- and bis-phenols from the marine red alga *Symphyocladia latiuscula* with radical-scavenging activity. *J. Nat. Prod.* **2007**, *70*, 1210–1213. [CrossRef]

23. Bjordal, M.V. Field Studies on the Abundance, Growth and Biofouling of the Harvestable Red Alga *Vertebrata lanosa* (Linnaeus) T.A.Christensen. Master's Thesis, University of Bergen, Bergen, Norway, 2018.

24. Woodin, S.A.; Lindsay, S.M.; Lincoln, E. Biogenic bromophenols as negative recruitment cues. *Mar. Ecol. Prog. Ser.* **1997**, *157*, 303–306. [CrossRef]

25. Kicklighter, C.E.; Kubanek, J.; Hay, M.E. Do brominated natural products defend marine worms from consumers? Some do, most don't. *Limnol. Oceanogr.* **2004**, *49*, 430–441. [CrossRef]

26. Oh, K.B.; Lee, J.H.; Chung, S.C.; Shin, J.; Shin, H.J.; Kim, H.K.; Lee, H.S. Antimicrobial activities of the bromophenols from the red alga *Odonthalia corymbifera* and some synthetic derivatives. *Bioorg. Med. Chem. Lett.* **2008**, *18*, 104–108. [CrossRef] [PubMed]

27. Jesus, A.; Correia-da-Silva, M.; Afonso, C.; Pinto, M.; Cidade, H. Isolation and Potential Biological Applications of Haloaryl Secondary Metabolites from Macroalgae. *Mar. Drugs* **2019**, *17*, 73. [CrossRef] [PubMed]

 *marine drugs* 

Article

Variation in Lipid Components from 15 Species of Tropical and Temperate Seaweeds

Eko Susanto [1,2,*], A. Suhaeli Fahmi [2], Masashi Hosokawa [1] and Kazuo Miyashita [1]

[1] Laboratory of Bio-resources Chemistry, Graduate School of Fisheries Sciences, Hokkaido University, Hakodate 041-8611, Japan; hoso@fish.hokudai.ac.jp (M.H.); kmiya@fish.hokudai.ac.jp (K.M.)
[2] Department of Fisheries Products Technology, Faculty of Fisheries and Marine Science, Diponegoro University, Jl. Prof. Soedarto SH Kampus Tembalang, Semarang 50275, Republic of Indonesia; suhaeli.fahmi@live.undip.ac.id
* Correspondence: eko.susanto@live.undip.ac.id; Tel.: +81-138-40-8804

Received: 26 September 2019; Accepted: 29 October 2019; Published: 6 November 2019

Abstract: The present study describes the variation in lipid components from 15 species of seaweeds belonging to the Chlorophyta, Ochrophyta, and Rhodophyta phyla collected in tropical (Indonesia) and temperate (Japan) areas. Analyses were performed of multiple components, including chlorophylls, carotenoids, *n-3* and *n-6* polyunsaturated fatty acids (PUFAs), and alpha tocopherol (α-Toc). Chlorophyll (Chl) and carotenoid contents varied among phyla, but not with the sampling location. Chl a and b were the major chlorophylls in Chlorophyta. Chl a and Chl c were the main chlorophylls in Ochrophyta, while Chl a was the dominant chlorophylls in Rhodophyta. β-Carotene and fucoxanthin were detected as major seaweed carotenoids. The former was present in all species in a variety of ranges, while the latter was mainly found in Ochrophyta and in small quantities in Rhodophyta, but not in Chlorophyta. The total lipids (TL) content and fatty acids composition were strongly affected by sampling location. The TL and *n-3* PUFAs levels tended to be higher in temperate seaweeds compared with those in tropical seaweeds. The major *n-3* PUFAs in different phyla, namely, eicosapentaenoic acid (EPA) and stearidonic acid (SDA) in Ochrophyta, α-linolenic acid (ALA) and SDA in Chlorophyta, and EPA in Rhodophyta, accumulated in temperate seaweeds. Chlorophylls, their derivatives, and carotenoids are known to have health benefits, such as antioxidant activities, while *n-3* PUFAs are known to be essential nutrients that positively influence human nutrition and health. Therefore, seaweed lipids could be used as a source of ingredients with health benefits for functional foods and nutraceuticals.

Keywords: seaweeds; chlorophylls; carotenoids; *n-3* PUFAs; EPA

---

1. Introduction

The consumption of seaweeds has existed for millennia, not only in Japan, Korea, and China, but also in some Southeastern Asian countries, such as Malaysia, the Philippines, and Indonesia [1]. Recently, global seaweed production has risen considerably with more than 291 exploited species for food, feed, paper, fertilizer, medicinal, and industrial product uses [2]. This might be because of the increasing interest in seaweeds as nutraceuticals, functional foods, cosmetics, and pharmaceuticals because of the presence of characteristic nutrients and bioactives in seaweeds [3].

Seaweeds, promising marine products owing to their sustainability, contain valuable bioactive compounds that possess potential benefits for human health, such as anti-obesity, anti-diabetes, anticancer, and cardioprotective activities [4–13]. The most abundant nutrients in seaweeds are non-starch polysaccharides and minerals. The lipid content of seaweeds is low (1%–5% of the weight of a dry weight (DW) sample); however, the lipids comprise many kinds of bioactives, such as *n-3* and *n-6* polyunsaturated fatty acids (PUFAs), chlorophylls, carotenoids, terpenoids, and sterols [14–18].

Seaweeds are marine algae that are rich sources of lipid soluble pigments, such as chlorophylls and carotenoids. These natural pigments from seaweeds show several biological activities. Specifically, there have been many studies on the nutritional properties of seaweeds' carotenoids, such as fucoxanthin (Fx) [19]. Fx is a specific carotenoid found in brown seaweeds and known to help manage obesity and type 2 diabetes mellitus [20]. Chlorophylls and their various derivatives have a long-established history of use in traditional medicine and for therapeutic purposes [21]. Antioxidant and antitumor activities have been reported as general physiological activities of chlorophylls and their derivatives [21,22]. In addition, chlorophylls from seaweeds exhibit therapeutic effects, such as anti-degranulation properties in RBL-2H3 cells [23,24], anti-inflammatory activity in RAW 264.7 cells [25,26], and neuroprotective activity in PC12 cells [27]. Because of their valuable health benefits, these natural pigments have attracted interest in their potential application in functional foods, cosmetics, and pharmaceuticals [28], while more efforts are needed to clarify the detailed biological activities of chlorophylls from seaweeds.

In addition to the pigments, seaweed lipids generally contain high levels of PUFAs, such as α-linolenic acid (18:3*n-3*, ALA), stearidonic acid (18:4*n-3*, SDA), and eicosapentaenoic acid (20:5*n-3*, EPA) as *n-3* PUFAs, and arachidonic acid (20:4*n-6*, ARA) as an *n-6* PUFAs [14,16,18,29]. Many studies, including substantial epidemiological, case-control, clinical, genetic, and nutrigenetic approaches, demonstrate a reduction in cardiovascular disease risk from intake of *n-3* PUFAs, such as EPA and docosahexaenoic acid (22:6*n-3*, DHA), which are the active forms of *n-3* PUFAs [16,30]. ARA also plays an important role in biological systems, such as in the immune response, thrombosis, and brain function [16]. ARA and DHA are used as supplements in commercial infant formulas because both PUFAs are essential for infant neurodevelopment. Furthermore, the combination of ARA and DHA has also been found to improve age-related disorders of the brain and cognitive function [31].

Tropical seaweed species have significantly lower lipid content than cold-water species [16]. Comparative studies have revealed that the total lipids (TL) content of a major brown seaweed family, Sargassaceae, was higher in subarctic zones (approximately 5% of the DW) than in tropical zones (0.9%–1.8% of the DW) [14,32–34]. In addition, different seaweed species have different TL contents, and some species have shown exceptionally high TL contents. For example, Gosch et al. [35] found higher TL contents in three tropical brown seaweed species (10.80%–11.91% of the DW) and two tropical green seaweed species (12.14%–13.04% of the DW) collected in tropical North Queensland, Australia. Higher TL contents have also been reported in brown seaweeds collected in the Indian Ocean (7%–8% of the DW) [36], as well as the Hawaiian coast (16%–20% of the DW) [37]. Therefore, seaweed lipids content varies greatly by species, geographical location, season, temperature, salinity, and light intensity, as well as interactions among these factors [14,34,38–42]. The content of lipid-soluble bioactive compounds in seaweeds also depends on several factors, such as species, sampling sites, environment, and seasonal variation [14,34,37–47].

Although many papers have reported variations in the lipid components from different species of seaweeds collected in various areas, most of them have focused on the fatty acids composition [48–51] and on specific bioactives, such as Fx [14,34,38]. In finding a potential target in seaweed as an ingredient for foods, animal feed, cosmetics, and nutraceuticals, information on the wide range of lipid and related compound profiles, especially on lipid soluble pigments, will be required. Among seaweed pigments, chlorophylls play a key role in photosynthesis, while carotenoids serve as antenna pigments by absorbing and passing light energy to chlorophyll. Both pigments are not only essential for photosynthesis, but also show various health benefits, as described above. However, little information is available on both lipid soluble pigment levels in seaweeds. The investigation of lipid-soluble compounds in seaweed grown near the coast of Indonesia and northern coast of Japan is limited [34].

Therefore, the primary aim of this study was to determine chlorophylls and carotenoids in 15 different species of seaweeds belonging to Ochrophyta (brown seaweeds), Chlorophyta (green seaweeds), and Rhodophyta (red seaweeds). In addition, the present study also revealed the fatty acids and α-tocopherol (α-Toc) profiles of these different seaweeds. Each sample was collected from tropical

(Indonesia) or temperate (Japan) waters. The present study will be useful for the comparison between the characteristics of lipid soluble compounds from seaweeds collected from the two different areas.

2. Results

2.1. Total Lipids

The TL of seaweeds from Indonesia (tropical area) and Japan (temperate area) are described in Table 1. The TL level varied among species, ranging from 6.91 to 62.48 mg·g^{-1} DW. On average, the highest TL content was recorded in Ochrophyta (42.44 mg·g^{-1} DW), followed by Chlorophyta (34.60 mg·g^{-1} DW), and Rhodophyta (17.75 mg·g^{-1} DW). In the present study, the lowest TL content was reported in *Chondrus yendoi* (6.91 mg·g^{-1} DW), while *Ulva australis* had the highest total lipids content among all seaweeds tested.

2.2. Pigments

The primary photosynthetic pigments in seaweed are comprised of chlorophylls and carotenoids. In this study, we investigated the variation of these pigment contents in different phyla of seaweeds, which were collected from two regions (Tables 2 and 3). Chlorophylls easily decompose to produce their derivatives, pheophytins, by elimination of magnesium. This change from chlorophylls to pheophytins is accelerated by heat treatment. In the dried samples from tropical seaweeds, chlorophyll a (Chl a) and chlorophyll b (Chl b) levels were very low or undetected, while their corresponding derivatives, pheophytin a (Phy a) and pheophytin b (Phy b), were major chlorophyll components in these seaweeds. The lower level of Chl a and Chl b in the tropical seaweeds was derived from the decomposition of these chlorophylls to corresponding pheophytin derivatives during the drying process.

Table 1. List of seaweeds examined and total lipids (TL) content of these seaweeds *.

Harvesting Location	Phylum	Family	Seaweeds	Local Name	Collection Date	Total Lipids (mg·g^{-1} DW)
Tual, Indonesia [a,d]	Chlorophyta	Caulerpaceae	*Caulerpa lentillifera*	*Lat*	Feb-17	15.75 ± 0.82
Hakodate, Japan [b,f]	Chlorophyta	Ulvaceae	*Ulva australis*	*Anaaosa*	Jun-17	62.48 ± 3.05
Hakodate, Japan [b,f]	Chlorophyta	Ulvaceae	*Ulva intestinalis*	*Bouaonori*	Jun-17	37.46 ± 7.56
Tual, Indonesia [a,d]	Chlorophyta	Ulvaceae	*Ulva reticulata*	*Lumut daun*	May-17	22.70 ± 3.01
Hakodate, Japan [b,f]	Ochrophyta	Agaraceae	*Costaria costata*	*Sujime*	May-17	33.71 ± 1.86
Hakodate, Japan [b,f]	Ochrophyta	Alariaceae	*Undaria pinnatifida*	*Wakame*	May-17	58.10 ± 4.56
Hakodate, Japan [b,f]	Ochrophyta	Laminariaceae	*Saccharina japonica*	*Konbu*	May-17	37.42 ± 6.23
Tual, Indonesia [a,c,d]	Ochrophyta	Sargassaceae	*Sargassum aquifolium* [a]	*Pama*	Feb-17	20.87 ± 0.70
Hakodate, Japan [b,f]	Ochrophyta	Sargassaceae	*Sargassum fusiforme*	*Hijiki*	May-17	48.54 ± 1.61
Iwate, Japan [b,g]	Ochrophyta	Sargassaceae	*Sargassum horneri*	*Akamoku*	May-15	55.97 ± 2.51
Hakodate, Japan [b,f]	Rhodophyta	Endocladiaceae	*Gloiopeltis furcata*	*Fukurofunori*	May-17	8.94 ± 1.45
Hakodate, Japan [b,f]	Rhodophyta	Gigartinaceae	*Chondrus yendoi*	*Kurohaginnansou*	Jun-17	6.91 ± 0.21
Hakodate, Japan [b,f]	Rhodophyta	Gigartinaceae	*Mazzaella japonica*	*Akabaginnansou*	Jun-17	14.25 ± 0.30
Jepara, Indonesia [a,e]	Rhodophyta	Gracilariaceae	*Gracilariopsis longissima*	*agar-agar*	May-17	8.86 ± 0.15
Hakodate, Japan [b,f]	Rhodophyta	Rhodomelaceae	*Chondria crassicaulis*	*Yuna*	Jun-17	49.77 ± 2.40

* The data value is expressed as the mean ± SD of three replicate measurements; [a] fresh samples collected, immediately washed, air dried, and kept in refrigerator (tropical seaweeds); [b] fresh samples collected, immediately kept in refrigerator (temperate seaweeds); [c] non-edible seaweeds; [d] latitudes/longitudes: 5° 37′27.3396″ S 132°43′22.3392″ E; [e] latitudes/longitudes: 6.5868° S, 110.6444° E; [f] latitudes/longitudes: 41°45′ N, 140°49′ E; [g] latitudes/longitudes: 39°28′46.5″ N 142°00′11.9″ E. DW, dry weight.

Although Chl a was detected in all seaweeds from the temperate area, its level widely ranged from 1.74 (*Chondria crassicaulis*) to 268.82 (*Sargassum horneri*) mg·100 g^{-1} DW. Chlorophyll b, the characteristic pigment in Chlorophyta, was present in small amounts and ranged from 16.83 (*Caulerpa lentillifera*) to 101.50 (*U. intestinalis*) mg·100 g^{-1} DW. A unique finding was the occurrence of chlorophyll c (Chl c), the characteristic accessory pigment in Ochrophyta. It was not only found in Ochrophyta, but also in Rhodophyta (*C. crassicaulis* and *Mazaella japonica*). Phy a was abundantly found in seaweeds of the Laminariaceae family and Rhodomelaceae family, while Phy b was only found in Chlorophyta.

Table 2. Composition of chlorophylls and pheophytins in 15 species of seaweeds (mg·100 g^{-1} DW) *.

Seaweeds	Chl a	Phy a	Chl b	Phy b	Chl c (c1 + c2) [b]	Total Chlorophylls
Caulerpa lentillifera	n.d.	118.68 ± 44.41	16.83 ± 4.60	90.28 ± 8.97	n.d.	225.79
Ulva australis	81.90 ± 4.94	26.17 ± 3.27	64.45 ± 9.36	n.d.	n.d.	172.52
Ulva intestinalis	115.57 ± 51.28	165.98 ± 14.70	101.50 ± 29.21	64.38 ± 11.30	n.d.	447.43
Ulva reticulata	n.d.	115.85 ± 21.17	18.04 ± 0.61	267.69 ± 158.86	n.d.	401.58
Costaria costata	16.28 ± 3.80	291.92 ± 67.84	n.d.	n.d.	21.78 ± 0.84	329.97
Undaria pinnatifida	54.67 ± 14.60	423.42 ± 57.72	n.d.	n.d.	38.58 ± 5.03	518.67
Saccharina japonica	26.13 ± 16.79	425.42 ± 19.17	n.d.	n.d.	17.11 ± 2.91	469.24
Sargassum aquifolium [a]	12.17 ± 4.02	149.64 ± 56.00	n.d.	n.d.	2.12 ± 0.21	163.93
Sargassum fusiforme	210.72 ± 43.63	107.59 ± 30.30	n.d.	n.d.	18.20 ± 0.34	336.51
Sargassum horneri	268.82 ± 59.93	220.86 ± 22.84	n.d.	n.d.	23.29 ± 4.9	512.97
Gloiopeltis furcata	2.98 ± 1.96	28.21 ± 4.92	n.d.	n.d.	n.d.	31.19
Chondrus yendoi	4.77 ± 2.10	29.29 ± 2.21	n.d.	n.d.	n.d.	34.06
Mazzaella japonica	6.70 ± 1.28	16.25 ± 1.85	n.d.	n.d.	7.63 ± 1.25	30.58
Gracilariopsis longissima	n.d.	14.66 ± 2.24	n.d.	n.d.	n.d.	14.66
Chondria crassicaulis	1.74 ± 1.19	322.72 ± 56.48	n.d.	n.d.	10.18 ± 1.29	334.64

* The data value is expressed as the mean ± SD of three replicate measurements; [a] non-edible seaweed; [b] Chl c composes of Chl c1 and Chl c2. However, both chlorophylls could not be separated on HPLC to be detected as one peak; therefore, Chl c1 and Chl c2 were expressed as Chl c altogether in the present study. Quantification of Chl c was done using the calibration curve of c1 standard, because Chl c1 has been reported to be the major chlorophyll c in brown seaweeds [37]. n.d. = not detected. Abbreviations: Chl a: chlorophyll a; Chl b: chlorophyll b; Chl c: chlorophyll c; Phy a: pheophytin a; Phy b: pheophytin b.

Carotenoids, accessory pigments in photosynthesis, also varied among seaweed species (Table 3). Provitamin A carotenoids, α-carotene (α-Car) and β-carotene (β-Car), exhibited different patterns. β-Car was present in all species with a range from 1.12 to 58.24 mg·100 g^{-1} DW, while α-Car was detected only in Chlorophyta and Rhodophyta. Fx was mainly found in Ochrophyta and found in a small amount in Rhodophyta, but not in Chlorophyta. In addition, Zeaxanthin (Zx) content was relatively higher in two *Sargassum* species (*S. horneri* and *S. fusiforme*) and *Undaria pinnatifida* collected from the temperate area. Lutein (Lut) was found in Rhodophyta and Chlorophyta with a range from 0.62 mg·100 g^{-1} DW (*M. japonica*) to 124.82 mg·100 g^{-1} DW (*U. australis*), but was not found in Ochrophyta. Violaxanthin (Vx) was only found in Ochrophyta and Chlorophyta collected from Japan with a range from 1.45 mg·100 g^{-1} DW (*Costaria costata*) to 23.30 mg·100 g^{-1} DW (*U. australis*). Neoxanthin, only present in Chlorophyta, also showed a wide range in its content from 30.63 mg·100 g^{-1} DW (*U. reticulata*) to 185.56 mg·100 g^{-1} DW (*U. australis*). Higher total carotenoid levels were found in Ochrophyta (*S. aquifolium, C. costata, S. japonica, U. pinnatifida, S. fusiforme, S. horneri*) from the two different areas. Chlorophyta species collected from the temperate area (*U. australis* and *U. intestinalis*) also showed higher total carotenoid levels, whereas Chlorophyta from the tropical area (*C. lentillifera* and *U. reticulata*) showed relatively low levels. The same trend was also found among Rhodophyta species collected from the temperate area (*C. crassicaulis, C. yendoi, Gloiopeltis furcata, M. japonica*) and those collected from the tropical area (*Gracilariopsis longissima*).

Table 3. Carotenoids content of 15 species of seaweeds (mg·100g^{-1} DW) *.

Seaweeds	β-Car	α-Car	Zx	Lut	Vx	Nx	Fx	Total Carotenoids
Caulerpa lentillifera	1.84 ± 0.70	17.15 ± 2.96	n.d.	1.02 ± 0.53	n.d.	n.d.	n.d.	20.01
Ulva australis	58.24 ± 6.52	63.01 ± 0.04	n.d.	124.82 ± 13.14	23.30 ± 0.74	185.56 ± 59.65	n.d.	454.93
Ulva intestinalis	39.91 ± 9.54	17.83 ± 6.66	n.d.	77.78 ± 18.14	13.15 ± 4.19	153.70 ± 84.20	n.d.	302.37
Ulva reticulata	3.89 ± 0.39	4.83 ± 0.35	n.d.	14.12 ± 2.85	n.d.	30.63 ± 7.84	n.d.	53.47
Costaria costata	7.31 ± 0.86	n.d.	n.d.	n.d.	1.45 ± 0.19	n.d.	97.60 ± 12.87	106.36
Undaria pinnatifida	25.58 ± 3.71	n.d.	12.03 ± 3.99	n.d.	3.51 ± 0.61	n.d.	169.48 ± 12.98	210.61
Saccharina japonica	18.10 ± 0.61	n.d.	3.04 ± 0.13	n.d.	1.47 ± 0.25	n.d.	154.71 ± 11.29	177.32
Sargassum aquifolium [a]	12.51 ± 1.24	n.d.	2.45 ± 0.32	n.d.	n.d.	n.d.	108.44 ± 9.17	123.40
Sargassum fusiforme	44.70 ± 7.52	n.d.	7.06 ± 1.65	n.d.	12.15 ± 2.45	n.d.	140.93 ± 12.98	204.84
Sargassum horneri	42.70 ± 6.70	n.d.	29.21 ± 2.72	n.d.	4.74 ± 0.98	n.d.	216.50 ± 31.97	293.15
Gloiopeltis furcata	8.99 ± 2.53	3.73 ± 0.70	0.47 ± 0.30	+(8.76 ± 2.31) [c]	n.d.	n.d.	3.43 ± 0.48	25.38
Chondrus yendoi	2.07 ± 0.19	4.60 ± 0.47	n.d.	1.78 ± 0.81	n.d.	n.d.	2.57 ± 0.07	11.02
Mazzaella japonica	2.23 ± 0.18	1.95 ± 0.27	n.d.	0.62 ± 0.12	n.d.	n.d.	7.26 ± 0.42	12.06
Gracilariopsis longissima	1.12 ± 0.25	n.d.	+(0.75 ± 0.16) [b]	n.d.	n.d.	n.d.	n.d.	1.87
Chondria crassicaulis	13.78 ± 2.40	8.19 ± 0.82	+(4.94 ± 1.66) [b]	n.d.	n.d.	n.d.	67.76 ± 9.63	94.67

*The data value is expressed as the mean ± SD of three replicate measurements; [a] non-edible seaweeds; [b] Zx and Lut were calculated from the calibration curve of Zx standard; [c] Lut and Zx were calculated from the calibration curve of Lut standard. n.d. = not detected. Abbreviations: β-Car: β-carotene; α-Car: α-carotene; Zx: zeaxanthin; Lut: lutein; Vx: violaxanthin; Nx: neoxanthin; Fx: fucoxanthin.

2.3. Fatty Acids

The details of fatty acids composition in seaweed lipids are listed in Table 4 and Tables S1a, S1b and S1c. All values are expressed as percentages of total fatty acids (FAs) content. The absolute content of total *n-3* PUFAs (mg·g^{-1} DW) is also shown in Table 4. The fatty acids composition varied widely among seaweed species and sampling locations.

In the three phyla, Ochrophyta, Chlorophyta, and Rhodophyta, palmitic acid (C16:0) was the predominant fatty acids (Tables S1a, S1b and S1c). The temperate seaweeds contained much more total PUFAs (28.15% (*C. crassicaulis*) to 51.28% (*S. japonica*)) than the tropical seaweeds (13.06% (*C. lentillifera*) to 25.75% (*U. reticulata*)) (Table 4). Specifically, the absolute amounts of total *n-3* PUFAs in the temperate seaweeds were much higher than those in the tropical seaweeds. Total saturated fatty acids (SFAs) contents in the tropical seaweeds were relatively higher than those in the temperate seaweeds, except for *M. japonica*. Furthermore, monounsaturated fatty acids (MUFAs), dominantly comprising C16:1*n-7* and C18:1*n-9*, were higher in temperate Rhodophyta, namely, *C. crassicaulis*. Overall, in the tropical seaweeds, SFAs accounted for 29.80% to 47.78%, followed by PUFAs (13.06% to 25.75%) and MUFAs (6.88% to 18.88%), while PUFAs (36.23% to 51.28%) were the most dominant FA class in the temperate seaweeds, except for *C. crassicaulis* (28.15%) and *M. japonica* (31.21%). Furthermore, Chlorophyta contained relatively less C20 FAs; however, C20 FAs were the major FAs in Ochrophyta (*S. horneri, S. fusiforme*) and in Rhodophyta, especially those collected in the temperate area (Table 4).

Total PUFAs levels varied among species, ranging from 13.06% (*C. lentillifera*) to 51.28% (*S. japonica*) of total fatty acids. These seaweeds contained EPA (C20:5*n-3*), ARA (C20:4*n-6*), ALA (C18:3*n-3*), and SDA (C18:4*n-3*) as the major PUFAs; however, there was significant variation in the level of these PUFAs in the TL content from each species. The main PUFAs in Ochrophyta collected from the temperate area were ARA (10.55% (*U. pinnatifida*) to 14.87% (*S. horneri*)) and EPA (8.36% (*C. costata*) to 13.04% (*S. japonica*)), whereas the main PUFAs in Ochrophyta collected from the tropical area (*S. aquifolium*) was ALA (10.40%), and the EPA level of this seaweed was very low (0.96%) (Table S1a). The same trend was found in Rhodophyta (Table S1c). Rhodophyta collected from the temperate area contained relatively high levels of EPA (13.08% (*C. crassicaulis*) to 35.81% (*G. furcata*)) and of ARA (3.33% (*C. crassicaulis*) to 17.26% (*C. yendoi*)); however, the main PUFAs of Rhodophyta collected in the tropical area (*G. longissima*) was ARA (11.41%), and the EPA level was 0.21%. Chlorophyta collected from the two geographical locations contained ALA as its major PUFAs (Table S1b). Among the 15 seaweeds analyzed, *G. furcata* exhibited the highest value of total *n-3* PUFAs (38.11%) with 35.81% EPA, whereas *U. intestinalis* exhibited the highest value of *n-3* PUFAs expressed in mg·g^{-1} DW (9.84 mg·g^{-1} DW) (Table 4).

Table 4. Fatty acids composition (weight % of total fatty acids (FAs)) *.

Seaweeds	Σ C16	Σ C18	Σ C20	Σ SFAs	Σ MUFAs	Σ PUFAs	Σ n-3 PUFAs	Σ n-6 PUFAs	Σ n-3 PUFAs [b]
Caulerpa lentillifera	33.69 ± 0.64	13.57 ± 0.91	2.84 ± 0.53	29.80 ± 1.65	9.08 ± 2.75	13.06 ± 0.32	7.31 ± 0.41	5.75 ± 0.10	0.01
Ulva australis	24.37 ± 0.35	35.41 ± 0.44	4.23 ± 0.08	25.67 ± 0.47	3.45 ± 0.24	36.23 ± 0.82	29.00 ± 0.66	7.22 ± 0.22	6.83
Ulva intestinalis	23.26 ± 1.27	36.61 ± 0.87	7.72 ± 2.31	24.37 ± 1.21	4.24 ± 0.87	39.16 ± 1.54	29.03 ± 2.02	10.14 ± 0.65	9.84
Ulva reticulata	44.66 ± 0.38	23.96 ± 0.38	6.01 ± 0.63	43.01 ± 0.46	6.88 ± 0.42	25.75 ± 0.52	23.25 ± 0.36	2.50 ± 0.43	0.02
Costaria costata	23.88 ± 1.12	30.84 ± 1.13	22.00 ± 1.92	37.46 ± 2.02	15.92 ± 1.46	36.59 ± 2.62	15.38 ± 2.15	21.21 ± 0.79	0.53
Undaria pinnatifida	25.73 ± 1.22	38.31 ± 4.44	23.01 ± 0.39	31.12 ± 1.75	14.07 ± 0.90	48.19 ± 5.72	28.94 ± 5.71	19.25 ± 0.30	4.66
Saccharina japonica	16.79 ± 0.75	38.83 ± 0.79	26.93 ± 0.18	24.35 ± 1.18	15.76 ± 0.64	51.28 ± 0.61	31.24 ± 1.91	20.04 ± 1.57	2.26
Sargassum aquifolium [a]	48.45 ± 2.70	21.83 ± 0.75	13.46 ± 0.30	46.80 ± 5.31	18.88 ± 1.40	22.90 ± 0.89	6.26 ± 0.37	16.64 ± 0.52	0.10
Sargassum fusiforme	24.80 ± 0.52	27.01 ± 0.43	30.38 ± 0.45	27.62 ± 0.58	11.95 ± 0.22	47.32 ± 1.08	30.22 ± 0.78	17.11 ± 0.27	2.66
Sargassum horneri	26.35 ± 1.30	30.25 ± 0.96	30.49 ± 1.14	26.98 ± 1.49	14.24 ± 0.25	49.00 ± 2.36	26.97 ± 1.64	22.03 ± 1.16	4.03
Gloiopeltis furcata	23.34 ± 1.31	19.36 ± 0.11	44.89 ± 2.24	26.97 ± 1.45	18.05 ± 0.23	45.33 ± 2.36	38.11 ± 2.12	7.22 ± 0.24	0.46
Chondrus yendoi	33.12 ± 0.48	12.60 ± 0.23	42.35 ± 0.89	37.28 ± 0.62	9.89 ± 0.19	43.91 ± 0.85	24.23 ± 0.81	19.68 ± 0.21	0.68
Mazzaella japonica	40.50 ± 0.38	16.42 ± 0.05	30.12 ± 0.29	45.29 ± 0.51	14.49 ± 0.61	31.21 ± 0.72	20.05 ± 0.50	11.15 ± 0.24	0.10
Gracilariopsis longissima	46.44 ± 0.91	10.16 ± 0.50	15.21 ± 0.69	47.78 ± 2.36	11.18 ± 1.73	13.90 ± 0.64	1.63 ± 0.26	12.28 ± 0.71	0.01
Chondria crassicaulis	38.41 ± 1.54	19.93 ± 1.20	18.50 ± 0.69	36.88 ± 1.23	20.49 ± 0.88	28.15 ± 1.37	19.82 ± 0.71	8.33 ± 0.82	2.73

* The data value is expressed as the mean ± SD of three replicate measurements; [a] non-edible seaweeds; [b] absolute content of total n-3 PUFAs (mg·g^{-1} DW). Abbreviations: SFAs: saturated fatty acids; MUFAs: mono unsaturated fatty acids; PUFAs: polyunsaturated fatty acids.

2.4. Nutritional Quality Index

The nutritional quality index based on the fatty acids composition was calculated in this study (Table 5). The *n-6/n-3* ratio was less than 3, except for *G. longissima* (7.69), and most of the ratios were less than 1.0. In general, the atherogenicity index (AI) and thrombogenicity index (TI) values increase with the decreasing degree of unsaturation of FAs, while the fatty acids hypocholesterolemic/hypercholesterolemic ratio (h/H) increases with the increasing unsaturated fatty acids composition. Therefore, AI and TI tended to be higher in the tropical seaweeds compared with those in temperate seaweeds. The h/H ratios of all temperate seaweeds were higher than those of the tropical seaweeds. In addition, the unsaturation index (UI) in all seaweeds was more than 3, and in Rhodophyta collected in the temperate area, it was more than 4.

Table 5. Nutritional quality index of different seaweeds judged from fatty acids composition.

Seaweeds	*n-6/n-3* PUFAs	AI	TI	h/H	UI
Caulerpa lentillifera	0.79 ± 0.05	1.53 ± 0.28	0.96 ± 0.13	0.44 ± 0.04	3.00 ± 0.07
Ulva australis	0.25 ± 0.01	0.64 ± 0.01	0.26 ± 0.01	1.04 ± 0.02	3.27 ± 0.01
Ulva intestinalis	0.35 ± 0.05	0.57 ± 0.04	0.23 ± 0.01	1.39 ± 0.13	3.19 ± 0.03
Ulva reticulata	0.11 ± 0.02	1.31 ± 0.05	0.35 ± 0.27	0.33 ± 0.02	3.67 ± 0.01
Costaria costata	1.40 ± 0.18	1.37 ± 0.16	0.54 ± 0.08	1.28 ± 0.12	3.63 ± 0.06
Undaria pinnatifida	0.68 ± 0.13	0.69 ± 0.08	0.28 ± 0.06	1.66 ± 0.23	3.71 ± 0.04
Saccharina japonica	0.65 ± 0.09	0.67 ± 0.04	0.20 ± 0.02	2.23 ± 0.08	3.88 ± 0.04
Sargassum aquifolium [a]	2.66 ± 0.07	1.42 ± 0.25	1.23 ± 0.20	0.67 ± 0.07	3.40 ± 0.01
Sargassum fusiforme	0.57 ± 0.01	0.62 ± 0.02	0.24 ± 0.01	1.76 ± 0.06	3.93 ± 0.00
Sargassum horneri	0.82 ± 0.05	0.53 ± 0.05	0.26 ± 0.03	1.82 ± 0.16	3.86 ± 0.02
Gloiopeltis furcata	0.19 ± 0.00	0.50 ± 0.05	0.19 ± 0.02	2.46 ± 0.26	4.75 ± 0.01
Chondrus yendoi	0.81 ± 0.03	0.79 ± 0.03	0.41 ± 0.02	1.48 ± 0.05	4.46 ± 0.01
Mazzaella japonica	0.56 ± 0.01	1.14 ± 0.03	0.58 ± 0.03	0.97 ± 0.03	4.49 ± 0.00
Gracilariopsis longissima	7.69 ± 1.58	2.03 ± 0.24	2.80 ± 0.25	0.44 ± 0.03	3.87 ± 0.02
Chondria crassicaulis	0.42 ± 0.03	1.19 ± 0.09	0.47 ± 0.03	0.86 ± 0.05	4.13 ± 0.03

[a] Non-edible seaweeds. Abbreviations: AI: atherogenicity index; TI: thrombogenicity index; h/H: fatty acids hypocholesterolemic/hypercholesterolemic ratio; UI: unsaturation index.

2.5. Alpha Tocopherol

α-Toc content varied with seaweed species (Table 6). Relatively lower α-Toc levels were found in Chlorophyta (*C. lentillifera*, *U. reticulata*, *U. australis*, *U. intestinalis*). Seaweeds belonging to Ochrophyta and Rhodophyta showed relatively higher levels of α-Toc, except for *C. crassicaulis*. Specifically, *C. yendoi* showed the highest α-Toc content (9.34 mg·100 g^{-1}), followed by two species of *Sargassum* (*S. fusiforme* and *S. horneri*).

Table 6. α-Tocopherols content (mg·100g^{-1} DW) *.

Seaweeds	α-Toc
Caulerpa lentillifera	0.87 ± 0.21
Ulva australis	0.44 ± 0.01
Ulva intestinalis	0.83 ± 0.06
Ulva reticulata	1.13 ± 0.08
Costaria costata	1.37 ± 0.10
Undaria pinnatifida	1.09 ± 0.05
Saccharina japonica	1.82 ± 0.02
Sargassum aquifolium [a]	2.40 ± 0.02
Sargassum fusiforme	3.56 ± 0.08
Sargassum horneri	3.65 ± 0.031
Gloiopeltis furcata	2.71 ± 0.34
Chondrus yendoi	9.34 ± 0.19
Mazzaella japonica	1.72 ± 0.009
Gracilariopsis longissima	2.58 ± 0.015
Chondria crassicaulis	0.54 ± 0.009

* The data value is expressed as the mean ± SD of three replicate measurements, [a] non-edible seaweeds.

2.6. Multivariate Analysis

A statistical analysis was performed using the 15 fatty acids composition measured in this study in order to establish the relationship between geographical location and phylum. The fifteen FA variables explained the variability presented in the data, with principle component 1 (PC-1) accounting for 33.1% and principle component 2 (PC-2) for 22.5% of the variation (Figure 1a). Although the total percentage was only 55.6%, the analysis revealed a correlation of each seaweed within the phyla. Axis I separated the Rhodophyta species from the Chlorophyta, while Ochrophyta was positioned intermediately between these two, based on the principle component analysis (PCA) shown in the bi-plot data. However, these results could not determine whether the geographical location affected the fatty acids composition. Therefore, the PCA on the FA groups was analyzed. The bi-plot of the FA groups' data matrix showed 81.5% of variance with PC-1 contributing 51.9% and PC-2 contributing 22.5% (Figure 1b). These data exhibited a broad diversity of fatty acids in the seaweed samples. The analysis revealed that the tropical seaweeds were rich in SFAs, while temperate seaweeds were dominated by PUFAs, including *n-3* and *n-6* PUFAs, excluding *C. costata*, *M. japonica*, and *C. crassicaulis*. When PCA was used for clustering of the fatty acids group, the seaweeds could be grouped within the sampling location.

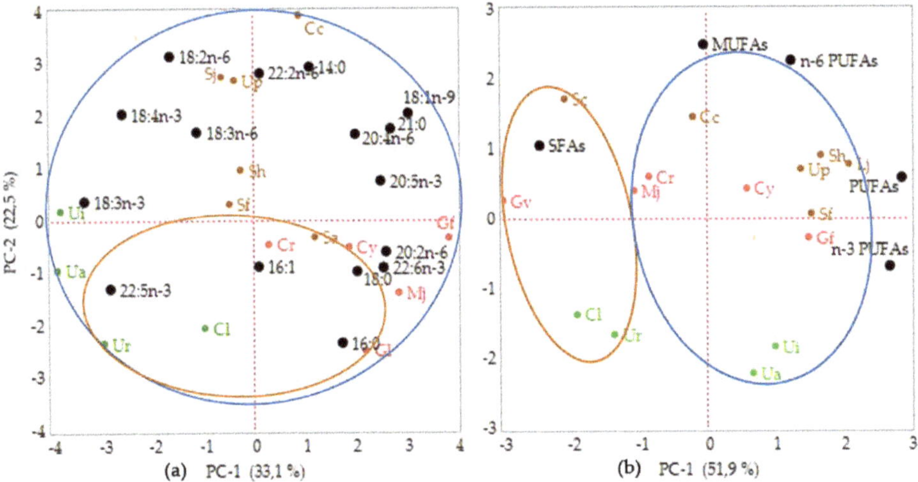

Figure 1. Bi-plot of the seaweeds' sample received from PCA analysis of (**a**) fatty acids (FAs) data composition (15 fatty acids), (**b**) FAs group (SFAs, MUFAs, PUFAs, *n-3* PUFAs, and *n-6* PUFAs). Abbreviations: Cl: *Caulerpa lentillifera*; Ua: *Ulva australis*; Ui: *Ulva intestinalis*; Ur: *Ulva reticulata*; Cc: *Costaria costata*; Up: *Undaria pinnatifida*; Sj: *Saccharina japonica*; Sa: *Sargassum aquifolium*; Sf: *Sargassum fusiforme*; Sh: *Sargassum horneri*, Cr: *Chondria crassicaulis*; Cy: *Chondrus yendoi*; Gf: *Gloiopeltis furcata*; Gl: *Gracilariopsis longissima*; Mj: *Mazzaella japonica*. The seaweeds species codes are colored based on the seaweeds' phyla (green: Chlorophyta; brown: Ochrophyta; red: Rhodophyta). The fatty acids composition and FAs group variable are colored in black. The circles were drawn based on geographical sampling location (orange circle: tropical seaweeds; blue circle: temperate seaweeds).

3. Discussion

Seaweeds have been an integral part of Japanese cuisine for the past one hundred years, similar to the diet of coastal communities in several Indonesian regions [52,53]. Industries use seaweeds as marine hydrocolloids and ingredients for foods and animal feed. Utilization of seaweeds as a dietary staple is because of their characteristic nutrients, which are shown to have therapeutic effects in humans [28,54–57]. Seaweeds are a source of novel bioactive compounds, such as certain polysaccharides and antioxidants, which are not found in terrestrial plants. The consumption of seaweeds has been linked to a lower incidence of chronic diseases, such as cancer, hyperlipidemia, and coronary heart disease [58]. Recently, seaweed lipids have drawn increased interest because of the presence of many kinds of bioactives, such as *n-3* and *n-6* PUFAs, carotenoids, and sterols [16]. Although seaweeds have a lower lipid content than marine fish, they are still a potential source of functional lipids because of their large stock in coastal waters.

Among bioactive seaweed lipids, much attention has been paid to the lipid soluble pigments, such as chlorophylls and carotenoids [28]. In addition, seaweed lipids are known to be rich in *n-3* and *n-6* PUFAs. However, little information has been available on the relationship between the multiple lipid bioactives from different species of seaweed. The present study analyzed the lipids from 15 different species of seaweed collected from two different regions, tropical (Indonesia) and temperate (Japan) regions, with an emphasis on the content of potentially bioactive components, chlorophylls, carotenoids, *n-3* and *n-6* PUFAs, and α-Toc. The results confirmed that the TL content and composition varied widely among species. Moreover, TL was sometimes affected by the location. As shown in Table 1, the TL level was higher in the temperate seaweeds than in the tropical seaweeds, although there were exceptions. It is generally accepted that the TL content of seaweeds varies with temperature, salinity, and light intensity. Specifically, the temperature strongly affects the TL level of seaweeds.

Several studies have revealed that the TL content in temperate seaweeds is much higher than that in tropical seaweeds. The TL variation observed in the present study is consistent with previous studies [14,33,34,46].

Although the major components of the seaweed TL content are glycerolipids, such as glycoglycerolipids [16], other kinds of lipid-related compounds, such as photosynthetic pigments, are also abundant [28]. Chlorophylls and carotenoids are representative and major photosynthetic pigments found in seaweeds [59]. Chlorophylls are the most abundant pigments on earth and allow seaweeds to convert light into biological energy [22]. Carotenoids also have an important function as light energy harvesters in photosynthesis by passing on light excitation to chlorophylls. In addition, carotenoids act as antioxidants that inactivate reactive oxygen species formed by exposure to light and air during photosynthesis. Both lipid soluble pigments have been known to show a variety of nutritional effects in humans [20,25,26,28,60].

In seaweeds, the most important chlorophyll is Chl a, which absorbs the energy from the wavelength of violet blue and orange-red light [61]. Chl b and Chl c are also important chlorophylls found in seaweeds. They are accessory pigments in the antenna system of Chl a [62–65]. The present study revealed that the variation of these three kinds of chlorophylls was mainly affected by phylum, namely, Ochrophyta, Chlorophyta, and Rhodophyta, but not by sampling location (Table 2). In addition, the study also showed less Chl a and Chl b in the tropical seaweeds compared with those in the temperate seaweeds, while considerable amounts of the corresponding pheophytin derivatives, Phy a and Phy b, were detected in the tropical seaweeds. Seaweeds from the tropical area were dried before the analysis; therefore, most of the Chl a and Chl b in the tropical seaweed samples would have decomposed to produce corresponding derivatives, Phy a and Phy b, respectively. The heating treatments on seaweeds cause the central magnesium ion losses in the chlorophylls, which convert into degradation products, i.e., pheophytins [21,66]. Therefore, considering the degradation of Chl a and Chl b during the treatment of the tropical seaweeds, the major chlorophylls may be considered to be Chl a and Chl c for Ochrophyta, Chl a and Chl b for Chlorophyta, and Chl a for Rhodophyta. This study is in accordance with another study by Chen et al. [43].

In the present study, Chl c was found in two species of Rhodophyta, *C. crassicaulis* and *M. japonica* (Table 2). This finding is noteworthy because Chl c has been regarded as a specific chlorophyll in Ochrophyta [43,67]. Wilhelm [68] detected Chl c in Chlorophyta, namely, *Mantoniella comigrates*; however, there has been no report on the presence of Chl c in Rhodophyta. Chl c is also found in diatoms as their major chlorophyll. Diatoms use the blue-green spectral region for photosynthetic energy. They have Chl a and Chl c as their major chlorophylls and Fx as their major carotenoid. In the diatoms, the fucoxanthin–chlorophyll protein complex is mainly responsible for light-harvesting [69,70]. The present study suggests that the fucoxanthin–chlorophyll protein complex may be the key molecular complex for light harvesting, not only in diatoms, but also in several kinds of seaweeds [69,71].

As described above, the content of chlorophylls and their derivatives mainly varied among the seaweed phyla, but not with the different locations. The same trend was also found in the carotenoids composition (Table 3). The carotenoid content in each phylum is related to the difference in the carotenoid biosynthesis of each species [59]. Ochrophyta species were rich in Fx, with a considerable amount of β-Car and Zx and with a small amount of Vx. The presence of Fx, Zx, and Vx in Ochrophyta suggests the involvement of the xanthophyll-cycle pathway in carotenoid biosynthesis in this phylum [14,59]. On the other hand, α-Car and Lut were detected as major carotenoids in Chlorophyta and Rhodophyta. The presence of both carotenoids is closely related to the presence of the α-Car pathway in Rhodophyta and Chlorophyta [72]. Two species (*C. yendoi* and *M. japonica*) of Rhodophyta analyzed in this study lacked Zx. The lack of Zx may be related to photo-acclimation in these species [73].

Chlorophylls and carotenoids have been considered as the most important biomolecules for photosynthesis. In addition, both pigments have attracted major interest from biochemists and nutritionists alike, because they are known to have significant biochemical and physiological effects, and primarily exhibit a positive influence on human nutrition and health [28]. Specifically, much attention has been paid to the antioxidant activity of chlorophylls and carotenoids. Carotenoids are one of the most famous natural antioxidants, together with tocopherols and polyphenols. They are regarded as the most efficient natural quenchers of singlet oxygen ($^1O_2$), and this effect has been attributed to a physical mechanism where the excess energy of singlet oxygen is transferred to conjugated double bonds of carotenoids [19]. In addition to the singlet oxygen quenching ability, carotenoids can scavenge free radicals. The reactions with free radicals are much more complex than those with singlet oxygen. Chlorophylls are also known to scavenge free radicals, and the effect may be strongly related to their chemical structure, especially the porphyrin ring, phytyl chain, and extended system of conjugated double bonds [28].

The strong ability of these seaweed pigments to quench singlet oxygen and/or to scavenge free radicals has been suggested as the main mechanism by which they afford their health benefits [19,22,28,74]. The seaweed lipids examined in the present study contained carotenoids and chlorophylls as major lipid soluble antioxidants. In addition, α-Toc, the most popular natural antioxidant, was detected in these lipids. Therefore, the seaweed lipids may be a potential source of natural antioxidants that can be used as cosmetic and food ingredients [70,75]. α-Toc is a major tocopherol analogue found in seaweeds [45]. In this study, the α-Toc content in two species of seaweeds from the Sargassaceae family (*S. fusiforme* and *S. horneri*) was higher than that of other seaweeds, except for *C. yendoi* (Table 6). This result is in accordance with previous studies [76,77]. The present study also suggests the relatively lower α-Toc level in Chlorophyta seaweeds (*C. lentillifera*, *U. reticulata*, *U. australis*, *U. intestinalis*) as compared with the other phyla, Ochrophyta and Rhodophyta.

Table 4 together with Tables S1a, S1b and S1c show the FAs composition of seaweed lipids. The composition was strongly affected by the sampling location, as seen in the TL analysis. As shown in Table 4, total PUFAs were the most dominant FAs class in the temperate seaweeds, and the level was higher than that in the tropical seaweeds. In the seaweeds collected from the tropical area, SFAs were a major FAs class. This result is consistent with other studies [14,48,49,78]. Gerasimenko and Logvinov [79] reported that FAs content and composition in seaweeds differed not only by species, but also by seawater temperature. The lower temperature of seaweed habitats results in the high accumulation of *n-3* PUFAs for accelerating cell metabolism, especially during the winter [38]. The present study revealed that three phyla grown in the temperate area greatly accumulated EPA and SDA in Ochrophyta, ALA and SDA in Chlorophyta, and EPA in Rhodophyta, while these *n-3* PUFAs levels were relatively lower in the corresponding phyla grown in the tropical area (Tables S1a, S1b and S1c). On the other hand, there was not much difference in the *n-6* PUFAs levels between the seaweeds collected from the temperate and the tropical areas. The higher level of total PUFAs in temperate seaweeds may be derived from the higher accumulation rate of *n-3* PUFAs.

It is known that TL content and fatty acids composition varies seasonally. For example, the seasonal changes of fatty acids composition have been determined in *Egregia menziesii* (Turner) [80], *S. japonica* [81,82], *C. costata* (Turner) [83], *Stephanocystis hakodatensis* [38], *S. horneri* [38], *S. oligocystum* [78], and *U. pinnatifida* (Harvey) [84]. In all cases, the higher level of total n-3PUFAs has been found during winter or spring; this is the growing period for these seaweeds. Seasonal variation was also found in photosynthetic pigments, such as chlorophylls and carotenoids [38,85–90]. In the present study, no concern has been given to the seasonal changes in the seaweed pigments. These pigments are strongly related to the growing rate of each seaweed species. Therefore, much attention has been paid to the relationship between the growth rate of seaweeds and the level of lipid compounds, especially focusing on the photosynthetic pigments such as chlorophylls and carotenoids.

The lower *n-6/n-3* PUFAs ratio in dietary lipids has been considered to reduce the risk for many kinds of chronic diseases [91]. Although, a balanced ratio between *n-6* and *n-3* PUFAs, usually 1:2 to 1:4 (*w/w*) [30], has been recommended. However, in modern society, intake of the *n-6* PUFAs is much lower than that of the *n-3* PUFAs, resulting in the increase in the *n-6/n-3* PUFAs ratio. Most of the seaweeds, except for *G. longissima*, showed a lower *n-6/n-3* ratio, mostly less than 1.0 (Table 5). In addition, TI and AI values in all seaweeds were less than 3, especially in the temperate seaweeds. The lower TI and AI values of temperate seaweeds are because of the considerable accumulation of *n-3* PUFAs in the temperate seaweeds. TI and AI values found in the present study were lower than those in Indian *macrolagae*, namely, *Gracillaria salicornia*, *Sarconema cinaioides*, *Hypnea spinella*, and *Laurencia dendroidea* [49]. Overall, all indices shown in Table 5 indicate that seaweed lipids analyzed in the present study, especially those from the temperate seaweeds, have a cardio-protective FA composition. Prabhasankar et al. [92] reported the change in the *n-6/n-3* ratio of pasta lipids after incorporation of the powder of brown seaweed collected from the temperate area. When 10% of the brown seaweed powder was incorporated in the pasta, the *n-6/n-3* ratio changed from 15.2 to 3.4. This result indicates that a drastic change could be found in the low-fat food by mixing seaweed powder rich in *n-3* PUFAs.

4. Materials and Methods

4.1. Materials

Standards for Chl a, Phy a, Phy b, and β-Car were obtained from Wako Pure Chemicals (Tokyo, Japan). Chl b and Chl c1 were purchased from Sigma-Aldrich Japan Co. (Tokyo, Japan) and DHI Laboratory Products (Hørsholm, Denmark), respectively. α-Carotene, Lut, Zx, Vx, and Nx were purchased from Carote *Nature* GmbH (Münsingen, Switzerland). dl-α-Tocopherol (α-Toc) was obtained from Kanto Chemical Co. Inc. (Tokyo, Japan). Fx standard for use in this study was purified from the TL content of *S. horneri*, as described previously [20]. Tricosanoic acid (C23:0), used as an internal standard in the gas chromatography (GC) analysis, was obtained from Sigma-Aldrich (Tokyo, Japan). High-performance liquid chromatography (HPLC) grade solvents were used for the HPLC analysis and purchased from Wako Pure Chemicals, Ltd. (Osaka, Japan). All other solvents and chemicals used in the study were of analytical grade.

4.2. Sample Collection and Handling

Seaweeds were collected from two different regions, Indonesia (tropic) and Japan (temperate). The 15 seaweeds including harvesting locations, phylum, family, scientific and local name, and date of collection are listed in Table 1. After collection, tropical seaweeds were washed, air-dried at room temperature (3–5 days; 27–30 °C) to reduce their water content, and transported to the laboratory. Temperate seaweeds were kept in plastic bags, stored on ice, and transported to the laboratory. These samples were then stored in a freezer (−30 °C).

4.3. Moisture Determination and TL Extraction

Before moisture determination, all temperate seaweed samples were thawed and washed several times with running tap water until there were no debris, sand, or other contaminants. Before extraction, the moisture content of 15 seaweeds species was estimated with an oven-drying method, as described by Gómez-Ordóñez et al. [93], and used for expressing TL, chlorophylls, carotenoids, *n-3* PUFAs, and α-Toc content in seaweed samples on a DW basis. The TL content from the seaweeds was obtained by overnight extraction with a ten-fold portion of ethanol (*w/v*). For the extraction, the samples were crudely cut into small pieces. In the case of tropical seaweeds, a nine-fold portion of water was added to the washed sample before the ethanol extraction. After immersion for one hour, the sample was subjected to the ethanol extraction, while the temperate raw seaweed sample was directly subjected to the ethanol extraction. The filtrate obtained by the extraction was filtered with Advantec No. 2 filter paper (Advantec Toyo Kaisha Ltd., Tokyo, Japan), and the residue was subjected to a further overnight extraction with the same solvent proportion. Both filtrates were combined and were vacuum-concentrated at 30 °C using a rotary evaporator to obtain the ethanol extract (dark green mass). The ethanol extract was dissolved in a chloroform/methanol/water solution with a ratio of 10:5:3 (*v/v/v*), and the solution was placed into a separatory funnel. After allowing the funnel to stand overnight, the solution was separated into two layers. The lower layer evaporated under reduced pressure in a rotary evaporator. The last traces of the solvent and water were removed using nitrogen and using a high-powered vacuum to obtain the seaweed TL content. The TL content was dissolved in ethanol and stored at -30 °C for further analysis.

4.4. Pigment Analysis

The contents of chlorophylls, their derivatives, and carotenoids in the seaweed TL content were analyzed using HPLC. HPLC was performed with a Hitachi HPLC La Chrome system (Hitachi Seisakusho Co., Tokyo, Japan) equipped with an auto sampler (L-2200), pump (L-2130), column oven (L-2300), and photodiode array detector (L-2455). An aliquot of the TL content was weighed and dissolved in acetone. The sample solution was filtered with a 0.45 µm membrane filter of polytetrafluoroethylene (PTFE) (Ekicrodisc 13CR; Nippon Genetics Co. Ltd., Tokyo, Japan) and subjected to HPLC analysis. The analysis was performed on an octadecylsilyl (ODS) column (TSK-gel ODS 80-Ts, 250 × 4.6 mm i.d., 5 µm particle size; Tosoh, Japan) protected with a guard column (15 × 3.2 mm) with the same stationary phase. The mobile phase consisted of methanol/acetonitrile/1 M ammonium acetate (5:3:2, *v/v/v*) (A) and acetonitrile/ethyl acetate (1:1, *v/v*) (B) [94] with slight modification. A gradient elution procedure was programmed as follows: 0–2 min, 100:0 (A/B, *v/v*); 2–28 min, a linearly elution gradient from A to B; and 28–37 min, 0:100 (A/B, *v/v*). The flow rate was kept at 1.0 mL min^{-1}. The UV/visible spectra absorption was recorded from 350 to 800 nm with a photodiode array detector, and sequential detection was carried out at 410, 430, 450, and 666 nm [43]. Quantification of chlorophylls, their derivatives, and carotenoids was performed with the corresponding calibration curves. A calibration curve for each pigment was prepared using an authentic standard.

4.5. Fatty Acids Analysis

The fatty acids composition of the TL content was determined by GC after conversion of fatty acyl groups in the lipid to their methyl esters. The fatty acid methyl esters (FAMEs) were prepared as per the method by Prevot and Mordret [95], with a slight modification. Briefly, 1 mL *n*-hexane containing an internal standard (tricosanoic acid, C23:0) and 0.2 mL 2 M NaOH in methanol were added to an aliquot of the TL content (*ca.* 10 mg), vortexed for 10 seconds, and incubated at 50 °C for 30 seconds. After the incubation, 0.2 mL 2 M HCl in methanol solution was added to the solution and vortexed for 60 seconds. The mixture was separated by centrifugation at 1000× *g* for five minutes. The upper hexane layer containing FAMEs was recovered and subjected to GC. The FAMEs analysis was performed on a Shimadzu GC-2014 (Shimadzu Seisakusho, Kyoto, Japan) equipped with a flame ionization detector and a capillary column (Omegawax-320; 30 m × 0.32 mm i.d.; Supelco, Bellefonte, PA, USA). The detector, injector, and column temperatures were 260, 250, and 200 °C, respectively. The carrier gas was helium at a flow rate of 1.0 mL s^{-1}. FA content was expressed as the relative weight percentage of the total FAs weight. Each *n-3* PUFAs level (mg·g^{-1} TL content) was quantified by comparing the peak ratio to that of the internal standard (23:0). Total *n-3* PUFAs (mg·g^{-1} DW) were calculated from the content of each *n-3* PUFAs in the TL content and the TL content g^{-1} DW. FAME peaks were identified by comparison of their retention time and log RRT with GLC-Reference standards by GC FID (Shimadzu GC-2014) after analysis. The fatty acids standard was prepared with GLC-Reference Standard, fatty acid methyl esters: GLC-462, and Echium oil, according to Susanto et al [34].

Several factors predicting the nutritional quality of the seaweed TL were calculated based on the formula as below [79,96,97]:

$$n-6/n-3 \text{ ratio} = \frac{\sum n-6\text{PUFAs}}{\sum n-3\text{PUFAs}} \tag{1}$$

$$\text{Atherogenicity index (AI)} = \frac{C12:0 + 4 \times C14:0 + C16:0}{\sum \text{MUFAs} + \sum n-6 \text{ PUFAs} + \sum n-3 \text{ PUFAs}} \tag{2}$$

$$\text{Thrombogenicity index (TI)} = \frac{\text{C14:0}+\text{C16:0}+\text{C18:0}}{(0.5\times\sum \text{MUFAs})+(0.5\times\sum n\text{-6 PUFAs})+(3\times\sum n\text{-3 PUFAs})+\frac{\sum n\text{-3 PUFAs}}{\sum n\text{-6 PUFAs}}} \quad (3)$$

$$\text{Fatty acids hypocholesterolemic/hypercholesterolemic ratio (h/H)} = \frac{\text{C18:1}n\text{-9}+\text{C18:2}n\text{-6}+\text{C20:4}n\text{-6}+\text{C18:3}n\text{-3}+\text{C20:5}n\text{-3}}{\text{C14:0}+\text{C16:0}} \quad (4)$$

$$\text{Unsaturation Index (UI)} = \sum (\% \text{ PUFAs} \times \text{DB}); \text{ DB is the number of double bonds.} \quad (5)$$

4.6. α-Toc Analysis

The content of α-Toc in the TL content was analyzed by HPLC with a silica column (Mightysil Si 60 silica, 250 × 4.6 mm i.d.; Kanto Chemical Co., Ltd., Tokyo, Japan). HPLC was carried out with the same system as described in the pigment analysis, except that the fluorescence detector (Hitachi L-7485) was used for the peak detection. The mobile phase was hexane/2-propanol (99.2:0.8, *v/v*) at a flow rate of 1.0 mL min^{-1}. The fluorescence detector was set at Ex. 298 nm and Em. 325 nm.

4.7. Data Analysis

All analytical data were conducted in triplicate and the mean values are presented. The PCA was performed using JMP 14.0 statistical analysis. Two principal components based on the fatty acids composition and group (SFAs, MUFAs, PUFAs, *n*-3 PUFAs, and *n*-6 PUFAs) were used to make clear the relationship between different sampling areas and seaweed phyla. Fatty acids composition (supplementary data) and FAs group composition (Table 4) were used for the multivariate analysis. However, an insignificant amount of fatty acids was not included in this analysis, these fatty acids were dodecanoic acid (C12:0), tridecanoic acid (C13:0), 7-tetradecenoic acid (C14:1*n*-7), pentadecanoic acid (C15:0), 8-pentadecenoic acid (C15:1*n*-7), heptadecanoic acid (C17:0), 10-heptadecenoic acid (C17:1*n*-7), 11,14,17-eicosatrienoic acid (C20:3*n*-3), docosanoic acid (C22:0), 13-docosaenoic acid (C22:1*n*-9), 13,16-docosadieonic acid (C22:2*n*-6), tetradocosanoic acid (C24:0), and 17-tetracosaenoic acid (C24:1*n*-7).

5. Conclusions

In conclusion, the present study clearly shows the major factors affecting the variation of seaweed lipid compounds. Liphophilic pigments were explained by seaweeds' phyla, while the geographical location influenced fatty acids composition. Lipid soluble pigments were one of the major compounds of interest in the present study. Among these pigments, relatively little attention has been paid to chlorophylls, despite the fact that chlorophylls are abundant in our diet and their biological significance has been recognized for a long time. The results for the compositions of chlorophylls and their derivatives will provide valuable information for the application of each seaweed's lipids as a good source of these chlorophyll compounds and other lipid soluble compounds, such as carotenoids and PUFAs.

Supplementary Materials: The following are available online at http://www.mdpi.com/1660-3397/17/11/630/s1, Table S1a: Fatty acids composition (weight % of total FAs) of Ochrophyta; Table S1b: Fatty acids composition (weight % of total FAs) of Chlorophyta; Table S1c: Fatty acids composition (weight % of total FAs) of Rhodophyta.

Author Contributions: E.S. designed and performed experiments, drafted the manuscript. A.S.F. provided samples and performed multivariate analysis. M.H. reviewed the manuscript and made suggestive revision. K.M. provided samples and developed manuscript.

Funding: This research received no external funding.

Acknowledgments: E.S. is an awardee of BUDI-LN Scholarship from Directorate Human Resources for Sciences and Higher Education, Ministry of Research Technology and Higher Education; and Indonesia Endowment Fund for Education (LPDP) Scholarship, Ministry of Finance, Republic of Indonesia. The authors would like to thank Bayu Kumayanjati and Nur Hamid who kindly collected seaweeds samples from Indonesia.

Conflicts of Interest: The authors declare no conflict of interest.

References

1. Dhargalkar, V.K.; Pereira, N. Seaweed: Promising plant of the millennium. *Source* **2005**, *71*, 60–66.
2. White, W.L.; Wilson, P. World seaweed utilization. In *Seaweed Sustainability*, 1st ed.; Tiwari, B.K., Troy, D.J., Eds.; Elsevier Inc.: San Diego, CA, USA, 2015; pp. 7–26. ISBN 9780124199583.
3. Buschmann, A.H.; Camus, C.; Infante, J.; Neori, A.; Israel, Á.; Hernández-González, M.C.; Pareda, S.V.; Gomez-Pinchetti, J.L.; Golberg, A.; Tadmor-Shalev, N.; et al. Seaweed production: Overview of the global state of exploitation, farming and emerging research activity. *Eur. J. Phycol.* **2017**, *52*, 391–406. [CrossRef]
4. Cardoso, S.M.; Pereira, O.R.; Seca, A.M.L.; Pinto, D.C.G.A.; Silva, A.M.S. Seaweeds as preventive agents for cardiovascular diseases: From nutrients to functional foods. *Mar. Drugs* **2015**, *13*, 6838–6865. [CrossRef] [PubMed]
5. Chater, P.I.; Wilcox, M.D.; Houghton, D.; Pearson, J.P. The role of seaweed bioactives in the control of digestion: Implications for obesity treatments. *Food Funct.* **2015**, *6*, 3420–3427. [CrossRef] [PubMed]
6. Sharifuddin, Y.; Chin, Y.X.; Lim, P.E.; Phang, S.M. Potential bioactive compounds from seaweed for diabetes management. *Mar. Drugs* **2015**, *13*, 5447–5491. [CrossRef] [PubMed]
7. Xu, S.Y.; Huang, X.; Cheong, K.L. Recent advances in marine algae polysaccharides: Isolation, structure, and activities. *Mar. Drugs* **2017**, *15*, 388. [CrossRef] [PubMed]
8. Admassu, H.; Gasmalla, M.A.A.; Yang, R.; Zhao, W. Bioactive peptides derived from seaweed protein and their health benefits: Antihypertensive, antioxidant, and antidiabetic properties. *J. Food Sci.* **2018**, *83*, 6–16. [CrossRef]
9. Alves, C.; Silva, J.; Pinteus, S.; Gaspar, H.; Alpoim, M.C.; Botana, L.M.; Pedrosa, R. From marine origin to therapeutics: The antitumor potential of marine algae-derived compounds. *Front. Pharmacol.* **2018**, *9*, 1–24. [CrossRef]
10. Circuncisão, A.R.; Catarino, M.D.; Cardoso, S.M.; Silva, A.M.S. Minerals from macroalgae origin: Health benefits and risks for consumers. *Mar. Drugs* **2018**, *16*, 400.
11. Gómez-Guzmán, M.; Rodríguez-Nogales, A.; Algieri, F.; Gálvez, J. Potential role of seaweed polyphenols in cardiovascular-associated disorders. *Mar. Drugs* **2018**, *16*, 250. [CrossRef]
12. Seca, A.M.L.; Pinto, D.C.G.A. Overview on the antihypertensive and anti-obesity effects of secondary metabolites from seaweeds. *Mar. Drugs* **2018**, *16*, 237. [CrossRef] [PubMed]
13. Viera, I.; Pérez-Gálvez, A.; Roca, M. Bioaccessibility of marine carotenoids. *Mar. Drugs* **2018**, *16*, 397. [CrossRef] [PubMed]
14. Terasaki, M.; Hirose, A.; Narayan, B.; Baba, Y.; Kawagoe, C.; Yasui, H.; Saga, N.; Hosokawa, M.; Miyashita, K. Evaluation of recoverable functional lipid components of several brown seaweeds (*Phaeophyta*) from Japan with special reference to fucoxanthin and fucosterol contents. *J. Phycol.* **2009**, *45*, 974–980. [CrossRef] [PubMed]
15. Jiménez-Escrig, A.; Gómez-Ordóñez, E.; Rupérez, P. Brown and red seaweeds as potential sources of antioxidant nutraceuticals. *J. Appl. Phycol.* **2012**, *24*, 1123–1132. [CrossRef]
16. Miyashita, K.; Mikami, N.; Hosokawa, M. Chemical and nutritional characteristics of brown seaweed lipids: A review. *J. Funct. Foods* **2013**, *5*, 1507–1517. [CrossRef]
17. Aryee, A.N.; Agyei, D.; Akanbi, T.O. Recovery and utilization of seaweed pigments in food processing. *Curr. Opin. Food Sci.* **2018**, *19*, 113–119. [CrossRef]
18. Ito, M.; Koba, K.; Hikihara, R.; Ishimaru, M.; Shibata, T.; Hatate, H.; Tanaka, R. Analysis of functional components and radical scavenging activity of 21 algae species collected from the Japanese coast. *Food Chem.* **2018**, *255*, 147–156. [CrossRef]
19. Miyashita, K.; Hosokawa, M. Health impact of marine carotenoids. *J. Food Bioact.* **2018**, *1*, 31–40. [CrossRef]
20. Maeda, H.; Hosokawa, M.; Sashima, T.; Funayama, K.; Miyashita, K. Fucoxanthin from edible seaweed, *Undaria pinnatifida*, shows antiobesity effect through UCP1 expression in white adipose tissues. *Biochem. Biophys. Res. Commun.* **2005**, *332*, 392–397. [CrossRef]

21. Ferruzzi, M.G.; Blakeslee, J. Digestion, absorption, and cancer preventative activity of dietary chlorophyll derivatives. *Nutr. Res.* **2007**, *27*, 1–12. [CrossRef]

22. Chen, K.; Roca, M. In-vitro bioavailability of chlorophyll pigments from edible seaweeds. *J. Funct. Foods* **2018**, *41*, 25–33. [CrossRef]

23. Yoshioka, H.; Kamata, A.; Konishi, T.; Takahashi, J.; Oda, H.; Tamai, T.; Toyohara, H.; Sugahara, T. Inhibitory effect of chlorophyll c2 from brown algae, *Sargassum horneri*, on degranulation of RBL-2H3 cells. *J. Funct. Foods* **2013**, *5*, 204–210. [CrossRef]

24. Yoshioka, H.; Ishida, M.; Nishi, K.; Oda, H.; Toyohara, H.; Sugahara, T. Studies on anti-allergic activity of *Sargassum horneri* extract. *J. Funct. Foods* **2014**, *10*, 154–160. [CrossRef]

25. Okai, Y.; Higashi-Okai, K. Potent anti-inflammatory activity of pheophytin a derived from edible green alga, *Enteromorpha prolifera (sujiao-nori)*. *Int. J. Immunopharmacol.* **1997**, *19*, 355–358. [CrossRef]

26. Islam, M.N.; Ishita, I.J.; Jin, S.E.; Choi, R.J.; Lee, C.M.; Kim, Y.S.; Jun, H.A.; Choi, J.S. Anti-inflammatory activity of edible brown alga *Saccharina japonica* and its constituents pheophorbide a and pheophytin a in LPS-stimulated RAW 264.7 macrophage cells. *Food Chem. Toxicol.* **2013**, *55*, 541–548. [CrossRef]

27. Ina, A.; Hayashi, K.I.; Nozaki, H.; Kamei, Y. Pheophytin a, a low molecular weight compound found in the marine brown alga *Sargassum fulvellum*, promotes the differentiation of PC12 cells. *Int. J. Dev. Neurosci.* **2007**, *25*, 63–68. [CrossRef]

28. Pangestuti, R.; Kim, S.K. Biological activities and health benefit effects of natural pigments derived from marine algae. *J. Funct. Foods* **2011**, *3*, 255–266. [CrossRef]

29. Kumar, M.; Kumari, P.; Trivedi, N. Minerals, PUFAs and antioxidant properties of some tropical seaweeds from Saurashtra coast of India. *J. Appl. Phycol.* **2011**, *23*, 797–810. [CrossRef]

30. Gebauer, S.K.; Psota, T.L.; Harris, W.S.; Kris-etherton, P.M. *n-3* Fatty acid dietary recommendations and food sources to achieve essentiality and cardiovascular benefits. *Am. J. Clin. Nutr.* **2006**, *83*, 1526–1535. [CrossRef]

31. Kiso, Y. Pharmacology in health foods: Effects of arachidonic acid and docosahexaenoic acid on the age-related decline in brain and cardiovascular system function. *J. Pharmacol. Sci.* **2011**, *115*, 471–475. [CrossRef]

32. Narayan, B.; Kinami, T.; Miyashita, K.; Park, S.B.; Endo, Y.; Fujimoto, K. Occurrence of conjugated polyenoic fatty acids in seaweeds from the Indian Ocean. *Z. Naturforsch. C.* **2004**, *59*, 310–314.

33. Narayan, B.; Miyashita, K. Comparative evaluation of fatty acid composition of different Sargassum (Fucales, Phaeophyta) species harvested from temperate and tropical waters. *J. Aquat. Food Prod. Tech.* **2008**, *13*, 53–70. [CrossRef]

34. Susanto, E.; Fahmi, A.S.; Abe, M.; Hosokawa, M.; Miyashita, K. Lipids, fatty acids, and fucoxanthin content from temperate and tropical brown seaweeds. *Aquat. Procedia* **2016**, *7*, 66–75. [CrossRef]

35. Gosch, B.J.; Magnusson, M.; Paul, N.A.; de Nys, R. Total lipid and fatty acid composition of seaweeds for the selection of species for oil-based biofuel and bioproducts. *GCB Bioenergy* **2012**, *4*, 919–930. [CrossRef]

36. Thinakaran, T.; Balamurugan, M.; Sivakumar, K. Screening of phycochemical constituents qualitatively and quantitatively certain seaweeds from Gulf of Mannar biosphere reserve. *Int. Res. J. Pharm.* **2012**, *3*, 261–265.

37. McDermid, K.J.; Stuercke, B. Nutritional composition of edible Hawaiian seaweeds. *J. Appl. Phycol.* **2003**, *15*, 513–524. [CrossRef]

38. Nomura, M.; Kamogawa, H.; Susanto, E.; Kawagoe, C.; Yasui, H.; Saga, N.; Hosokawa, M.; Miyashita, K. Seasonal variations of total lipids, fatty acid composition, and fucoxanthin contents of *Sargassum horneri* (Turner) and *Cystoseira hakodatensis* (Yendo) from the northern seashore of Japan. *J. Appl. Phycol.* **2013**, *25*, 1159–1169. [CrossRef]

39. Khotimchenko, S.V.; Yakovleva, I.M. Lipid composition of the red alga *Tichocarpus crinitus* exposed to different levels of photon irradiance. *Phytochemistry* **2005**, *66*, 73–79. [CrossRef]

40. Ganesan, M.; Mairh, O.P.; Eswaran, K.; Subba Rao, P.V. Effect of salinity, light intensity and nitrogen source on growth and composition of *Ulva fasciata* Delile (Chlorophyta, Ulvales). *Indian J. Mar. Sci.* **1999**, *28*, 70–73.

41. Zhao, Z.; Zhao, F.; Yao, J. Early development of germlings of *Sargassum thunbergii* (Fucales, Phaeophyta) under laboratory conditions. *J. Appl. Phycol.* **2008**, 925–931. [CrossRef]

42. Kumari, P.; Kumar, M.; Reddy, C.R.K.; Jha, B. Algal lipids, fatty acids and sterols. In *Functional Ingredients from Algae for Foods and Nutraceuticals*; Domínguez, H., Ed.; Woodhead Publishing Limited: Cambridge, UK, 2013; pp. 119–166, ISBN 978-0-85709-868-9.

43. Chen, K.; Ríos, J.J.; Pérez-Gálvez, A.; Roca, M. Comprehensive chlorophyll composition in the main edible seaweeds. *Food Chem.* **2017**, *228*, 625–633. [CrossRef] [PubMed]

44. Wong, K.H.; Cheung, P.C.K. Nutritional evaluation of some subtropical red and green seaweeds. *Food Chem.* **2000**, *71*, 475–482. [CrossRef]

45. Burtin, P. Nutritional value of seaweeds. *J. Environ. Agric. Food Chem.* **2003**, *2*, 498–503.

46. Sánchez-Machado, D.I.; López-Cervantes, J.; López-Hernández, J.; Paseiro-Losada, P. Fatty acids, total lipid, protein and ash contents of processed edible seaweeds. *Food Chem.* **2004**, *85*, 439–444. [CrossRef]

47. Marinho-Soriano, E.; Fonseca, P.C.; Carneiro, M.A.A.; Moreira, W.S.C. Seasonal variation in the chemical composition of two tropical seaweeds. *Bioresour. Technol.* **2006**, *97*, 2402–2406. [CrossRef]

48. Van Ginneken, V.J.T.; Helsper, J.P.F.G.; De Visser, W.; Van Keulen, H.; Brandenburg, W.A. Polyunsaturated fatty acids in various macroalgal species from north Atlantic and tropical seas. *Lipids Health Dis.* **2011**, *10*, 4–11. [CrossRef]

49. Kumari, P.; Bijo, A.J.; Mantri, V.A.; Reddy, C.R.K.; Jha, B. Fatty acid profiling of tropical marine macroalgae: An analysis from chemotaxonomic and nutritional perspectives. *Phytochemistry* **2013**, *86*, 44–56. [CrossRef]

50. Boulom, S.; Robertson, J.; Hamid, N.; Ma, Q.; Lu, J. Seasonal changes in lipid, fatty acid, α-tocopherol and phytosterol contents of seaweed, *Undaria pinnatifida*, in the Marlborough Sounds, New Zealand. *Food Chem.* **2014**, *161*, 261–269. [CrossRef]

51. Kumari, P.; Kumar, M.; Gupta, V.; Reddy, C.R.K.; Jha, B. Tropical marine macroalgae as potential sources of nutritionally important PUFAs. *Food Chem.* **2010**, *120*, 749–757. [CrossRef]

52. Sho, H. History and characteristics of Okinawan longevity food. *Asia Pac. J. Clin. Nutr.* **2001**, *10*, 159–164. [CrossRef]

53. McHugh, D.J. *A Guide to the Seaweed Industry*; Food and Agriculture Organization of the United Nations: Rome, Italy, 2003; pp. 1–105. ISBN 92-5-104958-0.

54. Mohamed, S.; Hashim, S.N.; Rahman, H.A. Seaweeds: A sustainable functional food for complementary and alternative therapy. *Trends Food Sci. Technol.* **2012**, *23*, 83–96. [CrossRef]

55. Moussavou, G.; Kwak, D.H.; Obiang-Obonou, B.W.; Maranguy, C.A.O.; Dinzouna-Boutamba, S.D.; Lee, D.H.; Pissibangnga, O.G.M.; Ko, K.; Seo, J.I.; Choo, J.K. Anticancer effects of different seaweeds on human colon and breast cancers. *Mar. Drugs* **2014**, *12*, 4898–4911. [CrossRef] [PubMed]

56. Pádua, D.; Rocha, E.; Gargiulo, D.; Ramos, A.A. Bioactive compounds from brown seaweeds: Phloroglucinol, fucoxanthin and fucoidan as promising therapeutic agents against breast cancer. *Phytochem. Lett.* **2015**, *14*, 91–98. [CrossRef]

57. Macartain, P.; Gill, C.I.R.; Brooks, M.; Campbell, R.; Rowland, I.R. Special article nutritional value of edible seaweeds. *Nutr. Rev.* **2007**, *65*, 535–543. [CrossRef]

58. Brown, E.M.; Allsopp, P.J.; Magee, P.J.; Gill, C.I.; Nitecki, S.; Strain, C.R.; McSorley, E.M. Seaweed and human health. *Nutr. Rev.* **2014**, *72*, 205–216. [CrossRef]

59. Takaichi, S. Distributions, biosyntheses and functions of carotenoids in algae. *Agro Food Ind. Hi Tech* **2013**, *24*, 55–58. [CrossRef]

60. Maeda, H.; Hosokawa, M.; Sashima, T.; Miyashita, K. Dietary combination of fucoxanthin and fish oil attenuates the weight gain of white adipose tissue and decreases blood glucose in obese/diabetic KK-A y mice. *J. Agric. Food Chem.* **2007**, *55*, 7701–7706. [CrossRef]

61. Holdt, S.L.; Kraan, S. Bioactive compounds in seaweed: Functional food applications and legislation. *J. Appl. Phycol.* **2011**, *23*, 543–597. [CrossRef]

62. Dougherty, R.C.; Strain, H.H.; Svec, W.A.; Uphaus, R.A.; Katz, J.J. The structure, properties, and distribution of chlorophyll c. *J. Am. Chem. Soc.* **1970**, *92*, 2826–2833. [CrossRef]

63. Katz, J.J.; Norris, J.R.; Shipman, L.L.; Thurnauer, M.C.; Wasielewski, M.R. Chlorophyll function in the photosynthetic reaction center. *Annu. Rev. Biophys. Bioeng.* **1978**, *7*, 393–434. [CrossRef]

64. Allakhverdiev, S.I.; Kreslavski, V.D.; Zharmukhamedov, S.K.; Voloshin, R.A.; Korol'kova, D.V.; Tomo, T.; Shen, J.R. Chlorophylls d and f and their role in primary photosynthetic processes of cyanobacteria. *Biochemistry* **2016**, *81*, 201–212. [CrossRef] [PubMed]

65. Fujiwara, T.; Nishida, N.; Nota, J.; Kitani, T.; Aoishi, K.; Takahashi, H.; Sugahara, T.; Hato, N. Efficacy of chlorophyll c2 for seasonal allergic rhinitis: Single-center double-blind randomized control trial. *Eur. Arch. Otorhinolaryngol.* **2016**, *273*, 4289–4294. [CrossRef] [PubMed]

66. Chen, K.; Roca, M. Cooking effects on chlorophyll profile of the main edible seaweeds. *Food Chem.* **2018**, *266*, 368–374. [CrossRef] [PubMed]

67. Jeffrey, S.W. Preparation and some properties of crystalline chlorophyll c1 and c2 from marine algae. *Biochim. Biophys. Acta* **1972**, *279*, 15–33. [CrossRef]

68. Wilhelm, C. Purification and identification of chlorophyll c1 from the green alga *Mantoniella squamata*. *Biochim. Biophys. Acta* **1987**, *892*, 23–29. [CrossRef]

69. Gelzinis, A.; Butkus, V.; Songaila, E.; Augulis, R.; Gall, A.; Büchel, C.; Robert, B.; Abramavicius, D.; Zigmantas, D.; Valkunas, L. Mapping energy transfer channels in fucoxanthin-chlorophyll protein complex. *Biochim. Biophys. Acta - Bioenerg.* **2015**, *1847*, 241–247. [CrossRef]

70. Kuczynska, P.; Jemiola-Rzeminska, M.; Strzalka, K. Photosynthetic pigments in diatoms. *Mar. Drugs* **2015**, *13*, 5847–5881. [CrossRef]

71. Apt, K.E.; Clendennen, S.K.; Powers, D.A.; Grossman, A.R. The gene family encoding the fucoxanthin chlorophyll proteins from the brown alga *Macrocystis pyrifera*. *Mol. Gen. Genet.* **1995**, *246*, 455–464. [CrossRef]

72. Mikami, K.; Hosokawa, M. Biosynthetic pathway and health benefits of fucoxanthin, an algae-specific xanthophyll in brown seaweeds. *Int. J. Mol. Sci.* **2013**, *14*, 13763–13781. [CrossRef]

73. Schubert, N.; García-Mendoza, E.; Pacheco-Ruiz, I. Carotenoid composition of marine red algae. *J. Phycol.* **2006**, *42*, 1208–1216. [CrossRef]

74. Cho, M.; Lee, H.S.; Kang, I.J.; Won, M.H.; You, S.G. Antioxidant properties of extract and fractions from *Enteromorpha prolifera*, a type of green seaweed. *Food Chem.* **2011**, *127*, 999–1006. [CrossRef] [PubMed]

75. Sathasivam, R.; Ki, J.S. A review of the biological activities of microalgal carotenoids and their potential use in healthcare and cosmetic industries. *Mar. Drugs* **2018**, *16*, 26. [CrossRef] [PubMed]

76. Miyashita, K.; Takagi, T. Tocopherol content of Japanese algae and its seasonal variation. *Agric. Biol. Chem.* **1987**, *51*, 3115–3118.

77. Santos, S.A.O.; Vilela, C.; Freire, C.S.R.; Abreu, M.H.; Rocha, S.M.; Silvestre, A.J.D. Chlorophyta and Rhodophyta macroalgae: A source of health promoting phytochemicals. *Food Chem.* **2015**, *183*, 122–128. [CrossRef] [PubMed]

78. Praiboon, J.; Palakas, S.; Noiraksa, T.; Miyashita, K. Seasonal variation in nutritional composition and anti-proliferative activity of brown seaweed, *Sargassum oligocystum*. *J. Appl. Phycol.* **2018**, *30*, 101–111. [CrossRef]

79. Gerasimenko, N.; Logvinov, S. Seasonal composition of lipids, fatty acids pigments in the brown alga *Sargassum pallidum*: The potential for health. *Open J. Mar. Sci.* **2016**, *06*, 498–523. [CrossRef]

80. Nelson, M.M.; Phleger, C.F.; Nichols, P.D. Seasonal lipid composition in macroalgae of the northeastern Pacific Ocean. *Bot. Mar.* **2002**, *45*, 58–65. [CrossRef]

81. Honya, M.; Kinoshita, T.; Ishikawa, M.; Mori, H.; Nisizawa, K. Seasonal variation in the lipid content of cultured *Laminaria japonica*: Fatty acids, sterols, β-carotene and tocopherol. *J. Appl. Phycol.* **1994**, *6*, 25–29. [CrossRef]

82. Sanina, N.M.; Goncharova, S.N.; Kostetsky, E.Y. Seasonal changes of fatty acid composition and thermotropic behavior of polar lipids from marine macrophytes. *Phytochemistry* **2008**, *69*, 1517–1527. [CrossRef]

83. Gerasimenko, N.I.; Busarova, N.G.; Moiseenko, O.P. Seasonal changes in the content of lipids, fatty acids, and pigments in brown alga *Costaria costata*. *Russ. J. Plant Physiol.* **2010**, *57*, 205–211. [CrossRef]

84. Gerasimenko, N.I.; Skriptsova, A.V.; Busarova, N.G.; Moiseenko, O.P. Effects of the season and growth stage on the contents of lipids and photosynthetic pigments in brown alga *Undaria pinnatifida*. *Russ. J. Plant Physiol.* **2011**, *58*, 885–891. [CrossRef]

85. Altamirano, M.; Flores-Moya, A.; Conde, F.; Figueroa, F.L. Growth seasonality, photosynthetic pigments, and carbon and nitrogen content in relation to environmental factors: A field study of *Ulva olivascens* (Ulvales, Chlorophyta). *Phycologia* **2000**, *39*, 50–58. [CrossRef]

86. Christaki, E.; Bonos, E.; Giannenasa, I.; Florou-Paneria, P. Functional properties of carotenoids originating from algae. *J. Sci. Food Agric.* **2013**, *93*, 5–11. [CrossRef] [PubMed]

87. Ismail, M.M.; Osman, M.E.H. Seasonal fluctuation of photosynthetic pigments of most common red seaweeds species collected from Abu Qir, Alexandria, Egypt. *Rev. Biol. Mar. Oceanogr.* **2016**, *51*, 515–525. [CrossRef]

88. Sampath-Wiley, P.; Neefus, C.D.; Jahnke, L.S. Seasonal effects of sun exposure and emersion on intertidal seaweed physiology: Fluctuations in antioxidant contents, photosynthetic pigments and photosynthetic efficiency in the red alga *Porphyra umbilicalis* Kützing (Rhodophyta, Bangiales). *J. Exp. Mar. Biol. Ecol.* **2008**, *361*, 83–91. [CrossRef]

89. Pereira, D.C.; Trigueiro, T.G.; Colepicolo, P.; Marinho-Soriano, E. Seasonal changes in the pigment composition of natural population of *Gracilaria domingensis* (Gracilariales, Rhodophyta). *Braz. J. Pharmacogn.* **2012**, *22*, 874–880. [CrossRef]

90. Zavodnik, N. Seasonal variations in rate of photosynthetic activity and chemical composition of the littoral seaweeds common to North Adriatic part II. *Wrangelia penicillata* C. AG. *Bot. Mar.* **1973**, *16*, 166–170. [CrossRef]

91. Simopoulos, A.P. The importance of the ratio of omega-6/omega-3 essential fatty acids. *Biomed. Pharmacother.* **2002**, *56*, 365–379. [CrossRef]

92. Prabhasankar, P.; Ganesan, P.; Bhaskar, N.; Hirose, A.; Stephen, N.; Gowda, L.R.; Hosokawa, M.; Miyashita, K. Edible Japanese seaweed, *wakame* (*Undaria pinnatifida*) as an ingredient in pasta: Chemical, functional and structural evaluation. *Food Chem.* **2009**, *115*, 501–508. [CrossRef]

93. Gómez-Ordóñez, E.; Jiménez-Escrig, A.; Rupérez, P. Dietary fibre and physicochemical properties of several edible seaweeds from the northwestern Spanish coast. *Food Res. Int.* **2010**, *43*, 2289–2294. [CrossRef]

94. Garrido, J.L.; Zapata, M. High performance liquid chromatography of chlorophylls c3, c1, c2 and a and of carotenoids of chromophyte algae on a polymeric octadecyl silica column. *Chromatographia* **1993**, *35*, 543–547. [CrossRef]

95. Prevot, A.F.; Mordret, F.X. Utilisation des colonnes capillaires de verre pour l'analyse des corps gras par chromotographie en phase gazeuse. *Rev. Fr. Corps Gras.* **1979**, *23*, 409–423.

96. Matos, Â.P.; Feller, R.; Moecke, E.H.S.; de Oliveira, J.V.; Junior, A.F.; Derner, R.B.; Sant'Anna, E.S. Chemical characterization of six microalgae with potential utility for food application. *J. Am. Oil Chem. Soc.* **2016**, *93*, 963–972. [CrossRef]

97. Poerschmann, J.; Spijkerman, E.; Langer, U. Fatty acid patterns in *Chlamydomonas* sp. as a marker for nutritional regimes and temperature under extremely acidic conditions. *Microb. Ecol.* **2004**, *48*, 78–89. [CrossRef] [PubMed]

 *marine drugs*

Article

Extraction and Characterization of Alginate from an Edible Brown Seaweed (*Cystoseira barbata*) Harvested in the Romanian Black Sea

Bogdan Trica [1,2,3], Cédric Delattre [2], Fabrice Gros [2], Alina Violeta Ursu [2], Tanase Dobre [3], Gholamreza Djelveh [2], Philippe Michaud [2] and Florin Oancea [1,*]

[1] Department of Bioresources, National Institute for Research & Development in Chemistry and Petrochemistry-ICECHIM Bucharest, Splaiul Independenței 202, 060021 Bucharest, Romania
[2] CNRS, SIGMA Clermont, Institut Pascal, Université Clermont Auvergne, F-63000 Clermont-Ferrand, France
[3] Department of Chemical and Biochemical Engineering, University "POLITEHNICA" of Bucharest, Splaiul Independenței 313, 060042 Bucharest, Romania
* Correspondence: florin.oancea@icechim.ro; Tel.: +40-21-316-3071

Received: 19 June 2019; Accepted: 5 July 2019; Published: 8 July 2019

Abstract: *Cystoseira barbata* is an edible brown seaweed, traditionally used in the Black Sea area as functional food. Both alginate and brown seaweed biomass are well known for their potential use as adsorbents for heavy metals. Alginate was extracted from *C. barbata* recovered from the Romanian coast on the Black Sea with a yield of 19 ± 1.5% (*w/w*). The structural data for the polysaccharide was obtained by HPSEC-MALS, ^{1}H-NMR. The M/G ratio was determined to be 0.64 with a molecular weight of 126.6 kDa with an intrinsic viscosity of 406.2 mL/g. Alginate beads were used and their adsorption capacity with respect to Pb^{2+} and Cu^{2+} ions was determined. The adsorption kinetics of *C. barbata* dry biomass was evaluated and it was shown to have an adsorption capacity of 279.2 ± 7.5 mg/g with respect to Pb^{2+}, and 69.3 ± 2 with respect to Cu^{2+}. Alginate in the form of beads adsorbs a maximum of 454 ± 4.7 mg/g of Pb^{2+} ions and 107.3 ± 1.7 mg/g of Cu^{2+} ions.

Keywords: alginate; *Cystoseira barbata*; Black Sea; heavy metals adsorption; diffusion model

1. Introduction

Edible seaweeds have significant potential as functional food, especially as they are active against various human non-communicable diseases such as cardiovascular diseases, cancers, type 2 diabetes/metabolic syndrome, auto-immune diseases, due to their high level of bioactive components—polysaccharides, peptides, polyunsaturated fatty acids, polyphenols, vitamins and minerals [1,2]. One of the main active components of edible brown seaweed is alginate, a polysaccharide composed of two different uronic acids, mannuronic and guluronic [3]. Due to the fact that alginates are not digested by human enzymes, they act as prebiotics, supporting the production of short-chain fatty acids, and as potential immunomodulators [4].

Traditionally, alginates have been used as tablet excipients and for the treatment of stomach ulcers, gastric reflux and heartburn [5]. Alginates' absorptive, swelling and haemostatic features are involved in their mode of action against such health conditions. These features also substantiate the use of alginate in wound treatment [6]. Their absorptive and swelling features have also been linked to many other health-related effects. For example, binding of glucose and α-amylase inhibition, which reduce post-prandial glucose levels [7,8]. Alginates (calcium) have the ability to absorb bile acids and lipids and therefore to lower cholesterol [9] and lipid levels [10].

The adsorption of heavy metal ions (biosorption) from the intestinal system is related to organism detoxification. Treatment with calcium alginate in a dose of 500 mg/kg removes the lead accumulated in rats after intoxication with lead acetate [11]. Simultaneous intoxication of lead acetate and treatment with calcium alginate significantly reduced lead accumulation in rat organs [12]. Administration of alginate in combination with modified citrus pectin reduces the total body burden of heavy metals [13].

Heavy metals are highly toxic to the environment and to humans. The recommended values for human consumption are in the very low ppm ranges: Pb, 0.01 ppm; Cu, 2 ppm; Hg, 0.001 ppm; As, 0.01 ppm; Ni, 0.02 ppm; etc. [14]. High levels of heavy metals have been reported in sources of drinking water [15], which have been contaminated from a variety of sources such as pipe corrosion [16,17] or industrial activities [18–20]. Other activities that have been associated with high contamination risks are petrochemistry [21] and electronics waste disposal (E-waste) [22]. In many cases, the recovery of heavy metal contaminants using adsorbents [15,23] is mainly due to their reduced costs as compared to other methods such as membrane filtration, chemical precipitation, ion exchange and others [15,24–26].

Biosorption is the process by which some type of biological material is used as an adsorbent to bind certain compounds [27]. It is regarded as a promising alternative to classical methods due to its cost-effectiveness and environment-friendly nature [21]. The biosorbents that are currently under development are mostly trying to take advantage of the adsorption properties of natural biomasses [28], or of composite materials based on natural sources [21,29]. Apart from bacteria [30–32] and fungi [21,33,34], agricultural [35] and algae-based adsorbents [28,36] have also been used. Among macroalgae, brown seaweeds are of particular interest because they contain alginate, which has a chemical affinity for divalent metals [37,38]. In fact, alginate chains form a tridimensional matrix in the presence of divalent metals by ionic crosslinking following the egg-box model [39]. Ca^{2+} ions are used for this purpose. However, divalent metals of higher atomic mass replace the calcium ions as the active sites in the alginate hydrogel have increased affinity for larger divalent metal ions [40]. For decades, alginate has been extracted from brown seaweed with a good yield at an industrial level to be used as a thickening and gelling agent in the food and cosmetics industry [41,42]. The main species that are currently exploited industrially are *Macrocystis pyrifera*, *Laminaria hyperborea*, *Laminaria digitata* and *Asphyllum nodosum* [43]. However, alternative uses are proposed for alginate in the biomedical field as a drug delivery system, for wound healing and tissue engineering [6], or as a biosorbent material. In fact, both brown seaweed biomass and alginate, in the form of gel beads have been extensively studied as biosorbents [28,37,44–47]. Recently, composite materials based on alginate have been derived with improved properties. For instance, including magnetite has enhanced the recovery of biosorbents. In this case, adsorption of Cd^{2+} ions was improved by also including activated carbon in the alginate matrix leading to an improved material in terms of adsorption capacity and costs [48]. Another approach is to obtain a compressive alginate sponge where the contaminant is recovered after adsorption simply by compressing the biosorbent before reuse. The efficiency of this system was proven in the case of treating methylene blue contaminated wastewaters [49]. The crosslinked alginate matrix has also been used as a coating for composite materials that are to be used as controlled drug release systems [50].

Cystoseira is a polyphyletic genus of brown seaweed, included in the Sargassaceae family, which is found extensively on the coasts of the Mediterranean Sea and the eastern Atlantic Ocean and also in the Black Sea [51,52] Most studies of *Cystoseira* sp. have shown a large variety of secondary metabolites with biological activity including phlorotannins, terpenoids and carbohydrates [53]. *Cystoseira barbata* recovered from the coasts of Tunisia has been proven to contain compounds that show biological activity, including laminarin, which has antioxidant, antibacterial and wound healing properties [54]; fucoxanthin, used as a color enhancer and oxidative stability enhancer of meat products [55]; and polyphenolic-protein-polysaccharide ternary conjugates, which are used as biopreservatives [56]. In the Black Sea, *C. barbata* and *Cystoseira crinita* are the only representatives of *Cystoseira* sp., although a *Cystoseira bosphorica* member has sometimes been reported [51,57]. *C. barbata* is traditionally used in the Black Sea area as functional food [2,58]. Recent studies have been based on ecological interest in the levels of heavy elements in certain parts of the Black Sea coast [59–61], as well as the structure of certain metabolites found in *C. barbata* and *C. crinita* [62]. The use of *Cystoseira* spp. biomass as a biosorbent has been shown in one study involving *C. amentacea* var. *stricta* (formerly *C. stricta*) [47]. Dried *C. barbata* biomass from the Turkish coast of the Black Sea has also been investigated as a biosorbent [46]. To the best of our knowledge, the structure of alginate extracted from *C. barbata* and its use as a biosorbent in the form of beads to adsorb heavy metals has never been studied.

The aim of this paper is to characterize alginate extracted from *C. barbata* recovered from the Romanian Black Sea and to prove that this bioactive component presents heavy metal adsorption properties compatible with its use as an adsorbent/biosorbent for heavy metal detoxification. The adsorption properties of the initial seaweed powder were also evaluated.

2. Results and Discussion

2.1. Extraction and Structural Characterization

In recent years, many studies on the characterization of polysaccharides extracted from brown algae, and more specifically on *Cystoseira* sp. have been conducted and described in the literature [63–65]. Nevertheless, alginate extracted from *C. barbata* recovered from the Romanian Black Sea coast has never been investigated from a structural point of view.

The extraction method (Figure 1) that was employed for the extraction of alginate had a yield of 19 ± 1.5% (*w/w*) and relative to the algae dry matter estimated at 95 ± 2%. This method is presented in detail in Section 3.2. The final alginate product is obtained as a fine powder (under 0.2 mm), which is referred to as CBA UF.

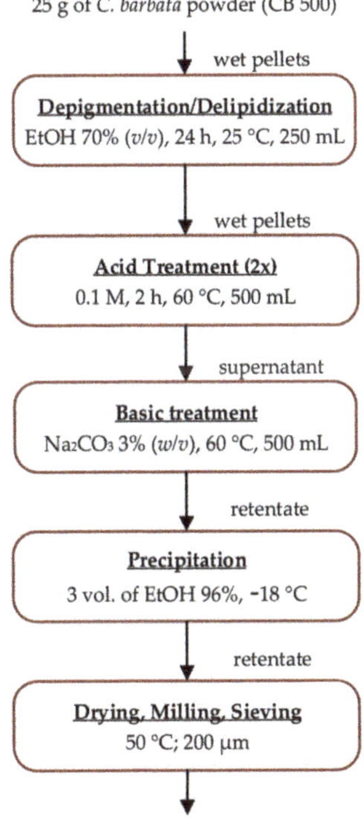

Figure 1. Processing steps for the extraction of alginate from *C. barbata* dry seaweed biomass.

2.2. HPSEC-MALS

The average molecular weight in mass (M_w), average molecular weight in number (M_n) and the intrinsic viscosity of alginate (CBA UF) extracted from *C. barbata* were determined by high performance size-exclusion chromatography equipped with a multi-angle light diffusion detector coupled to a differential refractometer and an in-line viscometer. The recovery rate of the sample was estimated at 90%.

The values of the determined average molecular weights and hydrodynamic radius are reported in Table 1. The M_n and M_w of alginate (CBA UF) were estimated at 85.2 (±2.7%) kDa and 126.6 (±1.0%) kDa, indicating a low polydispersity index (PDI = M_w/M_n =1.49). These results are similar to those for other alginates obtained from brown algae, and especially from *Cystoseira* sp. such as *Cystoseira sedoides* [66], *Cystoseira compressa* [63,66], *C. crinita* [66]. In fact, our present results for the M_w values were very consistent with earlier investigations on alginates from Fucales algae families, for example, *Fucus vesiculosus*, *A. nodosum*, *C. compressa* or *C. sedoides*, which have M_w values ranging from around 100 kDa to 200 kDa [66,67]. Nevertheless, compared to other brown algae from *Sargassum* species, where M_w values ranged from 300 to 1000 kDa, the M_w of alginate fractions (CBA UF) extracted from Romanian *C. barbata* were lower [67,68].

Table 1. Characterization of alginate (CBA UF) extracted from *C. barbata* collected from the Romanian Black Sea.

	Mn (kDa)	Mw (kDa)	PDI	R_h(w) (nm)	[η] (mL/g)
CBA UF	85.2	126.6	1.49	19.2	406.2

Mn—average molecular weight in number; Mw—Average molecular weight in mass (Mw); PDI—polydispersity index; R_h(w)—hydrodynamic radius; [η]—intrinsic viscosity.

Compared with other alginates from brown algae, and especially from *Cystoseiraceae* species, the PDI value of 1.49 is very close to the alginate from Tunisian *C. compressa* [63]. This PDI value of alginate from Romanian *C. barbata* confirms a good M_w polysaccharides distribution, and indicates that there is no depolymerization of polysaccharides during the extraction/purification process steps. Finally, the intrinsic viscosity, which signifies the hydrodynamic volume occupied by the macromolecules in a dilute solution, was estimated for CBA UF at 406 mL/g. This value is lower than the alginate [η] values from *Sargassum* (800–1300 mL/g) brown algae but is close to other intrinsic viscosity values of alginate extracted from *Cystoseiraceae* species [63,64,68]. Notably, a similar [η] was observed with alginate extracted from Tunisian brown algae *C. compressa* [63]. As generally described, these observed differences are related to the origin of the algae and the extraction / purification processes used, which affect molecular weight and intrinsic viscosity.

2.3. NMR Analysis

Alginate extracted from *C. barbata* (CBA UF) was analyzed by ^{1}H NMR. As observed in Figure 2, the 1D ^{1}H-NMR spectrum showed the specific signal characteristics of the sodium alginate fraction, revealing high purity [69,70].

As well defined in the literature [69,70], ^{1}H-NMR analysis identifies the alginate structure with three signals in the anomeric region: Signal I corresponds to the anomeric proton of the guluronic acid residue (G-1), Signal II corresponds to the overlap between the mannuronic acid anomeric proton (M-1) and the H-5 of alternating blocks (GM-5), and Signal III corresponds to proton H-5 guluronic acid from the GG-5G block (G-5). As a general rule, by using the NMR method we could estimate the proportions of each individual block of guluronic and mannuronic acids (F_G and F_M), the homogeneous (F_{GG} and F_{MM}) and heterogeneous (F_{GM} and F_{MG}) blocks of alginate [70] extracted from *C. barbata* (CBA UF) using the areas (A) of signals I, II and III and the Equations (1), (2) and (3):

$$F_G = A_I / A_{II} + A\_III \tag{1}$$

$$F_M = 1 - F_G \tag{2}$$

$$F_{GG} = A_{III} / A_{II} + A_{III} \tag{3}$$

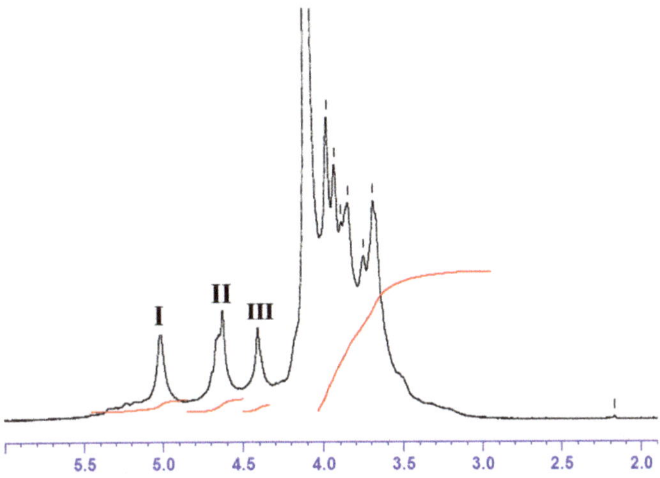

Figure 2. ^{1}H NMR analysis of CBA UF extracted from *C. barbata* with: signal I = guluronic acid anomeric proton (G-1), signal II = overlap between the mannuronic acid anomeric proton (M-1) and the H-5 of alternating blocks (GM-5), signal III = guluronic acid H-5 position (block GG-5G), G-6 the C-6 from guluronic acid residue and M-6 the C-6 from mannuronic residue.

Regarding the estimation of the block F_{GM}, F_{MM} and M/G molar ratio of alginate extracted from *C. barbata* (CBA UF), we used the following equations:

$$F_{GM} = F_G - F_{GG} \tag{4}$$

$$F_{MM} = F_M - F_{GM} \tag{5}$$

$$M/G = F_M / F_G \tag{6}$$

Consequently, Equations (4), (5) and (6) allow the complete structural characterization [70] of CBA UF, which is summarized in Table 2.

Table 2. Structural characterization of alginate (CBA UF) extracted from *C. barbata* from Romanian Black Sea.

Fraction	F_G [1]	F_M [2]	F_{GG} [3]	F_{GM} or F_{MG} [4]	F_{MM} [5]	M/G [6]
CBA UF	0.61	0.39	0.34	0.27	0.12	0.64

[1] F_G—fraction of individual blocks of guluronic acid units; [2] F_M—fraction of individual blocks of mannuronic acid units; [3] F_{GG}—fraction of homogeneous block of guluronic acid; [4] F_{GM} or F_{MG}—fraction of heterogeneous blocks of alternating mannuronic and guluronic acids; [5] F_{MM}—fraction of homogeneous block of mannuronic acid; [6] M/G—ratio between F_M and F_G.

The frequencies of structural blocks shown in Table 2 provide information about the alginate composition in *C. barbata*. For example, the ratio between mannuronic and guluronic acid gives information about the quality of Ca^{2+} reticulated gels which, in this case, has a strong and rigid quality [71]. This value is generally higher than those reported for other species from *Cystoseira* genus such as *Sirophysalis trinodis* (formerly *C. trinodis*) (0.59), *C. myrica* [68], although it is lower in some cases,

C. compressa (0.77) [63], *C. humilis* (1.46) [72]. Compared to other species from *Laminaria* or *Sargassum*, this value is higher in some cases (*L. hyperborea*, 0.41 [70]; *Sargassum filipendula*, 0.19 [73]), and lower in other cases (*S. vulgare*, 1.27 [69]; *L. digitata*, 1.12 [74]).

Structural information can be derived by evaluating the parameter $\eta = 1.13$ obtained using Equation (7). The value suggests that alginate extracted from *C. barbata* is of an alternate block type since it is greater than 1 [75]. Surprisingly, alginates extracted from other species from the *Cystoseira* genus present $\eta < 1$, a feature of alginates that have predominantly homopolymeric blocks [63,68,72].

$$\eta = \frac{F_{GM}}{F_M \cdot F_G} \tag{7}$$

2.4. Kinetics of Adsorption

Sodium alginate beads were obtained as described in Section 3.5. Photos (at 10× magnification) of the beads (not shown here) were taken and analyzed digitally in order to determine the diameter. Twenty beads were analyzed, and the mean diameter was determined to be 4413 ± 134 µm.

Initially, the kinetics of adsorption was studied by contacting sodium alginate beads and dry seaweed powder (<500 µm) with Cu^{2+} and Pb^{2+} solutions at 20 ppm and 74 ppm, respectively (C_i). [14] Four pairs of substrate/heavy metal were obtained. The instantaneous sorption capacity (q_t) was determined as defined in Equation (8) where C_t is the concentration of heavy metals at time t. This equation is deduced from mass balance. S_m represents the value of substrate mass divided by the volume of solution that was used.

$$q_t(mg/g) = \frac{C_i - C_t}{S_m} \tag{8}$$

The kinetic curves for copper and lead adsorption of each of the two substrates are shown in Figures 3 and 4.

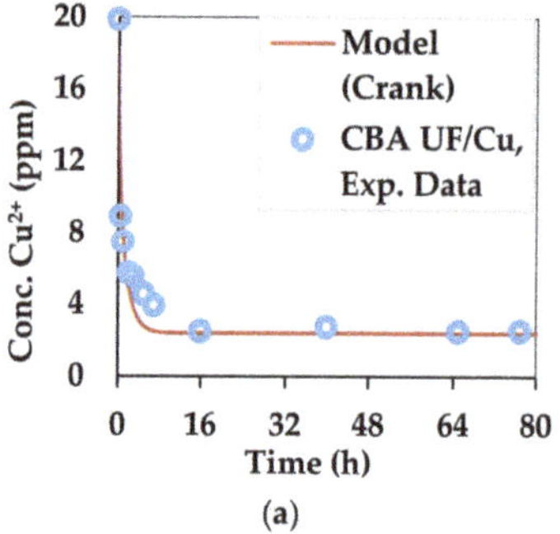

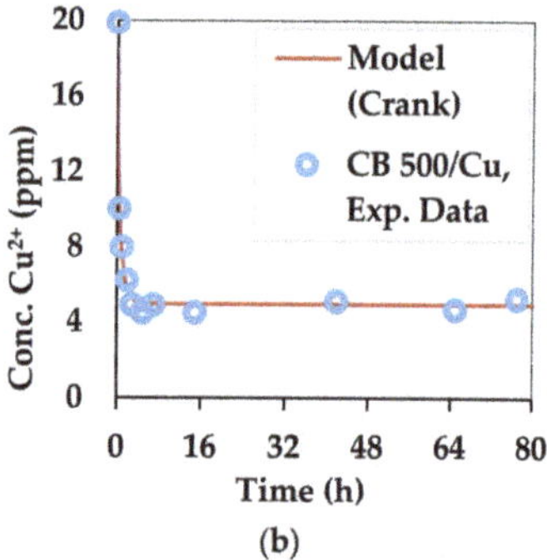

Figure 3. Kinetics of copper adsorption by the two substrates: (**a**) *C. barbata* alginate (CBA UF) and (**b**) *C. barbata* powder (CB 500).

The diffusion equation represented by Equation (9) [76] was used to simulate the experimental results of adsorption of Pb^{2+} and Cu^{2+} and is based on the diffusion of ions from a solution at a given ion concentration until the solution is solid free or has a constant negligible concentration of ions, which is the case of the present study. α in Equations (9) and (10) represents the ratio between the volume of the liquid and the volume of the solid. In the case of the alginate beads, the volume of solid is represented by the total volume of the beads used. When algal powder is used, the volume of the solute can be approximated by the equivalent volume of water and powder, which is taken up by the powder when swelling. This value is 4.14 times greater than the mass of the used powder. The ratio of a sphere equivalent in volume to the substrate needs to be found for both the alginate and the powder. This term is named a. The effective diffusivity of the solute in the solid is termed D_{eff} (m^2/s). q_n represents the six non-zero roots of Equation 10. q_e represents the equilibrium value of the adsorption while q_t is the value of the ion adsorption at time t.

$$\frac{q_t}{q_e} = 1 - \sum_{n=1}^{\infty} \frac{6\alpha(\alpha+1)\exp\left(-D_{eff}q_n{}^2 t/a^2\right)}{9 + 9\alpha + q_n{}^2\alpha^2} \tag{9}$$

$$\tan q_n = \frac{3q_n}{3 + \propto q_n{}^2} \tag{10}$$

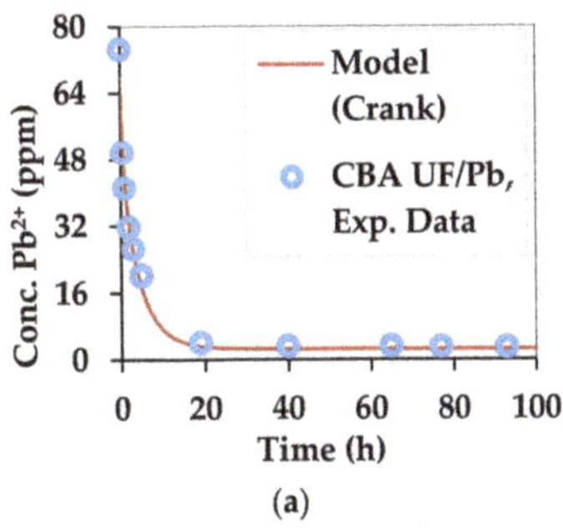
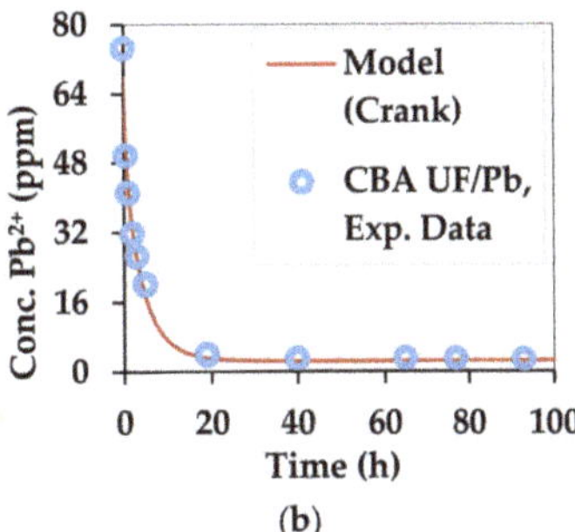

Figure 4. Kinetics of lead adsorption by the two substrates: (a) *C. barbata* alginate (CBA UF) and (b) *C. barbata* powder (CB 500).

Figures 3 and 4 compare the experimental and theoretical results from using Equation (9) by regression. The parameters that were optimized are q_e and D_{eff}. The optimization algorithm, GRG Nonlinear was applied in Excel 2013 by the Solver tool. The Multistart feature was used with a population size of 100, using central derivatives to converge towards the solution. Upper and lower bounds with physical significance were applied for each optimized parameter. The combinations of D_{eff} and q_e which best fitted the experimental data for all 4 combinations of substrate and divalent metal ions are summarized in Table 3.

Table 3. Crank diffusion model for a sphere. Model coefficients for 4 pairs of substrate / heavy metal.

Substrate/Metal Ion	$D_{eff} \times 10^{-9}$ (m^2/s)	q_e (mg/g)
CBA UF/Pb^{2+}	0.85	359.8
CBA UF/Cu^{2+}	3.98	43.8
CB 500/Pb^{2+}	1.39	172
CB 500/Cu^{2+}	1.79	37.3

D_{eff}—the effective diffusivity of the solute in the solid; q_e—the equilibrium value of the adsorption.

The values of D_{eff} in all cases are similar to the self-diffusivity of Cu^{2+} and Pb^{2+} in water: 0.71×10^{-9} m^2/s and 0.95×10^{-9} m^2/s [77]. This can be explained by the high content of water in the beads and the swollen biomass. Alginate gels are generally nanoporous [6,78], leading to high diffusion rates of small solutes [6]. However, in this case D_{eff} is accelerated by the affinity of divalent metal ions towards the egg-box structure of gels [39] formed by the G fractions in the alginate structure. The Ca^{2+} ions, which occupy this site initially, are replaced by the Pb^{2+} and Cu^{2+} ions, which have a higher affinity [40]. The q_e is the maximum adsorption capacity in the given setup, which was not meant to saturate the substrate. The diffusivities obtained for the substrates used in this work are higher than those obtained for materials used in a similar work for Pb^{2+} and Cu^{2+} [79].

2.5. Adsorption Isotherms

The experimental values for the maximum sorption capacities deduced from the adsorption isotherms are represented in Figures 5 and 6 and are presented in Table 4. Such levels of Pb^{2+} concentration are in the range of contaminated mine waters [80,81].

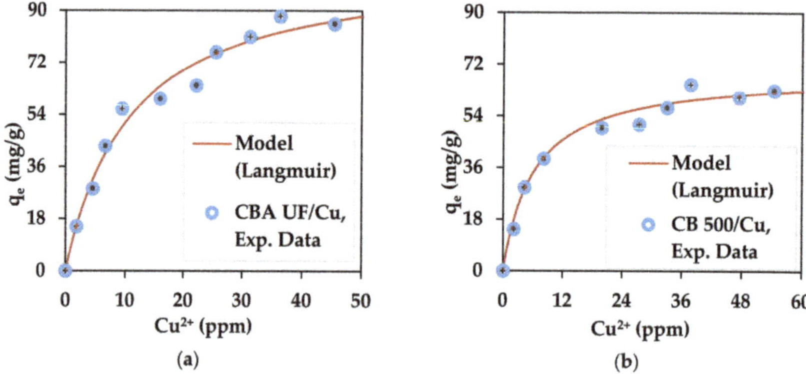

Figure 5. Copper adsorption isotherms for: (**a**) *C. barbata* alginate (CBA UF) and (**b**) *C. barbata* powder (CB500).

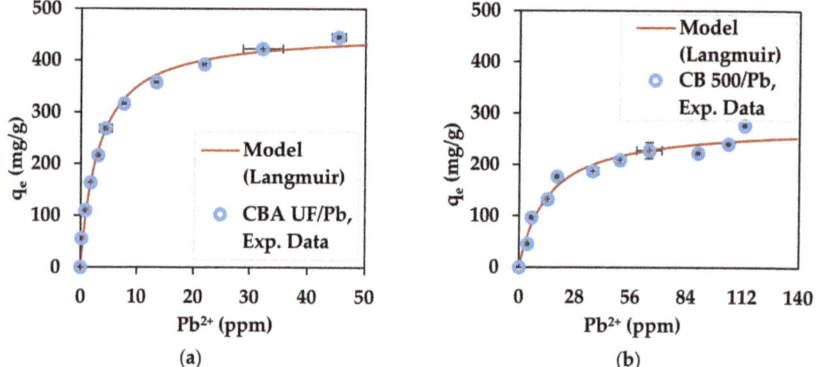

Figure 6. Lead adsorption isotherms for: (**a**) *C. barbata* alginate (CBA) and (**b**) *C. barbata* powder (CB500).

Table 4. Adsorption isotherms - model parameters (Langmuir).

Substrate/Metal Ion	q_{max} (mg/g)	K_L (mg/L)	U_t (mmol/g)
CBA UF/Pb^{2+}	454 ± 4.7	0.32 ± 0.04	0.77
CB 500/Pb^{2+}	279.2 ± 7.5	0.069 ± 0.005	0.15
CBA UF/Cu^{2+}	107.3 ± 1.7	0.092 ± 0.005	0.77
CB 500/Cu^{2+}	69.3 ± 2	0.16 ± 0.03	0.15

q_{max}—maximum theoretical adsorption capacity according to Langmuir, K_L—Langmuir constant, U_t—theoretical egg-box sites per mass of substrate.

On the same figures, the Langmuir adsorption model is shown for optimized parameters: q_{max} and K_L. This model has been used in many studies that involve the adsorption of pollutants such as heavy metals, dyes and phenol [24,82,83]. This model is represented by Equation (11). C_e represents the concentration of metal ions still in the solution.

$$q_e = \frac{q_{max} K_L C_e}{1 + K_L C_e} \tag{11}$$

The optimized parameters for each substrate/metal pair are shown in Table 4.

In all cases the capacity of the substrates to adsorb the heavy metals used in this work remains high and reliable because the standard deviation is very low. Interestingly, the seaweed powder without any treatment also had excellent capacity for the adsorption of heavy metals. This is an interesting result because it shows that the brown seaweed recovered from the Romanian Black Sea can be used in its native form for heavy metal adsorption. The values obtained for q_{max} show that the substrates used for both metal ions have good theoretical adsorption capacity according to Langmuir (infinite equilibrium time considered). These values are similar to results obtained in a previous work [79] where for Pb^{2+}, alginate beads were found to have a q_{max} of 390.3 mg/g, while brown seaweed biomass (*L. digitata*) was shown to have a q_{max} value of 264.2 mg/g [79]. For $Cu^{2,+}$ similar results can be found in the literature with, for example, values of 107.5 mg/g and 74.5 mg/g for alginate and algal biomass (*L. digitata*), respectively [79]. The algal biomass can also be compared to other agricultural wastes that have been investigated as heavy metal adsorbents, such as treated green coconut (*Cocos nucifera*) shells, Pb^{2+}: 54.62 mg/g, Cu^{2+}: 41.36 mg/g [84]; apricot stone activated carbon, Pb^{2+} 22.85 mg/g, Cu^{2+} 24.21 mg/g [85]; tea waste, Cu^{2+}: 48 mg/g [86]; and rose waste biomass, Pb^{2+}: 151.51 mg/g [87]. As can be observed, the algal powder has a better adsorption capacity than other dried biomasses. The heavy metals adsorption capacity of *C. barbata* from the Black Sea has been studied before on samples recovered from the Turkish Black Sea Coast [46], where dried *C. barbata* biomass was found to have a maximum sorption capacity of 253 mg/g for lead, which is only slightly lower than the value obtained in this study. Other seaweed from *Cystoseira* sp. have also shown to have heavy metal adsorption properties when used directly as dried biomass. *Cystoseira amentacea* var. *stricta* (formerly *C. stricta*) recovered on the Algerian coast proved to adsorb 64.5 mg/g of lead ions after undergoing several chemical treatments [47].

The good performance of alginate from *C. barbata* can be explained by the low value of the M/G ratio (0.64), which indicates the higher availability of G blocks [40,88]. This could explain the better performance compared to other studies where alginate from *L. digitata* was used [79]. *L. digitata* was shown to have an M/G value of 1.12 [74] or 1.63 [44].

The natural capacity of *C. barbata* to adsorb heavy metals is proven in this study and confirms the high adsorption capacity which was also proven in a previous study [46]. The affinity of brown seaweed for heavy metals can be explained by the presence of alginate, especially the G homogeneous block fractions, which chelate divalent metal ions. This is why alginate extracted from *C. barbata* has an increased capacity.

The good heavy metal absorption characteristics of the edible brown seaweed *C. barbata* and of its major bioactive component, alginic acid, offer a possible explanation regarding the traditional use of this seaweed as functional food. Further studies on biological systems are needed.

Beside its use as nutraceutical/functional food, the biomass of *C. barbata* and/or of its bioactive component, alginic acid, could be used for water treatment to recover heavy metals ions. In this case, this seaweed and the alginate extracted from it have the potential to be used to develop systems such as cartridges, which could facilitate their use in processes under continuous conditions [44]. Also, the behavior during sorption/desorption cycles needs to be determined to prove reusability.

3. Materials and Methods

3.1. Raw Material and Chemicals

C. barbata seaweed was recovered from the Romanian seashore of the Black Sea in the city of Mangalia (GPS coordinates at latitude: N 43° 49.2′, longitude: E 28° 35.4′). The biomass was thoroughly washed with tap water to remove sand particles and epiphytes and dried at room temperature for one week in the dark to prevent any possible degradation associated with sunlight. Afterwards, the dry biomass was milled and sieved at 500 µm to obtain the material referred to as CB 500. The powder recovered under the sieve was used for further manipulation. The chemicals used for this work were analytical grade and were purchased from Sigma-Aldrich (Sigma-Aldrich, Saint-Louis, MO, USA).

3.2. Extraction of Alginate

Alginate was extracted by adapting several methods used in the literature [64]. Initially, 25 g of dried seaweed powder underwent a mild depigmentation and defatting in EtOH (70% *v/v*, 24 h, 250 mL). The solid was removed by vacuum filtration (Whatman filter paper, 25 µm, Whatman, Maidstone, UK) and added to HCl (0.1 M, 2 h, 60 °C, 500 mL). After vacuum filtration, this operation was repeated for the recovered wet pellets. The excess acid was washed away with distilled water before extracting the alginate in a Na_2CO_3 solution (3% (*w/v*), 2 h, 60 °C). The sodium alginate extract was left to cool before separating it from the solid waste by centrifugation (15,000 g, 30 min, 4 °C). After neutralization with dilute HCl, the extract was treated by ultrafiltration on a Vivaflow 200 crossflow cassette module (100 kDa, polyethersulfone, Sartorius, Göttingen, Germany) fitted with a peristaltic pump. Following the treatment in diafiltration mode with 5 diavolumes, alginate remained in solution in the retentate while most of contaminants were removed in the permeate. This procedure was followed by a concentration process, which reduced the volume of the retentate 3 times before precipitating the alginate by adding 3 volumes of EtOH (96%, −18 °C). The alginate pellets were dehydrated twice with 20 mL of acetone at −18 °C. The pellets were then dried at 50 °C, milled and sieved (0.2 mm) to obtain a fine powder called CBA UF.

3.3. NMR Analysis

The sample was prepared and analyzed under conditions described in the literature [63]. CBA UF was dissolved in D_2O (99.9% D) at a concentration of 50 g/L. After dissolution the sample was freeze-dried resulting in CBA UF alginate with exchangeable protons replaced by deuterium. In total, this operation was done 3 times. Before analysis, the CBA UF lyophilized sample was dissolved again in D_2O (99.9% D) at a concentration of 40 g/L. NMR spectra were obtained at 80 °C using a 400 MHz Bruker Avance spectrometer (Bruker, Billerica, MA, USA), equipped with a BBFO probe. A spectral width of 3000 Hz was used for acquiring the data obtained under the following acquisition parameters: acquisition mode = 2 s, pulse 90° = 8 µsec, scans = 64, recovery = 5 s (for a complete return after the 90° pulse).

3.4. HPSEC-MALS

High pressure size exclusion chromatography (HPSEC) was used to determine the number average molar mass, (M_n), the mass average molar mass (M_w), intrinsic viscosity ([η]), hydrodynamic radius (R_h) and gyration radius (R_g) for CBA UF. Three detectors were used in parallel, and were coupled to the chromatograph: multi-angle light scattering (MALS, DAWN-EOS from Wyatt Technology Corp., Santa Barbara, CA, USA) with a Ga-As laser (690 nm) and a K5 cell (50 µL) (HELEOS II Wyatt Technology Corp., USA), a viscosimeter (Viscostar II, Wyatt Technology Corp., USA) and refractive index detector (RID, RID10A Shimadzu, Kyoto, Japan). One OHPAK SB-G, two OHPAK SB 804 and 806 HQ columns were used in series for the HPSEC line (Shodex Showa Denko K.K., Tokyo, Japan). The system was eluted with $LiNO_3$ 0.1 M, filtered using a 0.1 µm unit (Millipore, Merck Group, Darmstadt, Germany) and degassed (DGU-20A3 Shimadzu, Kyoto, Japan). The flow rate was set at a value of 0.5 mL/min (LC10Ai Shimadzu, Kyoto, Japan). The sample was diluted to 1 mg/mL in $LiNO_3$ 0.1M under stirring for 24h, filtered (0.45 µm, Millipore) before 500 µL was placed onto the analytical line of the instrument with an automatic injector (SIL-20A Shimadzu, Kyoto, Japan).

3.5. Preparation of Alginate Beads

CBA UF was dissolved in MilliQ water to obtain a solution with a concentration of 3% (*w/v*). The solution was then pumped through a capillary (d = 1 mm) with its other end above a 0.5 M $CaCl_2$ solution under stirring. Alginate beads are formed drop by drop as they contact the calcium solution. They were stored for 12 h at 4 °C in a 0.5 M $CaCl_2$ solution. A digital camera (Euromex CMEX-5000, Arnhem, The Netherlands) was used to take photos of the alginate beads. The diameter of the beads was measured using the ImageFocus software.

3.6. Adsorption Kinetics

The initial conditions for the kinetic experiments were adapted from the literature [79] and are shown in Table 5. The substrate mass represents the quantity of alginate used to produce the corresponding number of beads used or the mass of dried powder added. The initial concentration of heavy metal ions is also given in Table 5 and was obtained from $PbCl_2$ and $CuSO_4$. The corresponding quantity of substrate was added under stirring. The kinetics experiments were run for 100 h in order to achieve equilibrium. The concentration of metal ions was determined using MP-AES (microwave plasma-atomic emission spectrometry, Agilent 4200, Santa Clara, CA, USA).

Table 5. Initial conditions for adsorption kinetics experiments.

Substrate/Heavy Metal Pair	Substrate Mass/Solution Volume (g/L)	Initial Concentration of Heavy Metal Ions (ppm)	Heavy Metal Ions/Dry Substrate (*w/w*)
CBA UF/Pb^{2+}	0.2	74	0.37
CBA UF/Cu^{2+}	0.4	20	0.05
CB500/Pb^{2+}	0.4	74	0.185
CB500/Cu^{2+}	0.4	20	0.05

3.7. Adsorption Isotherms

Following the adsorption experiments, it was observed that the equilibrium was well reached after ~65 h. This value is used as the time necessary to achieve equilibrium in the adsorption isotherm experiments. Ten solutions of 50 mL each were obtained for each substrate/heavy metal pair with a concentration of heavy metal ions varying from 22.3 to 223 ppm for Pb^{2+} and from 8 to 79.6 ppm for Cu^{2+}(C_i). In this case, the mass of alginate contained in the beads with respect to the heavy metal solution (S_m) was fixed at 0.4 g/L of solution. The corresponding mass of substrate was added to the heavy metal solutions. The solutions are kept under agitation using a plate shaker until equilibrium was reached. Afterwards, the final concentration of heavy metals was determined by MP-AES.

Author Contributions: Conceptualization, B.T., C.D., A.-V.U., P.M., and G.D.; Investigation, B.T., F.G., P.M. and F.O.; Methodology, B.T., F.G., A.-V.U. and G.D.; Supervision, C.D., T.D. and G.D.; Writing—original draft, B.T., C.D. and G.D.; Writing—review & editing, B.T., C.D., T.D. and F.O.

Funding: This research was founded by a grant from the Hubert Curien (PHC) Brancusi 2017 Programme, project number 38365QK entitled "*On the advanced processing of marine algae from Romanian Black Sea edge*" and by the project PN19.23.01.01 "*Integrated platform for smart valorization of the biomass*" SmartBi, from Nucleu Programme ChemEmergent of NIRDCP-ICECHIM, founded by the Ministry of Research and Innovation, Romania. The APC was funded by the project PFE 31/2018, *Enhancing NIRDCP-ICECHIM research & innovation potential within the inter-disciplinary and cross-sectoral field of key enabling technologies* - TRANS-CHEM, founded by Ministry of Research and Innovation, Romania.

Acknowledgments: We would like to thank Didier Le Cerf for the HPSEC-MALS analysis (Normandie Univ, UNIROUEN, INSA Rouen, CNRS, PBS, 76000 Rouen, France), Jacques Desbrières (Université de Pau et des Pays de l'Adour (UPPA), IPREM, UMR 5254 CNRS/UPPA, Helioparc Pau Pyrénées, 2 avenue P. Angot, 64053 PAU CEDEX 09, France) for the ^{1}H-NMR and Diana Constantinescu-Aruxandei, for helpful discussion and English editing.

Conflicts of Interest: The authors declare no conflict of interests.

References

1. Tanna, B.; Mishra, A. Metabolites unravel nutraceutical potential of edible seaweeds: An emerging source of functional food. *Compr. Rev. Food Sci. Food Saf.* **2018**, *17*, 1613–1624. [CrossRef]
2. Pereira, L. *Therapeutic and Nutritional Uses of Slgae*; CRC Press: Boca Raton, FL, USA, 2018; p. 672. [CrossRef]
3. Tanna, B.; Mishra, A. Nutraceutical Potential of Seaweed Polysaccharides: Structure, Bioactivity, Safety, and Toxicity. *Compr. Rev. Food Sci. Food Saf.* **2019**, *18*, 817–831. [CrossRef]
4. Okolie, C.L.; Rajendran, S.R.C.K.; Udenigwe, C.C.; Aryee, A.N.A.; Mason, B. Prospects of brown seaweed polysaccharides (BSP) as prebiotics and potential immunomodulators. *J. Food Biochem.* **2017**, *41*, e12392. [CrossRef]
5. Draget, K.I.; Taylor, C. Chemical, physical and biological properties of alginates and their biomedical implications. *Food Hydrocoll.* **2011**, *25*, 251–256. [CrossRef]
6. Lee, K.Y.; Mooney, D.J. Alginate: Properties and biomedical applications. *Prog. Polym. Sci.* **2012**, *37*, 106–126. [CrossRef] [PubMed]
7. Idota, Y.; Kato, T.; Shiragami, K.; Koike, M.; Yokoyama, A.; Takahashi, H.; Yano, K.; Ogihara, T. Mechanism of suppression of blood glucose level by calcium alginate in rats. *Biol. Pharm. Bull.* **2018**. [CrossRef] [PubMed]
8. Cassidy, Y.M.; McSorley, E.M.; Allsopp, P.J. Effect of soluble dietary fibre on postprandial blood glucose response and its potential as a functional food ingredient. *J. Funct. Foods* **2018**, *46*, 423–439. [CrossRef]
9. Georg Jensen, M.; Pedersen, C.; Kristensen, M.; Frost, G.; Astrup, A. Review: Efficacy of alginate supplementation in relation to appetite regulation and metabolic risk factors: Evidence from animal and human studies. *Obes. Rev.* **2013**, *14*, 129–144. [CrossRef] [PubMed]
10. Kasahara, F.; Kato, T.; Idota, Y.; Takahashi, H.; Kakinuma, C.; Yano, K.; Arakawa, H.; Hara, K.; Miyajima, C.; Ogihara, T. Reduction Effect of Calcium Alginate on Blood Triglyceride Levels Causing the Inhibition of Hepatic and Total Body Accumulation of Fat in Rats. *Biol. Pharm. Bull.* **2019**, *42*, 365–372. [CrossRef]
11. Savchenko, O.V.; Sgrebneva, M.N.; Kiselev, V.I.; Khotimchenko, Y.S. Lead removal in rats using calcium alginate. *Environ. Sci. Pollut. Res.* **2015**, *22*, 293–304. [CrossRef]
12. Khotimchenko, M.; Serguschenko, I.; Khotimchenko, Y. Lead absorption and excretion in rats given insoluble salts of pectin and alginate. *Int. J. Toxicol.* **2006**, *25*, 195–203. [CrossRef] [PubMed]
13. Eliaz, I.; Weil, E.; Wilk, B. Integrative Medicine and the Role of Modified Citrus Pectin/Alginates in Heavy MetIntegrative Medicine and the Role of Modified Citrus Pectin/Alginates in Heavy Metal Chelation and Detoxification—Five Case Reports. *Complement. Med. Res.* **2007**, *14*, 358–364. [CrossRef] [PubMed]
14. Kopittke, P.M.; Asher, C.J.; Kopittke, R.A.; Menzies, N.W. Toxic effects of Pb2+ on growth of cowpea (Vigna unguiculata). *Environ. Pollut.* **2007**, *150*, 280–287. [CrossRef] [PubMed]
15. Chowdhury, S.; Mazumder, M.A.J.; Al-Attas, O.; Husain, T. Heavy metals in drinking water: Occurrences, implications, and future needs in developing countries. *Sci. Total Environ.* **2016**, *569–570*, 476–488. [CrossRef] [PubMed]
16. Alam, I.A.; Sadiq, M. Metal contamination of drinking water from corrosion of distribution pipes. *Environ. Pollut.* **1989**, *57*, 167–178. [CrossRef]
17. Al-Saleh, I.; Al-Doush, I. Survey of trace elements in household and bottled drinking water samples collected in Riyadh, Saudi Arabia. *Sci. Total Environ.* **1998**, *216*, 181–192. [CrossRef]
18. Simeonov, V.; Stratis, J.A.; Samara, C.; Zachariadis, G.; Voutsa, D.; Anthemidis, A.; Sofoniou, M.; Kouimtzis, T. Assessment of the surface water quality in Northern Greece. *Water Res.* **2003**, *37*, 4119–4124. [CrossRef]
19. Ahmad, M.K.; Islam, S.; Rahman, M.S.; Haque, M.R.; Islam, M.M. Heavy Metals in Water, Sediment and Some Fishes of Buriganga River, Bangladesh. *Int. J. Environ. Res.* **2010**, *4*, 321–332. [CrossRef]
20. Jane Wyatt, C.; Fimbres, C.; Romo, L.; Méndez, R.O.; Grijalva, M. Incidence of Heavy Metal Contamination in Water Supplies in Northern Mexico. *Environ. Res.* **1998**, *76*, 114–119. [CrossRef]
21. Jacob, J.M.; Karthik, C.; Saratale, R.G.; Kumar, S.S.; Prabakar, D.; Kadirvelu, K.; Pugazhendhi, A. Biological approaches to tackle heavy metal pollution: A survey of literature. *J. Environ. Manag.* **2018**, *217*, 56–70. [CrossRef]
22. Wu, Q.; Leung, J.Y.S.; Du, Y.; Kong, D.; Shi, Y.; Wang, Y.; Xiao, T. Trace metals in e-waste lead to serious health risk through consumption of rice growing near an abandoned e-waste recycling site: Comparisons with PBDEs and AHFRs. *Environ. Pollut.* **2019**, *247*, 46–54. [CrossRef] [PubMed]

23. Fu, H.-Z.; Wang, M.-H.; Ho, Y.-S. Mapping of drinking water research: A bibliometric analysis of research output during 1992–2011. *Sci. Total Environ.* **2013**, *443*, 757–765. [CrossRef] [PubMed]

24. Carolin, C.F.; Kumar, P.S.; Saravanan, A.; Joshiba, G.J.; Naushad, M. Efficient techniques for the removal of toxic heavy metals from aquatic environment: A review. *J. Environ. Chem. Eng.* **2017**, *5*, 2782–2799. [CrossRef]

25. World Health Organization. *Guidelines for Drinking-Water Quality*; World Health Organization: Geneva, Switzerland, 2004; Volume 1.

26. Pal, P. *Groundwater Arsenic Remediation: Treatment Technology and Scale UP*; Butterworth-Heinemann: Oxford, UK, 2015.

27. Holan, Z.R.; Volesky, B. Biosorption of lead and nickel by biomass of marine algae. *Biotechnol. Bioeng.* **1994**, *43*, 1001–1009. [CrossRef] [PubMed]

28. Yadanaparthi, S.K.; Graybill, D.; von Wandruszka, R. Adsorbents for the removal of arsenic, cadmium, and lead from contaminated waters. *J. Hazard. Mater.* **2009**, *171*, 1–15. [CrossRef] [PubMed]

29. Vakili, M.; Deng, S.; Cagnetta, G.; Wang, W.; Meng, P.; Liu, D.; Yu, G. Regeneration of chitosan-based adsorbents used in heavy metal adsorption: A review. *Sep. Purif. Technol.* **2019**, *224*, 373–387. [CrossRef]

30. Ullah, A.; Heng, S.; Munis, M.F.H.; Fahad, S.; Yang, X. Phytoremediation of heavy metals assisted by plant growth promoting (PGP) bacteria: A review. *Environ. Exp. Bot.* **2015**, *117*, 28–40. [CrossRef]

31. Puyen, Z.M.; Villagrasa, E.; Maldonado, J.; Diestra, E.; Esteve, I.; Solé, A. Biosorption of lead and copper by heavy-metal tolerant Micrococcus luteus DE2008. *Bioresour. Technol.* **2012**, *126*, 233–237. [CrossRef]

32. Jin, Y.; Yu, S.; Teng, C.; Song, T.; Dong, L.; Liang, J.; Bai, X.; Xu, X.; Qu, J. Biosorption characteristic of Alcaligenes sp. BAPb.1 for removal of lead(II) from aqueous solution. *3 Biotech.* **2017**, *7*, 123. [CrossRef]

33. Ucun, H.; Bayhana, Y.K.; Kaya, Y.; Cakici, A.; Algur, O.F. Biosorption of lead (II) from aqueous solution by cone biomass of Pinus sylvestris. *Desalination* **2003**, *154*, 233–238. [CrossRef]

34. Kariuki, Z.; Kiptoo, J.; Onyancha, D. Biosorption studies of lead and copper using rogers mushroom biomass 'Lepiota hystrix'. *S. Afr. J. Chem. Eng.* **2017**, *23*, 62–70. [CrossRef]

35. Abia, A.A.; Asuquo, E.D. Lead (II) and nickel (II) adsorption kinetics from aqueous metal solutions using chemically modified and unmodified agricultural adsorbents. *Afr. J. Biotechnol.* **2006**, *5*, 1475–1482.

36. Jalali, R.; Ghafourian, H.; Asef, Y.; Davarpanah, S.J.; Sepehr, S. Removal and recovery of lead using nonliving biomass of marine algae. *J. Hazard. Mater.* **2002**, *92*, 253–262. [CrossRef]

37. Zhao, L.; Wang, J.; Zhang, P.; Gu, Q.; Gao, C. Absorption of Heavy Metal Ions by Alginate. *In Bioact. Seaweeds Food Appl.* **2018**, 255–268. [CrossRef]

38. Liu, Y.; Cao, Q.; Luo, F.; Chen, J. Biosorption of Cd2+, Cu2+, Ni2+ and Zn2+ ions from aqueous solutions by pretreated biomass of brown algae. *J. Hazard. Mater.* **2009**, *163*, 931–938. [CrossRef] [PubMed]

39. Smidsrod, O.; Skjakbrk, G. Alginate as immobilization matrix for cells. *Trends Biotechnol.* **1990**, *8*, 71–78. [CrossRef]

40. Haug, A.; Bjerrum, J.; Buchardt, O.; Olsen, G.E.; Pedersen, C.; Toft, J. Affinity of some divalent metals to different types of alginates. *Acta Chem. Scand.* **1961**, *15*, 1794–1795. [CrossRef]

41. Draget, K.I.; Moe, S.T.; Skjak-Braek, G.; Smidsrod, O. *Alginates, Food Ppolysaccharides and Their Applications (Second Edition)*; CRC Press-Taylor & Francis Group: Boca Raton, FL, USA, 2006.

42. Hentati, F.; Ursu, A.V.; Pierre, G.; Delattre, C.; Tricǎ, B.; Abdelkafi, S.; Djelveh, G.; Dobre, T.; Michaud, P. *Production, Extraction and Characterization of Alginates from Seaweeds*; Université Clermont Auvergne: AUBIÈRE, France, 2018.

43. Kim, S.-K.; Chojnacka, K. *Processes, Products, and Applications*; Wiley-VCH Verlag GmbH & Co. KGaA: Weinheim, Germany, 2015.

44. Wang, S.; Vincent, T.; Faur, C.; Guibal, E. Algal Foams Applied in Fixed-Bed Process for Lead(II) Removal Using Recirculation or One-Pass Modes. *Mar. Drugs* **2017**, *15*, 315. [CrossRef]

45. Esteves, A.J.P.; Valdman, E.; Leite, S.G.F. Repeated removal of cadmium and zinc from an industrial effluent by waste biomass Sargassum sp. *Biotechnol. Lett.* **2000**, *22*, 499–502. [CrossRef]

46. Yalcin, S.; Sezer, S.; Apak, R. Characterization and lead(II), cadmium(II), nickel(II) biosorption of dried marine brown macro algae Cystoseira barbata. *Environ. Sci. Pollut. Res. Int.* **2012**, *19*, 3118–3125. [CrossRef]

47. Iddou, A.; Hadj Youcef, M.; Aziz, A.; Ouali, M.S. Biosorptive removal of lead (II) ions from aqueous solutions using Cystoseira stricta biomass: Study of the surface modification effect. *J. Saudi Chem. Soc.* **2011**, *15*, 83–88. [CrossRef]

48. De Castro Alves, L.; Yáñez-Vilar, S.; Piñeiro-Redondo, Y.; Rivas, J. Novel Magnetic Nanostructured Beads for Cadmium(II) Removal. *Nanomaterials* **2019**, *9*, 356. [CrossRef] [PubMed]

49. Shen, X.; Huang, P.; Li, F.; Wang, X.; Yuan, T.; Sun, R. Compressive Alginate Sponge Derived from Seaweed Biomass Resources for Methylene Blue Removal from Wastewater. *Polymers* **2019**, *11*, 961. [CrossRef] [PubMed]

50. Lisuzzo, L.; Cavallaro, G.; Parisi, F.; Milioto, S.; Fakhrullin, R.; Lazzara, G. Core/Shell Gel Beads with Embedded Halloysite Nanotubes for Controlled Drug Release. *Coatings* **2019**, *9*, 70. [CrossRef]

51. Marin, O.A.; Timofte, F. *Atlasul Macrofitelor de la Litoralul Romanesc*; Editura Boldas: Constanta, Romania, 2011.

52. Algae Base. Available online: http://www.algaebase.org (accessed on 24 May 2019).

53. Bruno de Sousa, C.; Gangadhar, K.N.; Macridachis, J.; Pavão, M.; Morais, T.R.; Campino, L.; Varela, J.; Lago, J.H.G. Cystoseira algae (Fucaceae): Update on their chemical entities and biological activities. *Tetrahedron Asymmetry* **2017**, *28*, 1486–1505. [CrossRef]

54. Sellimi, S.; Maalej, H.; Rekik, D.M.; Benslima, A.; Ksouda, G.; Hamdi, M.; Sahnoun, Z.; Li, S.; Nasri, M.; Hajji, M. Antioxidant, antibacterial and in vivo wound healing properties of laminaran purified from Cystoseira barbata seaweed. *Int. J. Biol. Macromol.* **2018**, *119*, 633–644. [CrossRef] [PubMed]

55. Sellimi, S.; Ksouda, G.; Benslima, A.; Nasri, R.; Rinaudo, M.; Nasri, M.; Hajji, M. Enhancing colour and oxidative stabilities of reduced-nitrite turkey meat sausages during refrigerated storage using fucoxanthin purified from the Tunisian seaweed Cystoseira barbata. *Food Chem. Toxicol.* **2017**, *107*, 620–629. [CrossRef] [PubMed]

56. Sellimi, S.; Benslima, A.; Barragan-Montero, V.; Hajji, M.; Nasri, M. Polyphenolic-protein-polysaccharide ternary conjugates from Cystoseira barbata Tunisian seaweed as potential biopreservatives: Chemical, antioxidant and antimicrobial properties. *Int. J. Biol. Macromol.* **2017**, *105*, 1375–1383. [CrossRef] [PubMed]

57. Berov, D.; Ballesteros, E.; Sales, M.; Verlaque, M. Reinstatement of Species Rank for *Cystoseira bosphorica* Sauvageau (Sargassaceae, Phaeophyceae). *BIOONE* **2015**, *36*, 65–80.

58. Milchakova, N. *Marine plants of the Black Sea. An Illustrated Field Guide*; DigitPrint: Sevastopol, Russia, 2011.

59. Nonova, T.; Tosheva, Z. Cesium and strontium in Black Sea macroalgae. *J. Environ. Radioact.* **2014**, *129*, 48–56. [CrossRef] [PubMed]

60. Jordanova, A.; Strezov, A.; Ayranov, M.; Petkov, N.; Stoilova, T. Heavy metal assessment in algae, sediments and water from the bulgarian black sea coast. *Water Sci. Technol.* **1999**, *39*, 207–212. [CrossRef]

61. Strezov, A.; Nonova, T. Influence of macroalgal diversity on accumulation of radionuclides and heavy metals in Bulgarian Black Sea ecosystems. *J. Environ. Radioact.* **2009**, *100*, 144–150. [CrossRef] [PubMed]

62. Milkova, T.; Talev, G.; Christov, R.; Dimitrova-Konaklieva, S.; Popov, S. Sterols and volatiles in Cystoseira barbata and Cystoseira crinita from the black sea. *Phytochemistry* **1997**, *45*, 93–95. [CrossRef]

63. Hentati, F.; Delattre, C.; Ursu, A.V.; Desbrières, J.; Le Cerf, D.; Gardarin, C.; Abdelkafi, S.; Michaud, P.; Pierre, G. Structural characterization and antioxidant activity of water-soluble polysaccharides from the Tunisian brown seaweed Cystoseira compressa. *Carbohydr. Polym.* **2018**, *198*, 589–600. [CrossRef] [PubMed]

64. Sellimi, S.; Younes, I.; Ayed, H.B.; Maalej, H.; Montero, V.; Rinaudo, M.; Dahia, M.; Mechichi, T.; Hajji, M.; Nasri, M. Structural, physicochemical and antioxidant properties of sodium alginate isolated from a Tunisian brown seaweed. *Int. J. Biol. Macromol.* **2015**, *72*, 1358–1367. [CrossRef] [PubMed]

65. Sellimi, S.; Kadri, N.; Barragan-Montero, V.; Laouer, H.; Hajji, M.; Nasri, M. Fucans from a Tunisian brown seaweed Cystoseira barbata: structural characteristics and antioxidant activity. *Int. J. Biol. Macromol.* **2014**, *66*, 281–288. [CrossRef] [PubMed]

66. Hadj Ammar, H.; Lajili, S.; Ben Said, R.; Le Cerf, D.; Bouraoui, A.; Majdoub, H. Physico-chemical characterization and pharmacological evaluation of sulfated polysaccharides from three species of Mediterranean brown algae of the genus Cystoseira. *Daru J. Fac. Pharm. Tehran Univ. Med. Sci.* **2015**, *23*. [CrossRef] [PubMed]

67. Fourest, E.; Volesky, B. Alginate Properties and Heavy Metal Biosorption by Marine Algae. *Appl. Biochem. Biotechnol.* **1997**, *67*, 215–226. [CrossRef]

68. Larsen, C.K.; Gåserød, O.; Smidsrød, O. A novel method for measuring hydration and dissolution kinetics of alginate powders. *Carbohydr. Polym.* **2003**, *51*, 125–134. [CrossRef]

69. Torres, M.R.; Sousa, A.P.; Silva Filho, E.A.; Melo, D.F.; Feitosa, J.P.; de Paula, R.C.; Lima, M.G. Extraction and physicochemical characterization of Sargassum vulgare alginate from Brazil. *Carbohydr. Res.* **2007**, *342*, 2067–2074. [CrossRef] [PubMed]

70. Grasdalen, H. High-field, 1H-n.m.r. spectroscopy of alginate: sequential structure and linkage conformations. *Carbohydr. Res.* **1983**, *118*, 255–260. [CrossRef]
71. Rioux, L.E.; Turgeon, S.L.; Beaulieu, M. Characterization of polysaccharides extracted from brown seaweeds. *Carbohydr. Polym.* **2007**, *69*, 530–537. [CrossRef]
72. Zrid, R.; Bentiss, F.; Ali, R.A.B.; Belattmania, Z.; Zarrouk, A.; Elatouani, S.; Eddaoui, A.; Reani, A.; Sabour, B. Potential uses of the brown seaweed Cystoseira humilis biomass: 1-Sodium alginate yield, FT-IR, H NMR and rheological analyses. *J. Mater. Environ. Sci.* **2016**, *7*, 613–620.
73. Davis, T.A.; Llanes, F.; Volesky, B.; Mucci, A. Metal Selectivity of Sargassum spp. and Their Alginates in Relation to Their α-l-Guluronic Acid Content and Conformation. *Environ. Sci. Technol.* **2003**, *37*, 261–267. [CrossRef] [PubMed]
74. Fertah, M.; Belfkira, A.; Dahmane, E.m.; Taourirte, M.; Brouillette, F. Extraction and characterization of sodium alginate from Moroccan Laminaria digitata brown seaweed. *Arab. J. Chem.* **2017**, *10*, S3707–S3714. [CrossRef]
75. Rioux, L.-E.; Turgeon, S.L.; Beaulieu, M. Rheological characterisation of polysaccharides extracted from brown seaweeds. *J. Sci. Food Agric.* **2007**, *87*, 1630–1638. [CrossRef]
76. Crank, J. Diffusion in a sphere. In *The Mathematics of Diffusion*; Oxford University Press: Oxford, UK, 1975; pp. 89–103.
77. Marcus, Y. *Ion Properties*; CRC Press: Boca Raton, FL, USA, 1997; Volume 1.
78. Boontheekul, T.; Kong, H.J.; Mooney, D.J. Controlling alginate gel degradation utilizing partial oxidation and bimodal molecular weight distribution. *Biomaterials* **2005**, *26*, 2455–2465. [CrossRef]
79. Wang, S.; Vincent, T.; Faur, C.; Guibal, E. Alginate and Algal-Based Beads for the Sorption of Metal Cations: Cu(II) and Pb(II). *Int. J. Mol. Sci.* **2016**, *17*, 1453. [CrossRef] [PubMed]
80. Peña, R.C.; Cornejo, L.; Bertotti, M.; Brett, C.M.A. Electrochemical determination of Cd(ii) and Pb(ii) in mining effluents using a bismuth-coated carbon fiber microelectrode. *Anal. Methods* **2018**, *10*, 3624–3630. [CrossRef]
81. Briso, A.; Quintana, G.; Ide, V.; Basualto, C.; Molina, L.; Montes, G.; Valenzuela, F. Integrated use of magnetic nanostructured calcium silicate hydrate and magnetic manganese dioxide adsorbents for remediation of an acidic mine water. *J. Water Process Eng.* **2018**, *25*, 247–257. [CrossRef]
82. Mahmoud, D.K.; Salleh, M.A.; Karim, W.A. Langmuir model application on solid–liquid adsorption using agricultural wastes: environmental application review. *J. Pur. Util. React. Environs.* **2012**, *1*, 170–199.
83. Wan Ngah, W.S.; Teong, L.C.; Hanafiah, M.A.K.M. Adsorption of dyes and heavy metal ions by chitosan composites: A review. *Carbohydr. Polym.* **2011**, *83*, 1446–1456. [CrossRef]
84. Sousa, F.W.; Oliveira, A.G.; Ribeiro, J.P.; Rosa, M.F.; Keukeleire, D.; Nascimento, R.F. Green coconut shells applied as adsorbent for removal of toxic metal ions using fixed-bed column technology. *J. Environ. Manage.* **2010**, *91*, 1634–1640. [CrossRef] [PubMed]
85. Kobya, M.; Demirbas, E.; Senturk, E.; Ince, M. Adsorption of heavy metal ions from aqueous solutions by activated carbon prepared from apricot stone. *Bioresour. Technol.* **2005**, *96*, 1518–1521. [CrossRef] [PubMed]
86. Amarasinghe, B.M.W.P.K.; Williams, R.A. Tea waste as a low cost adsorbent for the removal of Cu and Pb from wastewater. *Chem. Eng. J.* **2007**, *132*, 299–309. [CrossRef]
87. Javed, M.A.; Bhatti, H.N.; Hanif, M.A.; Nadeem, R. Kinetic and Equilibrium Modeling of Pb(II) and Co(II) Sorption onto Rose Waste Biomass. *Sep. Sci. Technol.* **2007**, *42*, 3641–3656. [CrossRef]
88. Persin, Z.; Stana-Kleinschek, K.; Foster, T.J.; van Dam, J.E.G.; Boeriu, C.G.; Navard, P. Challenges and opportunities in polysaccharides research and technology: The EPNOE views for the next decade in the areas of materials, food and health care. *Carbohydr. Polym.* **2011**, *84*, 22–32. [CrossRef]

Article

Sargassum muticum and *Osmundea pinnatifida* Enzymatic Extracts: Chemical, Structural, and Cytotoxic Characterization

Dina Rodrigues [1], Ana R. Costa-Pinto [1], Sérgio Sousa [1], Marta W. Vasconcelos [1], Manuela M. Pintado [1], Leonel Pereira [2], Teresa A.P. Rocha-Santos [3], João P. da Costa [3], Artur M.S. Silva [4], Armando C. Duarte [3], Ana M.P. Gomes [1,*] and Ana C. Freitas [1]

[1]   CBQF—Centro de Biotecnologia e Química Fina—Laboratório Associado, Escola Superior de Biotecnologia, Universidade Católica Portuguesa, Rua Diogo Botelho 1327, 4169-005 Porto, Portugal; drodrigues@porto.ucp.pt (D.R.); arpinto@porto.ucp.pt (A.R.C-P.); sdcsousa2@gmail.com (S.S.); mvasconcelos@porto.ucp.pt (M.W.V.); mpintado@porto.ucp.pt (M.M.P.); afreitas@porto.ucp.pt (A.C.F.)
[2]   Marine and Environmental Sciences Centre (MARE), Department of Life Sciences, Faculty of Sciences and Technology, University of Coimbra, 3000-456 Coimbra, Portugal; leonel.pereira@uc.pt
[3]   CESAM–Centre for Environmental and Marine Studies & Department of Chemistry, University of Aveiro, Campus Universitário de Santiago, 3810-193 Aveiro, Portugal; ter.alex@ua.pt (T.A.P.R.-S.); jpintocosta@ua.pt ((J.P.d.C.); aduarte@ua.pt (A.C.D.)
[4]   QOPNA–Organic Chemistry, Natural Products and Food Stuffs Research Unit & Department of Chemistry, University of Aveiro, Aveiro, 3810-193, Portugal; artur.silva@ua.pt
*   Correspondence: amgomes@porto.ucp.pt; Tel.: +0035-225-580-084.

Received: 27 February 2019; Accepted: 29 March 2019; Published: 3 April 2019

Abstract: Seaweeds, which have been widely used for human consumption, are considered a potential source of biological compounds, where enzyme-assisted extraction can be an efficient method to obtain multifunctional extracts. Chemical characterization of *Sargassum muticum* and *Osmundea pinnatifida* extracts obtained by Alcalase and Viscozyme assisted extraction, respectively, showed an increment of macro/micro elements in comparison to the corresponding dry seaweeds, while the ratio of Na/K decreased in both extracts. Galactose, mannose, xylose, fucose, and glucuronic acid were the main monosaccharides (3.2–27.3 mg/g$_{lyophilized\ extract}$) present in variable molar ratios, whereas low free amino acids content and diversity (1.4–2.7 g/100 g$_{protein}$) characterized both extracts. FTIR-ATR and 1H NMR spectra confirmed the presence of important polysaccharide structures in the extracts, namely fucoidans from *S. muticum* or agarans as sulfated polysaccharides from *O. pinnatifida*. No cytotoxicity against normal mammalian cells was observed from 0 to 4 mg$_{lyophilized\ extract}$/mL for both extracts. The comprehensive characterization of the composition and safety of these two extracts fulfils an important step towards their authorized application for nutritional and/or nutraceutical purposes.

Keywords: *Osmundea pinnatifida*; *Sargassum muticum*; enzymatic extracts; minerals; mono and polysaccharides; FTIR-ATR; NMR; cytotoxicity

1. Introduction

Seaweeds are extremely versatile organisms that are widely used for direct human consumption, being considered a food with high commercial value. Furthermore, they are also currently recognized as a huge source of new untapped ingredients, many of which, with biological activity, playing a positive role on health and with great potential to be exploited for food and/or nutraceutical applications. According to Singh and Reddy [1], a wide variety of products derived from seaweeds (including food products) are industrially produced, rendering an estimated total annual value of US $5.5 to 6 billion. Extraction plays a major role in supporting this substantial increase in the importance of seaweeds as a source of new bioactive compounds. Enzyme-assisted extraction (EAE) has gained attention as an effective tool to improve the extraction yield of bioactive compounds from different organisms containing a cell wall, such as seaweeds, while being able to maintain the bioactive properties of the derived extracts [2,3].

Among the edible seaweeds, *Sargassum muticum* (Phaeophyceae) is a brown seaweed of an invasive nature in Europe, containing a high content of antioxidant compounds, such as carotenoids and phenols [4], and therefore its exploitation as a potential food ingredient would be an added value helping to overcome its invasive character. In turn, *Osmundea pinnatifida* (Rhodophyta), an edible red macroalgae, is also found in several parts of European coasts, and recent studies have revealed its huge potential as a food ingredient [5]. Enzymatic extracts from *O. pinnatifida* and from *S. muticum* have demonstrated important biological properties, such as antioxidant, antidiabetic, and prebiotic activities [3]. Therefore, further insights into their chemical and structural properties as well as a potential cytotoxicity evaluation would consolidate the knowledge and safety criteria required for these seaweeds' enzymatic extracts, with demonstrated biological potential, to be used for food or nutraceutical applications.

To the best of our knowledge, the chemical and structural characterization of *S. muticum* and of *O. pinnatifida* seaweeds' enzymatic extracts, and the consecutive correlation with previously observed bioactivities [3], has not been performed. Hence, after identifying this important need, the main aim of the research was to determine: i) The elemental, as well as the amino acid and monosaccharide composition for each extract; ii) structural characterization based on FTIR-ATR and ^{1}H NMR analysis; and iii) safety validation by assessment of cytotoxicity by testing the metabolic activity of cells when in contact with the extracts. Safety validation as well as structural and chemical elucidation of the underlying multifunctional *O. pinnatifida* and *S. muticum* enzymatic extracts' roles in nutritional and supplement use were achieved.

2. Results and Discussion

2.1. Elemental Inorganic and Organic Composition of Seaweeds' Enzymatic Extracts

The elemental inorganic composition of the *S. muticum* and *O. pinnatifida* enzymatic extracts are shown in Table 1. For the majority of the macro and micro elements evaluated, increments were observed in comparison to the concentrations found in the corresponding dry seaweeds as supported by the respective ratios. Despite the higher contents in minerals, concentration factors thereof, represented by the calculated ratio, differed between both the macro element and extract origin.

The main constituent in both extracts was indeed K and it was also the macro element with the maximum associated ratio in the case of the *S. muticum* extract (ratio of 7.1, Table 1) and the second highest in the case of the *O. pinnatifida* extract (ratio of 6.7, Table 1). In the latter case, the macro element with the largest associated ratio was Mg (ratio of 7.7, Table 1), albeit it had almost a 5-fold lower content (compared to the K content—174 mg/mg$_{\text{lyophilized extract}}$). Magnesium levels were in the same range of values for both extracts (29.3–36.8 mg/mg$_{\text{lyophilized extract}}$) and represented the third major macro element found in both extracts (after K and Na); magnesium is known as an important mineral for cardiovascular function. The *S. muticum* extract also enabled a significant concentration of the P macro element with a ratio of 5.1 (Table 1). Phosphorus is essential because it is part of the skeletal structure and teeth, but it also has other important functions, such as its contribution to the control of the acid-base balance in the blood and to carbohydrate metabolism where it, contributes to the intestinal absorption of glucose by the process of phosphorylation [6].

Table 1. Elemental inorganic and organic composition of enzymatic seaweeds' extracts.

			EA_*S. muticum* Alcalase		EA_*O. pinnatifida* Viscozyme	
			(mg/mg$_{lyophilized\ extract}$)	Ratio [1]	(mg/mg$_{lyophilized\ extract}$)	Ratio [1]
Inorganic	Macro elements	K	407 ± 13	7.1	174 ± 3	6.7
		Na	66 ± 2	1.8	222 ± 3	2.4
		Ca	2.31 ± 0.06	0.3	8.34 ± 0.04	1.5
		Mg	29.3 ± 0.3	2.0	36.8 ± 0.8	7.7
		P	11.6 ± 0.5	5.1	6.30 ± 0.09	3.6
	Micro elements	Zn	0.033 ± 0.002	1.3	0.34 ± 0.02	5.9
		B	0.319 ± 0.004	3.0	0.28 ± 0.01	2.2
		Mn	0.045 ± 0.004	4.1	0.34 ± 0.06	29.2
		Fe	0.17 ± 0.02	0.9	1.8 ± 0.1	4.9
		Al	<LOD	-	0.085 ± 0.07	0.6
		Cu	0.065 ± 0.002	14.4	0.026 ± 0.001	5.3
		Ni	<LOD	-	0.20 ± 0.01	-
		Pb	0.040 ± 0.001	-	0.020 ± 0.001	-
Organic		%N	2.5	-	1.3	-
		%C	17.1	-	16.8	-
		%H	2.5	-	3.5	-
		%S	0.7	-	1.7	-

[1] Ratio = content in lyophilized extract/content in dry seaweed; values of organic elements are presented as average of triplicate samples. LOD: Limit of Detection.

Despite the widespread increase in the content of the various macro elements, the ratio of Na/K diminished in both extracts in comparison to the corresponding dry seaweeds. In the case of the *S. muticum* extract, the ratio of Na/K diminished four-fold from 0.65 to 0.16 whereas the three-fold decrease for *O. pinnatifida* extracts was of a larger amplitude, from 3.6 to 1.3. Extracts with a low ratio of Na/K are good candidates to be used as added-value multifunctional salt replacers contributing to the important current trend of salt reduction in food formulation and production.

In both enzymatic extracts, variations in terms of microelements content were observed. In general, *S. muticum* extract was poorer in microelement composition than the *O. pinnatifida* counterpart, namely in Zn, Fe, and Mn contents, which were 10-fold higher in the *O. pinnatifida* extract. Concerning the other microelements, higher variability was observed in both extracts. Noteworthy, there was a high increment of Cu and Mn in *S. muticum* and in *O. pinnatifida* extracts, respectively (Table 1), in comparison to the corresponding dried seaweeds' contents [5]. The Fe ratio increased 5 times in the *O. pinnatifida* extract (ratio of 4.9, Table 1) in comparison to the dried seaweed, being the major microelement found in this extract. In turn, in the *S. muticum* extract, the Fe ratio decreased slightly relative to the dried seaweed. The importance of Fe for human beings is well-known; Fe is a natural component of several enzymatic systems, being crucial for the transport of oxygen, and its deficiency has been reported as one of the most common nutritional disorders worldwide that may cause anaemia [7].

Regarding the elemental inorganic composition of the enzymatic *S. muticum* and *O. pinnatifida* extracts, some evidences are highlighted: i) Enzymatic extract of *O. pinnatifida* obtained with Viscozyme could be a good contributor of K, Mg, Zn, and Mn (added value) to recommended daily intakes (RDIs) as well as of Fe (similarly to the dried seaweed) whereas *S. muticum* obtained with Alcalase could be a good contributor of K, Mg, and P to RDIs (added value of extract given the concentration factor for P); ii) in general, enzymatic aqueous extraction enables concentrations of the majority of the macro and micro element in both extracts.

Observing the organic elemental data of both seaweed extracts (Table 1), slightly higher contents of %N and %C were observed for the *S. muticum* enzymatic extract, which was correlated with the higher nitrogen content found in this extract in comparison with the *O. pinnatifida* enzymatic extract [3]. In turn, slightly higher values of %H and %S were observed in *O. pinnatifida* enzymatic extract, which was correlated with the higher contents of polysaccharides, including sulfated polysaccharides, present in this extract in comparison to the *S. muticum* enzymatic extract [3].

2.2. Monosaccharides and Free Amino Acids

The composition of monosaccharides, uronic acids, and amino-monosaccharides in seaweed extracts is displayed in Table 2.

Table 2. Composition of monosaccharides, uronic acids, and amino-monosaccharide in enzymatic extracts of the seaweeds, *S. muticum* and *O. pinnatifida*.

		*S. muticum*_Alcalase (mg/g$_{\text{lyophyzed extract}}$)	*O. pinnatifida*_Viscozyme (mg/g$_{\text{lyophyzed extract}}$)
Monosaccharides	Glucose	<LOD	<LOD
	Galactose	19.1 ± 0.3	25.3 ± 0.2
	Mannose	7.8 ± 0.2	11.4 ± 0.2
	Arabinose	0.10 ± 0.01	0.16 ± 0.01
	Xylose	3.23 ± 0.02	4.8 ± 0.1
	Rhamnose	0.27 ± 0.01	0.52 ± 0.01
	Fucose	4.3 ± 0.1	5.60 ± 0.09
Uronic acids	Glucuronic acid	17.4 ± 0.3	27.3 ± 0.2
	Galacturonic acid	1.07 ± 0.01	1.50 ± 0.02
Amino-mon.	Glucosamine	7.9 ± 0.1	12.7 ± 0.4

All these compounds play important physiological roles in the original source as well as in the host ingesting these extracts. The most abundant monosaccharides in both seaweed extracts were galactose, mannose, glucuronic acid, and glucosamine.

Brown marine seaweeds are recognized as a source of complex polysaccharides, such as fucoidans, laminaram, and alginate. For example, fucoidans are made up of glucose, xylose, fucose, mannose, galactose, uronic acids, and acetyl groups, which can also contain some protein components as well as sulfate substituents [8].

According to Dore et al., (2013) [9], fucan sulfated polysaccharides extracted from the brown seaweed, *Sargassum vulgare*, were composed of galactose, fucose, xylose, mannose, and glucuronic acid; proportions varied according to the fractions of the fucans extracted. In the *S. muticum* extract obtained with Alcalase, the relative proportions of these monosaccharides found were 1.0:0.22:0.17:0.41:0.91 (Table 2), which provides evidence of the presence of fucoidans in the extract.

Studies on the *Sargassum* genus concerning fucans shows that they are generally composed of mannose, galactose, and glucuronic acid residues with partially sulfated-chains consisting of fucose, xylose, and galactose [9]. For example, alginate-free aqueous extracts of *S. muticum* had variable relative proportions of glucose, galactose, fucose, xylose, mannose, and uronic acids [8]. In addition, other factors, such as environmental conditions, geographic location, harvest season, species, and life-cycle stage, as well as extraction can have impacts on fucoidans' composition, structure, and molecular mass [8].

Fucoidans from brown seaweeds do not always have the same backbone: In some cases, a backbone of 3-linked α-L-fucopyranose is present, whereas in other cases, the backbone has alternating 3- and 4-linked α-L-fucopyranose residues and sulfated galactofucans [10]. The latter are mainly found in various *Sargassum* species [11]. These are built of (1→2)-β-D-mannose and/or (1→6)-β-D-galactose units with branching points formed by (1→4)-α-D-glucuronic acid, (1→4) and/or (1→3)-α-L-fucose, terminal β-D-xylose, and sometimes (1→4)-α-D-glucose [11].

The presence of glucosamine (Table 2) in the enzymatic extract of *S. muticum* (7.9 mg/g$_{\text{lyophlized extract}}$) indicates the presence of a proteoglycan-like material, such as that reported for *Sargassum filipendula* (Phaeophyceae) by García-Ríos et al. [12]. Glucosamine has been studied as an important amino-sugar with beneficial physiological roles in joint health [13]. Although experts have mentioned the need for more research to explore possible beneficial effects of glucosamine in healthy subjects or on risk factors of osteoarthritis, prophylactic evidence has been shown in some animal models. According to a survey of supplements available in the market [14], *S. muticum* extract may be used to develop glucosamine rich powder formulations.

Osmundea pinnatifida extract showed higher contents of amino-monosaccharides, uronic acids, and monosaccharides than the *S. muticum* counterpart (Table 2), which parallels the total sugars and sulfated sugars quantified therein. These uphold the potential role of prebiotic activity confirmed by high counts (>10^7 Log cfu/mL) of viable cells of *Bifidobacterium animalis* subsp. *lactis* BB-12 [3]. The richness of the *O. pinnatifida* extract in galactose and the lower quantities of xylose are in agreement with agaran polysaccharides. Agarans are galactans biosynthesized by red seaweeds constituted by 3-linked β-D-galactose alternating with 4-linked α-L-galactose units presenting different degrees of cyclization of the α-L-galactose residues to give 3,6-anhydro-α-L-galactose [15]. Agarans also present a certain degree of substitution with methyl ethers, pyruvate ketals, sulfate ester groups, D-xylose, and/or 4-O-methyl-L-galactose side chains, and different percentages of 3,6-anhydrogalactose [16]. Higher galactose contents with variable proportions of glucose, mannose, xylose, and fucose were reported by Canelón et al. [15] for aqueous extracts of *Laurencia* spp. (red seaweed). No data was found in the literature for red seaweed *O. pinnatifida* extracts.

Very low contents (1.4–2.7 g/100$g_{protein}$) and a low diversity of free amino acids (aspartic acid, glutamic acid, and methionine) characterized both seaweed extracts. Apparently, the higher nitrogen content in the *S. muticum* extract obtained with Alcalase [3] was not reflected in a higher content of free amino acids, which are probably present in the form of peptides.

2.3. Structural Characterization of Seaweed Extracts

The FTIR-ATR spectra of *S. muticum* and its enzymatic extract are displayed in Figure 1a. Both spectra reveal high similarity with most of the bands being common between *S. muticum* seaweed and corresponding extract. Practically, differences were only observed in the absorption intensity. N-H stretching vibrations at 3700–2900 cm^{-1} as well as from amide I and amide II at 1700–1420 cm^{-1} are present in both spectra and could be related to proteins; this is in agreement with the previous results for the nitrogen content [3].

A broad band at 3280–3350 cm^{-1} and a weaker signal at 2870–2960 cm^{-1} could be assigned to O-H and C-H stretching vibrations, but also to N-H stretching vibrations, respectively [17]. The two characteristic absorptions, a band at around 1630 cm^{-1} (C-O asymmetric stretching vibration) and a band of 1410 cm^{-1} (C-O symmetric stretching vibration) [18], indicates the presence of carboxyl groups in both *S. muticum* seaweed and in its extract. In the *S. muticum* Alcalase extract spectrum, there is a noteworthy intensity increment of the 1410 cm^{-1} band in comparison to other bands. This could indicate the higher presence of protein or peptides in the extract due to the role of the endopeptidase, Alcalase. This enzyme was, in fact, responsible for a significantly higher nitrogen content ($p < 0.05$) in the *S. muticum* extract [3], but lower contents of free amino acids were detected in this extract.

A focus on the 700–1400 cm^{-1} region is important because it is related to the seaweeds' polysaccharides, namely carrageenan and agar in red seaweeds and alginates and fucoidans in brown seaweeds. Alginate is a polysaccharide, which has been found in brown seaweeds, such as *S. muticum*, known to be a linear copolymer of β-D-mannuronic acid and α-L-guluronic acid (1-4)-linked residues arranged in heteropolymeric and/or homopolymeric blocks. The presence of these acids can be evidenced especially by the bands around 1030 and 1060 cm^{-1} assigned to guluronic acid and at 1320 cm^{-1} assigned to mannuronic acid, all present both in *S. muticum* seaweed and in its extract obtained by Alcalase. In terms of the broad band around 1220–1260 cm^{-1} in the FTIR-ATR spectrum, assigned to the presence of sulfate ester groups (S=O), which is a characteristic component in fucoidan and other sulfated polysaccharides that can be found in some brown seaweeds [17,19], it is particularly observable in the *S. muticum* seaweed spectrum and less in its extract obtained by

Alcalase (Figure 1a). These results are in agreement with the monosaccharide compositions observed previously for *S. muticum* extract, which evidences the presence of fucoidans (Table 2). Fucoidans have been reported as being responsible for some α-glucosidase inhibitory activity from fractions of *Sargassum duplicatum* [20]. The α-glucosidase inhibitory activity was also observed for *S. muticum* extracts obtained with Alcalase [3].

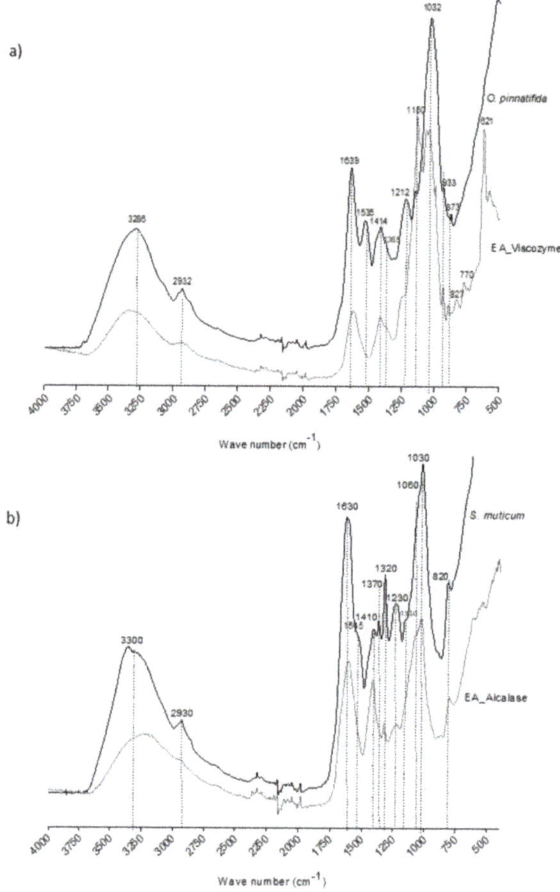

Figure 1. FTIR-ATR spectra of the red edible seaweed, *O. pinnatifida*, and of its enzymatic extract obtained by Viscozyme (**a**) and of the brown edible seaweed, *S. muticum*, and of its enzymatic extract obtained by Alcalase (**b**).

In terms of the *O. pinnatifida* seaweed and its enzymatic extract obtained by Viscozyme, a multi-enzyme complex of carbohydrases (arabanase, cellulase, β-Glucanase, hemicellulase, and xylanase), the two spectra presented some qualitative differences in the region of 1100 to 1600 cm^{-1} (Figure 1b). In the extract spectrum, the bands at 1535 and 1212 cm^{-1} almost disappeared whereas the band at 1150 cm^{-1} increased in comparison to the *O. pinnatifida* seaweed spectrum.

The absorbance bands at 1222 and 1150 cm^{-1} are characteristic of less sulfated polysaccharides, such as agar. Strong absorption at 1220 to 1260 cm^{-1} have been reported by Yu et al., (2012) [21] for agaran-type polysaccharides isolated from *Grateloupia filicina* (Rhodophyta), and according to Rodrigues et al. [5], *O. pinnatifida* was considered a red seaweed agar-like producer. The increment in the band at 1150 cm^{-1} may be related to the possible role of the multi-enzyme complex of carbohydrases on matrix polysaccharides (agar) and on cellulose, xylan, and manan fibrils of the complex composite cell walls of red seaweeds. The extract of *O. pinnatifida* was characterized by the highest content of sulfated sugars [5]. The absorbance band at 930 cm^{-1} in both spectra was assigned to the presence of 3,6-anhydro-D-galactose found in carrageenan and agar polysaccharides [22]. These results are in agreement with the monosaccharide compositions observed previously for *O. pinnatifida* extract, which evidences the presence of agaran polysaccharides (Table 2).

Since the region of 1700 to 1420 cm^{-1} was attributed to amide I and amide II, which in turn could be related to protein, the high reduction of the absorbance band at 1535 cm^{-1} is probably due to some loss of nitrogen content during the extraction process. This correlates well with the fact that enzymatic extracts of *O. pinnatifida* were characterized by low nitrogen contents [5] and low contents of free amino acids.

The ^{1}H NMR spectra of both seaweeds' enzymatic extracts were quite similar (Figure 2a,b).

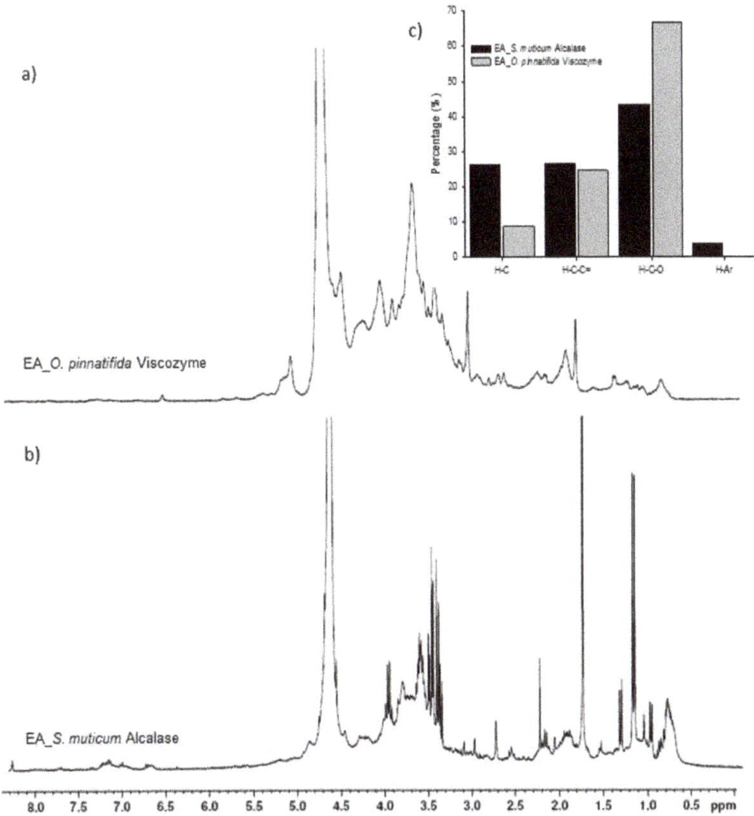

Figure 2. ^{1}H NMR spectra of enzymatic extracts of *Osmundea pinnatifida* obtained by Viscozyme (**a**) and *Sargassum muticum* obtained by with Alcalase (**b**) (the peak at 4.7 ppm indicates the water signal), as well as the relative abundance of each type of proton (**c**) estimated as the partial integrals of the spectra reported in Figure 2a,b for the enzymatic extracts of seaweeds where H-C: purely alkylic hydrogen atoms; H-C-C=: allylic (H-C$_\alpha$-C=), carbonyl or imino (H-C$_\alpha$-C=O or H-C$_\alpha$-C=N) groups; H-C-O: aliphatic C-H directly bound to an oxygen atom; H-Ar: aromatic hydrogen atoms.

These spectra reveal distinct peaks overlaying much broader bands, an expected observation given the nature of the NMR spectra of complex mixtures of organic compounds [23]. Despite the large variety of overlapping resonances, each ^{1}H NMR spectrum was investigated on the basis of the chemical shift assignments described in the literature for organic compounds [23,24]. Accordingly, four main regions of chemical shifts were considered in each spectrum:

1) δ_H = 0.6–1.8 ppm: Aliphatic protons, H–C; -CH > -CH$_2$ > -CH$_3$.
2) δ_H = 1.8–3.2 ppm: Protons bound to carbon atoms in the alpha position to unsaturated groups in allylic (H-C$_\alpha$-C=), carbonyl, or imino (H-C$_\alpha$-C=O or H-C$_\alpha$-C=N) groups, and protons in secondary and tertiary amines (H-C-NR$_2$ and NR$_3$, respectively).
3) δ_H = 3.2–4.1 ppm: Aliphatic protons on carbon atoms singly bound to oxygen atoms (H-C-O-CO-R > H-C-OH or H-C-O-C) in alcohols, polyols, ethers, and esters.
4) δ_H = 6.5–8.5 ppm (aromatic protons).

In order to further understand the ^{1}H NMR data, each spectral region was quantitatively integrated so that the abundance of each of the different types of protons in the different extracts could be assessed; the results are depicted in Figure 2c. Anomeric protons of glycosidic structures [25] related with the fifth region (δ_H = 4.1–6.0 ppm) were also considered, but not integrated since the wide and intense peak at 4.7 ppm is due to the water signal.

In accordance with the spectra, the relative abundance of each type of protons is, in general, relatively similar to the different extracts, but some points are worthy of being highlighted. The higher percentages of protons belong to the group of aliphatic H-C directly bound to an oxygen atom (H-C-O) probably due to the presence of nonaromatic ring structures, such as sugars [24]. Higher values (67%) are observed for the enzymatic extract of *O. pinnatifida* obtained with Viscozyme than for the enzymatic extract of *S. muticum* (43%) obtained with Alcalase. These values are in agreement with previously discussed trends for the FTIR-ATR spectra and sugars: Total sugar and sulfated sugar contents found in the *O. pinnatifida* enzymatic extract were 2.3 and 8-fold higher than in the *S. muticum* enzymatic extracts, respectively [3]. According to Bubb [26], ^{1}H spectra of carbohydrates do contain some well-resolved signals, including those of anomeric protons (δ_H = 4.4–5.5 ppm), acetyl (~δ_H = 2.0–2.1 ppm) and methyl (~δ_H = 1.2 ppm) groups as well as other protons that are influenced by specific functionality, including amino groups, phosphorylation, sulfatation, glycosylation, and acetylation, or lack of functionality as in deoxy-sugars. Signals due to polyols were described by Tanniou et al. [27], who studied the biochemical composition of *S. muticum* populations by ^{1}H HRMAS (high resolution magic angle spinning) NMR and assigned chemical shifts to polyols (mannitol) at δ_H = 3.5–4.0 ppm.

Not many differences were observed for protons in the alpha position to unsaturated groups in allylic (H-C$_\infty$-C=), carbonyl, or imino (H-C$_\infty$-C=O or H-C$_\infty$-C=N) groups for both seaweed extracts, but the slightly higher value for the enzymatic extract of *S. muticum* (27%) could be related to Alcalase activity, which lead to a 2.4 times higher content of nitrogen compounds in comparison to the enzymatic extract of *O. pinnatifida* obtained with Viscozyme [3]. According to Gonzaga et al. [25], a chemical shift at 2.78 ppm could be assigned to protein groups, which was visible in the spectrum of the *S. muticum* enzymatic extract, but not in the corresponding *O. pinnatifida* enzymatic extract (Figure 2a,b). Proteins were described as being associated to cell wall polysaccharides, being part of the structure of the seaweed cell walls [28].

In terms of purely alkylic hydrogen atoms (H-C; 26%) and of aromatic hydrogen atoms (H-Ar; 4%), the enzymatic extract of *S. muticum* was richer (26 and 4%, respectively) in these atoms than the enzymatic extract of *O. pinnatifida* (8.9 and 0.2%, respectively) (Figure 2c). Signals at δ_H = 7.2–7.3 ppm and δ_H = 7.6–7.8 ppm (Figure 2a,b) are consistent with the presence of aromatic units containing both electron-donor (e.g., phenolic groups) and electron-acceptor (e.g., carbonyl and carboxyl) substituents. The content of phenolic contents in aqueous extracts was shown to be low; albeit contents of 290 and 123 μg$_{cathecol\ equiv}$/g$_{lyoph\ extract}$ were determined in the enzymatic extracts of *S. muticum* and *O. pinnatifida* with some potential antioxidant activity, respectively [3]. Signals due to the presence of unsaturated fatty acids were described by Tanniou et al. [27] in *S. muticum* at δ_H = 1.0–1.5 ppm.

In the anomeric spectral region (δ_H = 4.1–6.0 ppm), different patterns were observed for both seaweed extracts (Figure 2a,b). In the case of the enzymatic extract of *O. pinnatifida*, signals in the region of δ_H = 4.2–4.5 and ~δ_H = 5.1 ppm could be assigned to α and β reducing end units whereas no particular signals were well resolved in this region for the enzymatic extract of *S. muticum*. According to Barros et al. [29], the signal from an anomeric proton at δ_H = 5.13 was assigned to 3,6-α-L-anhydrogalactose while the signal at δ_H = 4.56 was attributed to β-D-galactose for polysaccharides of a red seaweed, *Crassiphycus caudatus* (formerly *Gracilaria caudata*). No literature references were found for the ^{1}H NMR spectra *O. pinnatifida* extracts. According to Llanes et al. [30], the absence of signals in the regions of δ_H = 5.0–5.3 ppm and δ_H = 4.7–4.9 ppm expected for anomeric protons of α and β reducing end units that are released (D-anomeric protons of mannuronosyl and L-guluronosyl residues from hydrolysis of sodium alginate from *Sargassum* sp.) are an indication of limited hydrolysis of the *Sargassum* alginate or resulted from overlapping resonance with the water signal at 4.7 ppm.

2.4. Cytotoxicity Evaluation

In order to consolidate the safety of both seaweed extracts, the cytotoxicity of *O. pinnatifida* and *S. muticum* extracts was assessed by measuring the cellular metabolic activity of mammalian mouse fibroblasts cultured with increasing concentrations of both extracts. A range of concentrations from 0 to 4 mg$_{lyophilized\ extract}$/mL was tested. The results are summarized in Figure 3, where cell metabolic activity was evaluated by the resazurin assay. Resazurin (7-Hydroxy-3*H*-phenoxazin-3-one 10-oxide) is a blue and non-fluorescent compound, which is reduced by viable cells in the presence of mitochondrial NADPH dehydrogenases into pink and highly fluorescent resorufin [31]. After applying the resazurin assay, L929 cells were able to produce high amounts of the resorufin compound upon incubation with different concentrations of both seaweed extracts, which reveals normal cell metabolism and mitochondrial integrity/activity that can be inferred as a direct measure of cell viability.

Upon exposure to increasing concentrations of seaweed extracts, cells did not present cytotoxic behavior (Figure 3), indicating that both extracts do not provoke a detrimental effect on the metabolic activity of cells.

In contrast, the supplementation of the culture medium with the extracts seemed to promote an increase of cell viability, which was especially noticeable for the *O. pinnatifida* extracts (Figure 3a). When compared to the viability of cells cultured in the absence of extracts (0 mg$_{lyophilized\ extract}$/mL), cells cultured in the presence of the different concentrations of these extracts demonstrated equal or higher metabolic activity in comparison to the negative control (culture medium only), which can be attributed to the supplementation of the culture media with nutrients from the extracts. For the *O. pinnatifida* extracts, the time of exposure increased the cell metabolic activity for all tested concentrations (Figure 3a), while for *S. muticum*, a slight decrease was observed after 72 h of exposure (Figure 3b).

Consumer oriented applications, such as biomedical, pharmaceutical, or food applications, are highly demanding in terms of the cytotoxic evaluation for new compounds [32]. Cytotoxic evaluation of seaweed-based compounds [33–35] is under-exploited in the literature, and given their significance, this study is of key importance. The anti-tumorigenic effects of compounds extracted from seaweeds were reported [36,37], but it is also highly important to study the effects of these ingredients on normal mammalian cells, an added-value of this study.

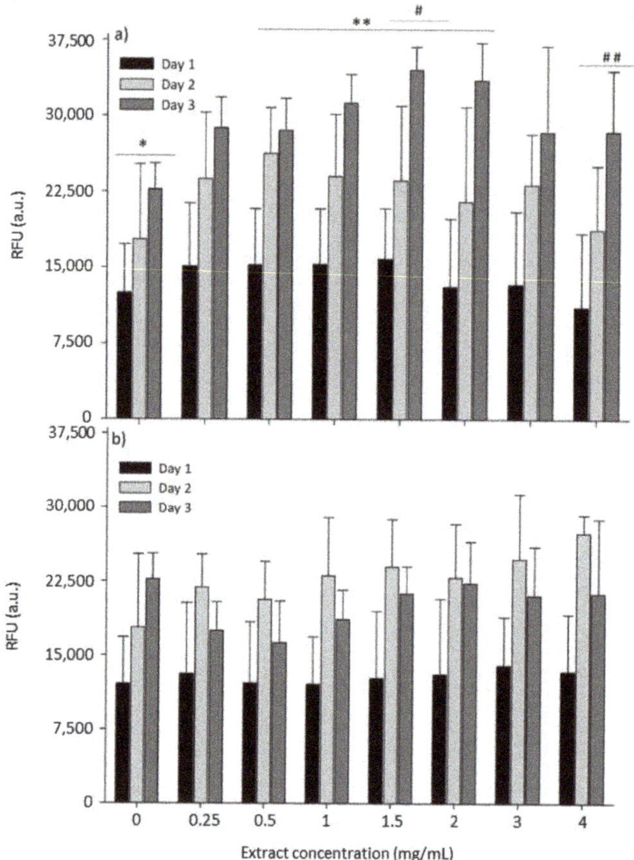

Figure 3. Metabolic activity by the resazurin assay on mouse lung fibroblast cells (L929) cells exposed to different concentrations of enzymatic extracts of *Osmundea pinnatifida* (**a**) and *Sargassum muticum* (**b**). Time of exposure was statistically significant ($p < 0.05$) for each concentration of extract of both seaweeds. Concentration of extract was not statistically significant ($p > 0.05$) for *S. muticum* extracts; for *O. pinnatifida* extracts, statistically significant differences were obtained between 0 mg/mL (*) and 0.5–2.0 mg/mL (* *) and between 1.5 mg/mL (#) and 4.0 mg/mL (# #), respectively.

3. Materials and Methods

3.1. Seaweed Extracts

Specimens of the red seaweed (Rhodophyta, Florideophyceae), *Osmundea pinnatifida* (Ceramiales) Rhodomelaceae family, and of the brown seaweed (Heterokontophyta, Phaeophyceae), *Sargassum muticum* (Fucales) Sargassaceae family, were harvested in Buarcos bay (Figueira da Foz, Portugal), and cleaned and dried according to Rodrigues et al., (2015b) [5]. Enzymatic extraction of *O. pinnatifida* by Viscozyme L (Sigma-Aldrich, St. Louis, MO, USA) and of *S. muticum* by Alcalase (Sigma-Aldrich) was performed according to procedures described by Rodrigues et al., (2015a) [3]. In this study, the authors used lyophilized aliquots of the same extracts obtained by Rodrigues et al., (2015a) [3] in order to enable accurate comparison and correlation analyses.

3.2. Chemical Characterization

3.2.1. Elemental Analysis

Determination of the inorganic elements, Mo, B, Zn, P, Cd, Co, Ni, Mn, Fe, Mg, Ca, Cu, Na, Al, and K, in lyophilized extracts was performed in two steps: Microwave-assisted digestion followed by determination of the 15 elements using an inductively coupled plasma (ICP) optical emission spectrometer (OES) with radial plasma configuration according to Rodrigues et al., (2015b) [5]. Three replicates were performed for each sample as well as blanks.

The organic elements, C, H, S, and N, in lyophilized extracts were quantified using a Truspec 630-200-200 Elemental Analyser (Mönchengladbach, Germany). Triplicate samples up to 3 mg for each extract were placed under combustion at 1075 °C. Carbon, H, and S were detected by infrared absorption whereas N was detected by thermal conductivity.

3.2.2. Analysis of Monosaccharide Composition

Monosaccharide composition was analyzed by high performance liquid chromatography coupled to a UV detector (HPLC-UV, Agilent 1100, Waldbronn, Germany,) after acid hydrolysis. For each lyophilized extract, 2.5 mg of sample was hydrolyzed with 2 mL of 2 M trifluoroacetic acid at 110 °C for 4 h. The hydrolysate was then dried by vacuum evaporation at 50 °C and re-dissolved in 2 mL of deionized water. The hydrolysate solution (450 µL) was mixed with 450 µL of 1-phenyl-3-methyl-5-pyrazolone solution (0.5 M in methanol) and 450 µL of NaOH solution (0.3 M) and then reacted at 70 °C for 30 min. The reaction was stopped by neutralizing with 450 µL of 0.3 M HCl, and the product was then partitioned with chloroform three times. The aqueous layer was collected and filtered through a 0.45 µM membrane and was applied to HPLC.

The HPLC analysis was performed using a ZORBAX ECLIPSE XDB-C18 column (Agilent) (4.6 × 150 mm, 5 micro) at 25 °C with potassium phosphate buffer saline (0.05 M, pH 6.9) with 15% (solvent A) and 40% acetonitrile (solvent B) as mobile phases and detected by a UV detector at 250 nm. All analyses were made in quintuplicate and quantified using a calibration curve built with monosaccharide standards (Sigma Aldrich, St. Louis, MO, USA) and expressed as mg/g$_{\text{lyophilized extract}}$. Glucose, galactose, arabinose, fucose, mannose, xylose, rhamnose, glucuronic acid, galacturonic acid, and glucosamine-6-phosphate were used as the standards. Recovery ranged between 93% and 99% with an LOD of 0.095 mg/g.

3.2.3. Analysis of Amino Acids

The free amino acid contents of each extract was performed by pre-column derivatization with the orthophthalaldehyde (OPA) methodology. Isoindole-type fluorescent derivatives were formed in an alkaline solution (borate buffer pH 10.4) from OPA, 2-sulfanylethanol, and the primary amine group of the amino acid. The derivatives were separated by HPLC (Beckman Coulter, California USA) coupled to a fluorescence detector (Waters, Milford. MA, USA) according to the procedure of Proestos et al., (2008) [38]. Of each sample, 100 µL, at concentration of 10 mg mL^{-1}, was derivatized according to the OPA method and the injection volume of derivatives was 20 µL. All analyses were made in triplicate and quantified using a calibration curve built with amino acid pure standards (Sigma Aldrich, St. Louis MO, USA) and expressed as g/100 g of protein content. Recovery ranged between 92% and 99% with an LOD of 0.02 g/100 g of protein content.

3.2.4. FTIR-ATR Analysis

Samples of lyophilized extracts were analyzed by Fourier Transform Infrared Spectroscopy with attenuated total reflectance (FTIR-ATR) (Spectrum 100, PerkinElmer, Shelton, CT, USA) according to procedures described in Rodrigues et al., (2015b) [5].

3.2.5. ^{1}H NMR Analysis

Twenty milligrams of each lyophilized extract were suspended in 700 µL of D_2O and homogenized for 10 min in a vortex. Then, an amount of dissolved lyophilized extract (650 µL) was placed in 5 mm NMR tubes (Sigma-Aldrich, 528-PP, 5 mm).

All spectra were acquired on a Bruker Advance 300 spectrometer (Karlsruhe, Germany) with an operating frequency of 300.13 MHz. Spectra were acquired with a spinning rate of 20 Hz, a contact time of 4.75 s, and with the pulse program, ZG30. The recycle delay was 1 s and the length of the proton 90 pulses was 9.00 µs. About 56 scans were collected for each spectrum. A 0.3 Hz line broadening weighting function and a baseline correction were applied. The identification of functional groups in the NMR spectra was based on their chemical shift (δ_H) relative to that of the water (4.7 ppm).

3.3. Cytotoxicity Assessment

Mouse lung fibroblast cell line (L929) was used to assess cytotoxicity. Cells were maintained in Dulbecco's Modified Eagle's Medium (DMEM) (Lonza, Verviers, Belgium) supplemented with 10% fetal bovine serum (FBS) (Lonza, Verviers, Belgium) and 1% of antibiotic–antimycotic mixture (Biowest, Nuaillé, France). The cells were grown at 37 °C and 5% CO_2 in a humidified incubator. Exponentially growing cells were used throughout the experiment. TrypLE™ Express Enzyme (Gibco, New York, NY, USA) was used to detach the cells.

For each cell culture assay, the cells were seeded in 48-well plates ($n = 3$) at a concentration of 2×10^4 cells/well and allowed to attach for 24 h at 37 °C and 5% CO2. After that time, cells were washed with phosphate buffered saline (PBS) and exposed for 24, 48, and 72 h to enzymatic extracts of *O. pinnatifida* and *S. muticum* containing different concentrations (0, 0.25, 0.5, 1, 1.5, 2, 3, and 4 mg/mL). All extracts were previously prepared using DMEM and filtered using Vacuum Filtration rapid-Filtermax (PES 0.22 µL membranes) (Trasadingen, Switzerland). Dimethyl sulfoxide (DMSO 20%) was used as a positive control of cell death, due to its strong cytotoxic effect. Culture medium was used as a negative control of cytotoxicity, considered to be the ideal situation for cell growth. At each time point, exposed and control wells were washed with PBS and cell metabolic activity was assessed using the resazurin assay [31]. Cells with 20% *v/v* of stock resazurin solution (1 mg/mL, Sigma) in culture medium were incubated for 2 h at 37 °C. Afterwards, 100 µL of supernatants were transferred to a 96-well black plate (Greiner) and the relative fluorescence units (RFU) were measured using a microplate reader, SynergyTM Mx HM550 (BioTek® Instruments, Vermont, NH, USA), set at 530/590 nm (excitation/emission wavelength, respectively). The obtained results were normalized by subtraction of the negative control (without cells). Samples were measured in triplicate and experiments repeated at least three times.

3.4. Statistical Analysis

Two-way analysis of variance (ANOVA) was carried out for cytotoxicity assessment of both extracts, with SigmaStat™ (Systat Software, Chicago, IL, USA) to assess if the concentration of the extract and time of exposure were significant sources of variation for metabolic activity (resazurin assay) of L929 fibroblasts. For each ANOVA, it was verified that both the normality and equal variance tests were non-significant ($p > 0.05$) as well as interactions between the concentration and time ($p > 0.05$). The Holm-Sidak method was used for pair-wise comparisons both for concentration and time effects.

4. Conclusions

The extensive chemical, structural, and cytotoxicity characterization of *S. muticum* and *O. pinnatifida* enzymatic extracts enabled a deeper understanding and justification for their previously identified multifunctionality. Analysis of the elemental inorganic composition of the seaweeds showed that enzymatic aqueous extraction enabled an important concentration effect of almost all macro and micro elements in comparison to the dry seaweeds' contents, and in some cases, the nutritional value (an extract containing at least 15% of the mineral RDI value) was enhanced, in particular for K and P in the *S. muticum* enzymatic extract and for K, Mg, Zn, and Mn in the *O. pinnatifida* extract.

Overall, the higher contents of monosaccharides, uronic acids, and glucosamine, which were observed in the *O. pinnatifida* extract obtained with Viscozyme in comparison to the *S. muticum* extract, were in agreement with the total sugars and sulfated sugars previously quantified in these extracts, and were further well correlated with the structural analysis obtained from the FTIR-ATR and ^{1}H NMR spectra. The results highlight the relevance of such characterization since both seaweed extracts showed a diversity of sugars in variable molar ratios. On the other hand, amino acids were not representative both qualitatively and quantitatively.

According to chemical and structural analysis by FTIR-ATR and ^{1}H NMR, both extracts, with prebiotic and antidiabetic potential, are composed of important polysaccharide structures, confirming, for example, fucoidans in *S. muticum* extract or agarans as sulfated polysaccharides in *O. pinnatifida* extract among the main representative polysaccharides. No cytotoxicity against normal mammalian cells was observed, making these seaweed extracts very interesting functional ingredients, which could be explored as a food ingredient (salt replacer, nutrient vector) or nutraceutical supplement.

Author Contributions: Experiment conceptualization by D.R., A.R.C.-P., A.C.D., A.C.F. and A.M.P.G., Methodology by D.R., A.R.C.-P., S.S., J.P.d.C., Data and statistical analysis by D.R., A.R.C.-P., A.C.F., T.A.P.R.-S. and S.S., M.W.V., L.P., M.M.P., J.P.d.C., A.M.S.S., A.D. and A.M.G. accomplished data interpretation. Original draft preparation by D.R., A.R.C.-P. and A.C.F., Review and editing by A.M.P.G., L.P., T.R.S. and A.D., All authors read and approved the final manuscript.

Funding: This work was supported by national funds through Fundação para a Ciência e a Tecnologia (FCT)/Ministério da Educação (MEC) (Programa de Investimentos e Despesas de Desenvolvimento da Administração Central—PIDDAC), through project references IF/00588/2015 and IF/00407/2013/CP1162/CT0023 as well by Fundo Europeu de Desenvolvimento Regional (FEDER) through project NORTE-01-0145-FEDER-000012 (Structured Programme on Bioengineering Therapies for Infectious) funded by NORTE2020 program. This work was also supported by the Fundação para a Ciência e a Tecnologia (FCT), through the strategic project UID/MAR/04292/2019 granted to MARE. Thanks, are also due, for the financial support, to CESAM (UID/AMB/50017) through FCT/MEC national funds.

Acknowledgments: Authors would like to thank the scientific collaboration of CBQF under the FCT project UID/Multi/50016/2013.

Conflicts of Interest: The authors declare no conflict of interest.

References

1. Singh, R.P.; Reddy, C.R.K. Unraveling the functions of the macroalgal microbiome. *Front. Microbiol.* **2016**, *6*, 1488. [CrossRef] [PubMed]

2. Gil-Chávez, G.J.; Villa, J.A.; Ayala-Zavala, J.F.; Heredia, J.B.; Sepulveda, D.; Yahia, E.M.; González-Aguilar, G.A. Technologies for extraction and production of bioactive compounds to be used as nutraceuticals and food ingredients: An Overview. *Compr. Rev. Food Sci. Food Saf.* **2013**, *12*, 5–23. [CrossRef]

3. Rodrigues, D.; Sousa, S.; Silva, A.G.; Amorim, M.; Pereira, L.; Rocha-Santos, T.A.P.; Gomes, A.M.; Duarte, A.C.; Freitas, A.C. Impact of Enzyme- and Ultrasound-Assisted Extraction Methods on Biological Properties of Red, Brown, and Green Seaweeds from the Central West Coast of Portugal. *J. Agric. Food Chem.* **2015**, *63*, 3177–3188. [CrossRef]

4. Milledge, J.J.; Nielsen, B.V.; Bailey, D. High-value products from macroalgae: The potencial uses of invasive brown seaweed, *Sargassum muticum*. *Rev. Environ. Sci. Biotechnol.* **2016**, *15*, 67–88. [CrossRef]

5. Rodrigues, D.; Freitas, A.C.; Pereira, L.; Rocha-Santos, T.A.P.; Vasconcelos, M.W.; Roriz, M.; Rodríguez-Alcalá, L.M.; Gomes, A.M.P.; Duarte, A.C. Chemical composition of red, brown and green macroalgae from Buarcos bay in Central West Coast of Portugal. *Food Chem.* **2015**, *183*, 197–207. [CrossRef]

6. Pérez, F.; Garaulet, M.; Gil, A.; Zamora, S. Calcio, fósforo, magnesio y flúor. Metabolismo óseo y su regulación. In *Tratado de Nutrición, Vol. I*; Gil, A., Ed.; Grupo Acción Médica: Madrid, Spain, 2005; pp. 897–925.

7. Allen, L.; de Benoist, B.; Dary, O.; Hurrel, R. *Guidelines on Food Fortification with Micronutrientes*; World Health Organization and Food and Agriculture Organization of the United Nations: Geneva, Switzerland, 2006; ISBN 92-4-159401-2.

8. Balboa, E.M.; Conde, E.; Moure, A.; Falque, E.; Dominguez, H. In vitro antioxidant properties of crude extracts and compounds from brown algae. *Food Chem.* **2013**, *138*, 1764–1785. [CrossRef] [PubMed]

9. Dore, C.M.; Alves, M.G.F.; Will, L.S.; Costa, T.G.; Sabry, D.A.; de Souza, R.L.A.; Accardo, C.M.; Rocha, H.A.; Filgueira, L.G.; Leite, E.L. A sulfated polysaccharide, fucans, isolated from brown algae *Sargassum vulgare* with anticoagulant, antithrombotic, antioxidant and anti-inflammatory effects. *Carbohydr. Polym.* **2013**, *91*, 467–475. [CrossRef]

10. Bilan, M.I.; Usov, A.I. Structural analysis of fucoidans. *Nat. Prod. Commun.* **2008**, *3*, 1639–1648. [CrossRef]

11. Duarte, M.E.R.; Cardoso, M.A.; Noseda, M.D.; Cerezo, A.S. Structural studies on fucoidans from the brown seaweed *Sargassum stenophyllum*. *Carbohydr. Res.* **2001**, *333*, 281–293. [CrossRef]

12. García-Ríos, V.; Ríos-Leal, E.; Robledo, D.; Freile-Pelegrin, Y. Polysaccharides composition from tropical brown seaweeds. *Phycol. Res.* **2012**, *60*, 305–315. [CrossRef]

13. Henrotin, Y.; Chevalier, X.; Herrero-Beaumont, G.; McAlindon, T.; Mobasheri, A.; Pavelka, K.; Biesalski, H. Physiological effects of oral glucosamine on joint health: Current status and consensus on future research priorities. *BMC Res. Notes* **2013**, *6*, 115. [CrossRef]

14. Zhou, J.Z.; Waszkuc, T.; Mohammed, F. Determination of glucosamine in raw materials and dietary supplements containing glucosamine sulfate and/or glucosamine hydrochloride by high-performance liquid chromatography with FMOC-Su derivatization: Collaborative Study. *J. AOAC Int.* **2005**, *88*, 1048–1058.

15. Canelón, D.J.; Ciancia, M.; Suárez, A.I.; Compagnone, R.S.; Matulewicz, M.C. Structure of highly substituted agarans from the red seaweeds *Laurencia obtusa* and *Laurencia filiformis*. *Carbohydr. Polym.* **2014**, *30*, 705–713. [CrossRef]

16. Usov, A.I. Polysaccharides of the red algae. *Adv. Carbohydr. Chem. Biochem.* **2015**, *65*, 115–217.

17. Gómez-Ordóñez, E.; Rupérez, P. FTIR-ATR spectroscopy as a tool for polysaccharide identification in edible brown and red seaweeds. *Food Hydrocoll.* **2011**, *25*, 1514–1520. [CrossRef]

18. Zheng, L.; Zhai, G.; Zhang, J.; Wang, L.; Ma, Z.; Jia, M.; Jia, L. Antihyperlipidemic and hepatoprotective activities of mycelia zinc polysaccharide from *Pholiota nameko*. *Int. J. Biol. Macromol.* **2014**, *70*, 523–529. [CrossRef]

19. Pereira, L.; Gheda, S.F.; Ribeiro-Claro, P.J.A. Analysis by vibrational spectroscopy of seaweed polysaccharides with potential use in food, pharmaceutical and cosmetic industries. *Int. J. Carbohydr. Chem.* **2013**, *2013*, 537202. [CrossRef]

20. Hardoko, S.T.; Eveline, Y.M.; Olivia, S. An in vitro of antidiabetic activity of *Sargassum duplicatum* and *Turbinaria decurens* seaweed. *Int. J. Pharm. Sci. Invent.* **2014**, *3*, 13–18.

21. Yu, Q.; Yan, J.; Wang, S.; Ji, L.; Ding, K.; Vella, C.; Wang, Z.; Hu, Z. Antiangiogenic effects of GFP08, an agaran-type polysaccharide isolated from *Grateloupia filicina*. *Glycobiology* **2012**, *22*, 1343–1352. [CrossRef]

22. Pereira, L. Identification of phycocolloids by vibrational spectroscopy. In *World Seaweed Resources—An Authoritative Reference System*; Critchley, A.T., Ohno, M., Largo, D.B., Eds.; Hybrid Windows and Mac DVD-ROM; ETI Information Services Ltd.: Amsterdam, The Netherlands, 2006; ISBN 90-75000-80-4.

23. Santos, P.S.M.; Santos, E.B.H.; Duarte, A.C. First spectroscopic study on the structural features of dissolved organic matter isolated from rainwater in different seasons. *Sci. Total Environ.* **2012**, *426*, 172–179. [CrossRef]

24. Clarke, C.J.; Haselden, J.N. Metabolic profiling as a tool for understanding mechanisms of toxicity. *Toxicol. Pathol.* **2008**, *36*, 140–147. [CrossRef]

25. Gonzaga, M.L.C.; Ricardo, N.M.P.S.; Heatley, F.; Soares, S.A. Isolation and characterization of polysaccharides from *Agaricus blazei* Murill. *Carbohydr. Polym.* **2005**, *60*, 43–49. [CrossRef]

26. Bubb, W.A. NMR spectroscopy in the study of carbohydrates; Characterizing the structural complexity. *Concepts Magn. Reson. Part A* **2003**, *19*, 1–19. [CrossRef]

27. Tanniou, A.; Vandanjon, L.; Gonçalves, O.; Kercvarec, N.; Stiger, P.V. Rapid geographical differentiation of the European spread brown macroalga *Sargassum muticum* using HRMAS NMR and Fourier-Transform Infrared spectroscopy. *Talanta* **2015**, *132*, 451–456. [CrossRef]

28. Robic, A.; Rondeau-Mouro, C.; Sassi, J.-F.; Lerat, Y.; Lahaye, M. Structure and interactions of ulvan in the cell wall of the marine green algae *Ulva rotundata* (Ulvales, Chlorophyceae). *Carbohydr. Polym.* **2009**, *77*, 206–216. [CrossRef]

29. Barros, F.C.N.; Silva, D.C.; Sombra, V.G.; Maciel, J.S.; Feitosa, J.P.A.; Freitas, A.L.P.; de Paula, R.C. Structural characterization of polysaccharide obtained from red seaweed *Gracilaria caudata* (J Agardh). *Carbohydr. Polym.* **2013**, *92*, 598–603. [CrossRef] [PubMed]

30. Llanes, F.; Sauriol, F.; Morin, F.G.; Perlin, A.S. An examination of sodium alginate from Sargassum by NMR spectroscopy. *Can. J. Chem.* **1997**, *75*, 585–590. [CrossRef]

31. O'Brien, J.; Wilson, I.; Orton, T.; Pognan, F. Investigation of the Alamar Blue (resazurin) fluorescent dye for the assessment of mammalian cell cytotoxicity. *Eur. J. Biochem.* **2000**, *267*, 5421–5426. [CrossRef] [PubMed]

32. Krugera, C.L.; Mann, S.W. Safety evaluation of functional ingredients. *Food Chem. Toxicol.* **2003**, *41*, 793–805. [CrossRef]

33. Alves, A.; Sousa, R.A.; Reis, R.L. In Vitro Cytotoxicity Assessment of Ulvan, a Polysaccharide Extracted from Green Algae. *Phytother. Res.* **2013**, *27*, 1143–1148. [CrossRef]

34. Heussner, A.H.; Mazija, L.; Fastner, J.; Dietrich, D.R. Toxin content and cytotoxicity of algal dietary supplements. *Toxicol. Appl. Pharmacol.* **2012**, *265*, 263–271. [CrossRef]

35. Süzgeç-Selçuk, S.; Meriçli, A.H.; Güven, K.C.; Kaiser, M.; Casey, R.; Hingley-Wilson, S.; Lalvani, A.; Tasdemir, D. Evaluation of Turkish Seaweeds for Antiprotozoal, Antimycobacterial and Cytotoxic Activities. *Phytother. Res.* **2011**, *25*, 778–783. [CrossRef]

36. Ayyad, S.N.; Ezmirly, S.T.; Basaif, S.A.; Alarif, W.M.; Badria, A.F.; Badria, F.A. Antioxidant, cytotoxic, antitumor, and protective DNA damage metabolites from the red sea brown alga *Sargassum* sp. *Pharmacogn. Res.* **2011**, *3*, 160–165. [CrossRef]

37. Khanavi, M.; Gheidarloo, R.; Sadati, N.; Ardekani, M.R.S.; Nabavi, S.M.B.; Tavajohi, S.; Ostad, S.N. Cytotoxicity of fucosterol containing fraction of marine algae against breast and colon carcinoma cell line. *Pharmacogn. Mag.* **2012**, *8*, 60–64.

38. Proestos, C.; Loukatos, P.; Komaitis, M. Determination of biogenic amines in wines by HPLC with precolumn dansylation and fluorimetric detection. *Food Chem.* **2008**, *106*, 1218–1224. [CrossRef]

 *marine drugs* 

Article

Phlorofucofuroeckol A from Edible Brown Alga *Ecklonia Cava* Enhances Osteoblastogenesis in Bone Marrow-Derived Human Mesenchymal Stem Cells

Jung Hwan Oh [1], Byul-Nim Ahn [1], Fatih Karadeniz [1], Jung-Ae Kim [1], Jung Im Lee [1], Youngwan Seo [2,3] and Chang-Suk Kong [1,4,*]

[1] Marine Biotechnology Center for Pharmaceuticals and Foods, Silla University, Busan 46958, Korea; wjdghks0171@naver.com (J.H.O.); icetwig@naver.com (B.-N.A.); karadenizf@outlook.com (F.K.); jale8469@gmail.com (J.-A.K.); think3433@daum.net (J.I.L.)

[2] Division of Marine Bioscience, College of Ocean Science and Technology, Korea Maritime and Ocean University, Busan 49112, Korea; ywseo@kmou.ac.kr

[3] Department of Convergence Study on the Ocean Science and Technology, Ocean Science and Technology School, Korea Maritime and Ocean University, Busan 49112, Korea

[4] Department of Food and Nutrition, College of Medical and Life Sciences, Silla University, Busan 46958, Korea

* Correspondence: cskong@silla.ac.kr; Tel.: +82-51-999-5429

Received: 14 August 2019; Accepted: 17 September 2019; Published: 21 September 2019

Abstract: The deterioration of bone formation is a leading cause of age-related bone disorders. Lack of bone formation is induced by decreased osteoblastogenesis. In this study, osteoblastogenesis promoting effects of algal phlorotannin, phlorofucofuroeckol A (PFF-A), were evaluated. PFF-A was isolated from brown alga *Ecklonia cava*. The ability of PFF-A to enhance osteoblast differentiation was observed in murine pre-osteoblast cell line MC3T3-E1 and human bone marrow-derived mesenchymal stem cells (huBM-MSCs). Proliferation and alkaline phosphatase (ALP) activity of osteoblasts during differentiation was assayed following PFF-A treatment along extracellular mineralization. In addition, effect of PFF-A on osteoblast maturation pathways such as Runx2 and Smads was analyzed. Treatment of PFF-A was able to enhance the proliferation of differentiating osteoblasts. Also, ALP activity was observed to be increased. Osteoblasts showed increased extracellular mineralization, observed by Alizarin Red staining, following PFF-A treatment. In addition, expression levels of critical proteins in osteoblastogenesis such as ALP, bone morphogenetic protein-2 (BMP-2), osteocalcin and β-catenin were stimulated after the introduction of PFF-A. In conclusion, PFF-A was suggested to be a potential natural product with osteoblastogenesis enhancing effects which can be utilized against bone-remodeling imbalances and osteoporosis-related complications.

Keywords: alkaline phosphatase; *Ecklonia cava*; phlorofucofuroeckol A; osteoblast; huBM-MSC

---

1. Introduction

Prevalence of bone related complications is steadily increasing while osteoporosis and linked syndromes are causing low life quality and mortality for more of today's elder population compared to the previous decade [1,2]. Among these complications, osteoporosis is one of the most commonly seen bone disorders and is characterized by bone formation imbalance. Bone fractures of elderly patients are mostly diagnosed with osteoporosis. Osteoporosis is being treated via different approaches, including but not limited to synthetic and natural-origin drugs which are targeting different mechanisms involved in bone formation [3–5]. The bone formation imbalance in osteoporosis is caused by bone volume loss while the formation of new bone tissue diminishes through inhibited osteogenesis. Hence, the stimulation of osteoblast formation is one of the main approaches to treat bone mass imbalance [6]. Additionally, links between osteogenesis and pathways related to onset of diabetes and obesity raises complex problems for elderly patients who are diagnosed with other metabolic syndromes along with osteoporosis [7,8]. Natural product research is gaining attention for the treatment or prevention of various metabolic syndromes including osteoporosis due their biocompatibility and fewer side effects compared to on the market drugs. In this context, several studies have reported the in vitro osteoblastogenesis inducing effects of marine-based nutraceuticals for the relief of osteoporotic complications [9,10].

Ecklonia cava (order Laminariales, family Lessoniaceae) is an edible brown alga growing abundantly on the shores of Japan and Korea where it is consumed as a part of daily diet. Numerous studies were conducted reporting its potential uses as rich source of bioactive secondary metabolites [11–13]. Phlorotannins are phloroglucinol oligomers [14] found richly in *E. cava* and credited for diverse health benefits including but not limited to antioxidant [15], antibacterial [16], anti-inflammation [17], anti-allergy [18], and anti-metastasis [19] activities. In this context, current study focused on the possible osteogenesis enhancing effect of phlorofucofuroeckol A (PFF-A) isolated from *E. cava*, as a part of ongoing research to elucidate natural compounds from marine sources with important bioactivities.

2. Results

PFF-A was obtained from *E. cava* and its chemical structure (Figure 1) was confirmed by comparison of ^{1}H NMR and ^{13}C NMR spectral data with previously published reports as described earlier [20]. PFF-A was tested for its potential activity on enhancing osteoblastic differentiation in two different cell lines; MC3T3-E1 murine pre-osteoblasts and human bone marrow-derived mesenchymal stem cells (huBM-MSCs). Cells were induced to differentiate into osteoblasts with or without PFF-A treatment to observe the enhancing effects of the PFF-A on osteoblastogenesis.

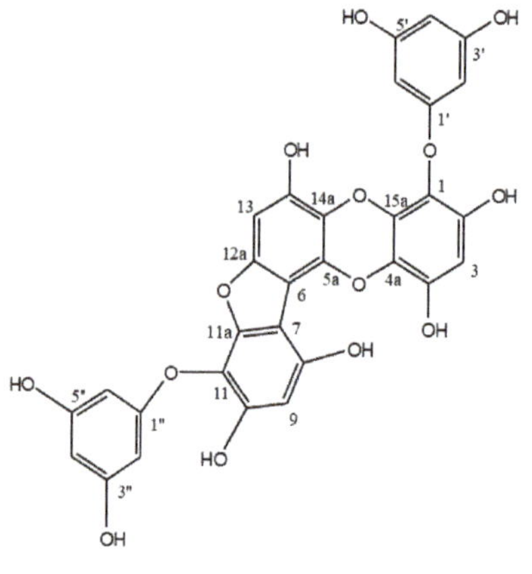

Figure 1. Chemical structure of phlorofucofuroeckol A (PFF-A).

2.1. Effect of PFF-A on the Differentiation of Murine Pre-osteoblasts

MC3T3-E1 cells are murine calvarial osteoblast precursor cells widely used in studies for their ability to differentiate into mature osteoblasts. Therefore, any effect of PFF-A on osteoblastic differentiation was tested first in differentiating MC3T3-E1 cells. Measurement of viable cells, alkaline phosphatase (ALP) activity and extracellular mineralization as calcium deposits were used as markers for maturation into osteoblasts.

Presence of PFF-A slightly stimulated the viable cell count of MC3T3-E1 osteoblasts at 5 µM (Figure 2a) compared to the untreated control. Differentiated osteoblasts showed elevated ALP activity which was significantly enhanced after PFF-A treatment at the concentration of 5 µM (Figure 2b). Enhanced ALP activity and osteoblastogenesis was also confirmed by extracellular mineralization assessed by Alizarin Red staining. The extracellular matrix calcium deposition levels were increased during differentiation of MC3T3-E1 cells into mature osteoblasts as a marker of bone tissue formation.

Treatment of MC3T3-E1 osteoblasts with PFF-A during differentiation exhibited elevated extracellular mineralization shown by quantification of the calcium staining (Figure 3a). PFF-A treatment at 5 and 20 µM resulted in enhanced calcium deposition compared to untreated control osteoblasts. Effect of PFF-A on the osteoblast differentiation pathways was analyzed by the investigation of the transcription pathways of osteoblastogenesis-related proteins. Cells treated with increasing doses of PFF-A expressed enhanced levels of ALP and osteocalcin mRNA expression (Figure 3b). In addition, protein expression levels of ALP, bone morphologic protein (BMP)-2 and osteocalcin were strongly elevated in PFF-A treated osteoblasts compared to untreated control (Figure 3c). Osteocalcin protein levels were enhanced in higher levels than its mRNA expression.

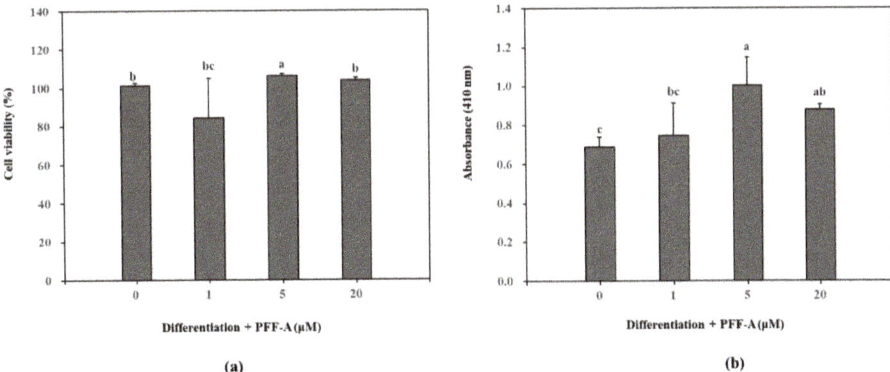

Figure 2. Effect of PFF-A on the cell viability (**a**) and alkaline phosphatase (ALP) activity (**b**) of MC3T3-E1 osteoblasts. Viability of cells and ALP activity were analyzed at 5 and 7 days after MC3T3-E1 cells were induced to differentiate, respectively. Cell viability was expressed as a percentage of osteoinduced untreated control group (0 µM). ALP activity was given as absorbance (410 nm) values of the colorimetric quantification of enzymatic activity. Values are means ± SD of three different experiments run in triplicate ($n = 3$). [a–c] Means with the different letters are significantly different ($p < 0.05$) by Duncan's multiple range test.

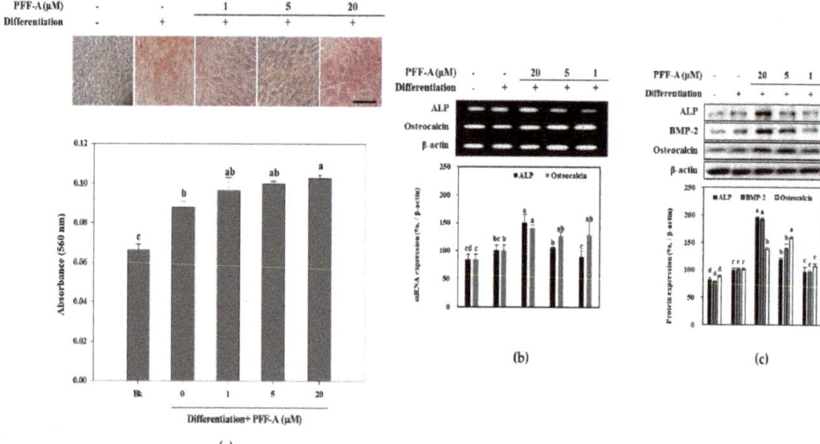

Figure 3. Images of MC3T3-E1 osteoblasts treated with PFF-A and stained with Alizarin Red for extracellular calcium deposits, and absorbance values (560 nm) of eluted dye retained in the cells (**a**). MC3T3-E1 osteoblasts were stained with Alizarin Red 14 days after the cells were induced to differentiate. (Bk: Non-differentiated blank group which was given growth medium instead of differentiation cocktail). Effect of PFF-A on the expression of mRNA (**b**) and protein (**c**) levels of osteoblast differentiation markers. Cells were harvested for reverse transcription-polymerase chain reaction (RT-PCR) and Western blot analysis 12 days after the cells were induced to differentiate. Expression levels quantified by densiometric analysis of bands were given as percentage of osteoinduced untreated control group after normalization using internal control β-actin. Non-induced untreated blank group was fed growth medium instead of differentiation cocktail. Values are means ± SD of three separate experiments ($n = 3$). [a–d] Means with the different letters are significantly different ($p < 0.05$) by Duncan's multiple range test. Scale bar: 25 μm.

2.2. Effect of PFF-A on the Osteogenic Differentiation of huBM-MSCs

Following the confirmation of its potential osteoblast differentiation enhancing effect in pre-osteoblasts, the mechanism behind this effect of PFF-A was investigated in bone marrow derived mesenchymal stem cells. The cytotoxicity assay showed that PFF-A treatment did not cause any significant toxicity in the huBM-MSCs up to concentration of 10 μM (Figure 4a). However, 20 μM PFF-A treatment caused a decrease in cell viability of the huBM-MSCs. Stem cells induced for osteogenic differentiation exhibited stimulated cell proliferation and PFF-A treatment (20 μM) enhanced the viable cell amount by increasing cell viability 24.53% compared to untreated control (Figure 4b). Enhancing effect of PFF-A was also confirmed on the ALP activity and extracellular mineralization of the osseous differentiated MSCs. At the highest concentration treated (20 μM), ALP activity in the osteoinduced huBM-MSCs was calculated 40.45 U/mL compared to 31.60 U/mL of untreated osteoblasts (Figure 5a). Enhancing of ALP activity was also observed as elevated extracellular mineralization. PFF-A treatment (20 μM) enhanced the extracellular calcium deposits by 20.52% compared to osseous differentiated MSCs without PFF-A treatment (Figure 5b).

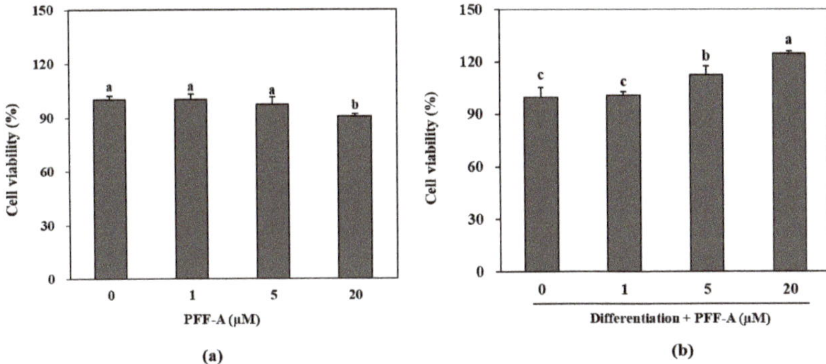

Figure 4. Effect of PFF-A on the viability of non-induced huBM-MSCs (**a**) and the viability of osteoinduced huBM-MSCs (**b**). Viability of osteoinduced huBM-MSCs was analyzed 5 days after inducement of differentiation. Cell viability was expressed as percentage of untreated (**a**) and osteoinduced untreated (**b**) control group (0 μM). Values are means ± SD of three separate experiment run in triplicates ($n = 3$). [a–c] Means with the different letters are significantly different ($p < 0.05$) by Duncan's multiple range test.

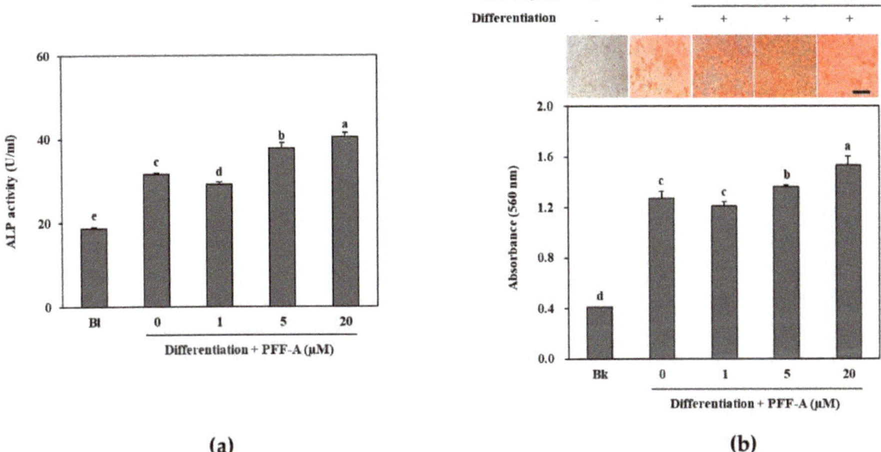

Figure 5. ALP activity of osteoinduced huBM-MSCs treated with PFF-A (**a**). Images of osteoinduced huBM-MSCs stained with Alizarin Red for extracellular calcium deposits and absorbance (560 nm) values of eluted dye retained in the cells (**b**). Cells were analyzed after 7 days of incubation with differentiation medium for ALP activity and 14 days of incubation with differentiation medium for Alizarin Red staining. Values are means ± SD of three separate experiments (run in triplicates for ALP activity) ($n = 3$). [a–e] Means with the different letters are significantly different ($p < 0.05$) by Duncan's multiple range test (Bk: Non-differentiated untreated blank group which was given growth medium instead of differentiation cocktail). Scale bar: 50 μm.

Osteoblastogenesis enhancing mechanism of PFF-A was investigated by analyzing the expression of osseous differentiation inducing and regulatory pathways during the osteoinduced huBM-MSC differentiation. Osteoinduced huBM-MSCs expressed high levels of ALP mRNA as a marker of differentiation, and Runx2 and osteocalcin mRNA as a marker of osseous differentiated MSCs. Treatment with PFF-A dose-dependently increased the mRNA expression levels of ALP, osteocalcin and Runx2 compared to untreated control (Figure 6a). Consequently, protein levels of same markers

(ALP, osteocalcin and Runx2) were observed to be enhanced following PFF-A treatment in a dose dependent manner (Figure 6b). In addition, osterix protein levels were investigated as osterix is a transcription factor activated by Wnt/β-catenin pathway and responsible for the osseous differentiation of huBM-MSCs. PFF-A treatment was also enhanced the levels of osterix in a dose-dependent manner (Figure 6b).

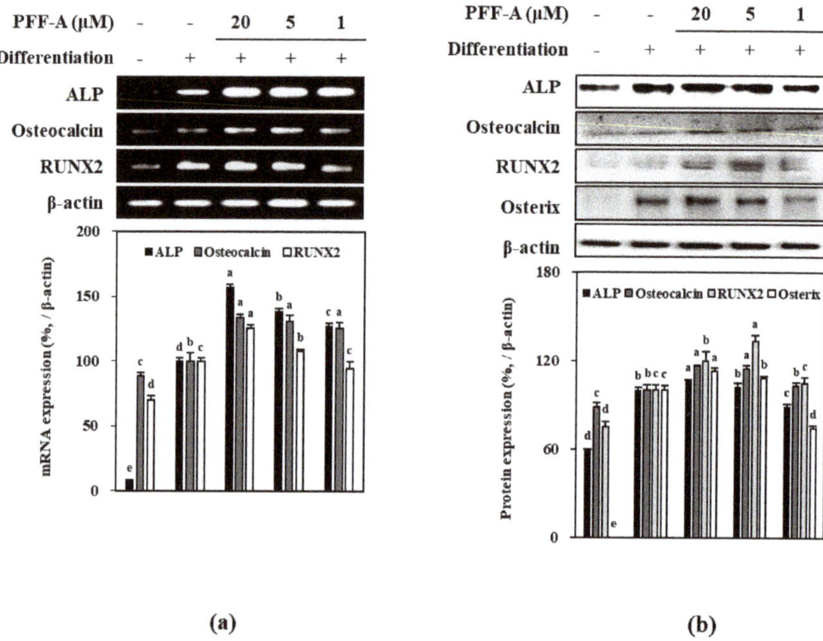

(a)   (b)

Figure 6. Effect of PFF-A on the expression of mRNA (a) and protein (b) levels of osteoblast differentiation markers in osteoinduced huBM-MSCs analyzed by RT-PCR and Western blotting, respectively. Cells were harvested for RT-PCR and Western blot analysis 12 days after the cells were induced to differentiate. Expression levels quantified by densiometric analysis of bands were given as percentage of osteoinduced untreated control group after normalization using internal control β-actin. Non-induced untreated blank group was fed growth medium instead of differentiation cocktail. Values are means ± SD of three separate experiments ($n = 3$). $^{a-e}$ Means with the different letters are significantly different ($p < 0.05$) by Duncan's multiple range test.

Effect of PFF-A on BMP and Wnt/β-Catenin Pathway

Mechanism of the enhancing ability of PFF-A was further analyzed by protein levels of the osteoblast differentiation ignitor pathways. Osteoinduced huBM-MSCs showed elevated levels of Smad1/5/8 complex, and β-catenin (Figure 7). Two main intracellular pathways that further activate the expression of osteoblast-specific genes. Activation of β-catenin and Smad1/5/8 complex were increased seen as the phosphorylated Smad1/5 and β-catenin were significantly higher than non-differentiated cells. Introduction of PFF-A enhanced the levels of this phosphorylation along with the expression of BMP2 and Wnt10a. However, axin, a negative regulator of the osteoblast differentiation was observed to be highly expressed during the differentiation and was not affected by PFF-A treatment (Figure 7).

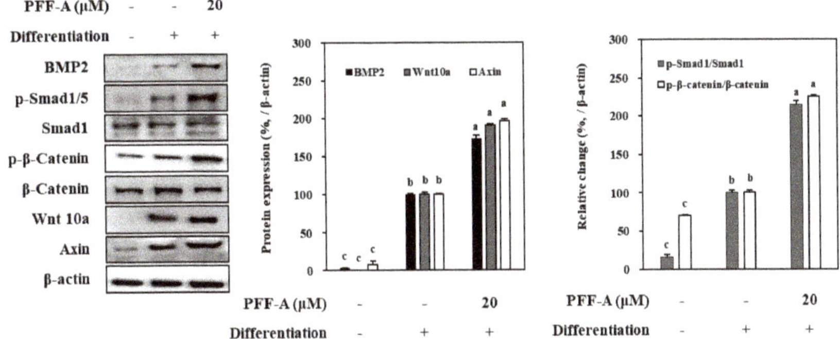

Figure 7. Effect of PFF-A on the levels of BMP2, Wnt 10a, and the inactive and phosphorylated (p-) Smad1 and β-catenin in osteoinduced huBM-MSCs analyzed by Western blotting. Cells were harvested for Western blot analysis 12 days after the cells were induced to differentiate. Expression levels of BMP2, Wnt 10a and Axin quantified by densiometric analysis of bands were given as percentage of osteoinduced untreated control group after normalization using internal control β-actin. Changes in p-Smad1 and p-β-catenin levels relative to their unphosphorylated forms were given as percentage of osteoinduced untreated control group after normalization using internal control β-actin. Values are means ± SD of three separate experiments ($n = 3$). [a-c] Means with the different letters are significantly different ($p < 0.05$) by Duncan's multiple range test.

3. Discussion

Regulation of bone remodeling is a highly critical process for a healthy bone metabolism, and any disorders of bone mass is evidently linked with variety of diseases such as arthritis, tumor growth and osteoporosis [21,22]. As the bone formation during the remodeling is dependent on the osteogenic differentiation in bone tissue, controlling the osteoblastogenesis is hence an important target to prevent or treat bone mass complications, especially osteoporosis. Bone marrow stroma derived osteoblasts are differentiated cells from mesenchymal stem cells, and they produce ALP for mineralization of the bone, and osteocalcin, which is an important hormone taking crucial roles in energy metabolism [23].

Osteoblastic differentiation of bone marrow mesenchymal stem cells is regulated by more than one signaling pathway mainly involving BMP growth factors and activation of Wnt/β-catenin/TCF cascade along with transforming growth factor (TGF)-β and mitogen-activated protein kinase (MAPK) pathways [24]. Among them, BMP pathway accompanied by β-catenin activation controls the osteoblastogenesis and subsequent formation of bone tissue by mineralization and required protein secretion.

Bone formation is carried out by increased proliferation of osteoblasts which is followed by elevated ALP activity. Results showed that PFF-A enhanced the viable cell count of osteoinduced pre-osteoblasts and huBM-MSCs. However, at 20 μM treatment PFF-A decreased the cell viability of non-induced huBM-MSCs while showing an opposite effect in osteoinduced huBM-MSCs. Although PFF-A was considered slightly cytotoxic for huBM-MSCs at 20 μM, osteoinducement was speculated to hinder or negate this effect. Increased ALP activity is a marker of osteoblast maturation and is needed for the last step of bone formation, extracellular mineralization [25]. Current data showed that PFF-A was able to increase the ALP activity in the osteoblasts differentiated from both pre-osteoblasts and stem cells. This also was confirmed by the elevated levels of osterix in huBM-MSCs treated with PFF-A. These results suggested that PFF-A had an enhancing effect on the osteogenic differentiation of cells.

Runx2 is the key transcription factor for osteoblastogenesis and has important roles in the maturation of osteoblasts. It is translocated into nucleus as a downstream factor in the Wnt and BMP pathways [26,27]. Both pathways stimulate the translocation of Runx2 and subsequent maturation of osteoblasts. Results showed that PFF-A increased the mRNA and protein expression levels of Runx2 in osteoinduced huBM-MSCs indicating that mechanism of PFF-A mediated osteoblast differentiation enhancement might occurred via enhancing the Wnt and/or BMP pathways. Therefore, it was suggested that PFF-A regulated the Wnt/β-catenin and BMP pathways which in turn induced Runx2 translocation all of which consequently elevated ALP activity and committed the stem cell differentiation to osteogenic lineage.

Mechanism of PFF-A mediated enhanced osteoblastogenesis was further analyzed via activation and expression levels of Wnt and BMP pathways. The pathway that starts with activation of BMP receptors stimulates and induces the commitment of stem cells to osteogenic lineage instead of adipogenic [28]. This signaling cascade includes phosphorylation of Smad1/5/8 complex in order to facilitate its interaction with Smad4 and their translocation to nucleus for further expression of osteoblast-specific genes via Runx2 [29,30]. Various plant-based metabolites such as flavonoids, phlorotannins, polysaccharides and similar phenol-based compounds were shown to stimulate osteoblast differentiation via BMP pathway by inducing the expression of BMP or any other activator protein in the cascade [31–33]. Current results showed that PFF-A presence not only increased BMP-2 expression levels for both mRNA and protein, but also notably stimulated the phosphorylation of Smad1/5/8 protein. Elevated levels of both BMP-2 and phosphorylated Smad1/5/8 compared to untreated control cells further confirmed that PFF-A-induced increase in osteoblast differentiation is related to BMP pathway.

Aside BMP pathway, Wnt/β-catenin pathway is also a key regulator for the stem cell commitment to osteogenic lineage [34]. It is heavily involved in new bone tissue remodeling and new bone formation. Activation of Wnt receptors in stem cells initiates a signaling cascade which involves activation and translocation to nucleus of β-catenin and stimulates the transcription of osteoblast-specific genes such as ALP and osterix [26]. The results showed that huBM-MSCs treated with PFF-A exhibited increased levels of Wnt-10a and phosphorylation of β-catenin. Significant increase in the activation ratio of β-catenin clearly suggested that Wnt/β-catenin pathway was also involved in the enhancing stem cell osteoblastogenesis by PFF-A treatment.

Although there are not any reports on specific side-effects of PFF-A, it could be speculated that it shares the common disadvantages of phlorotannins. Phlorotannins are known to be non-specific inhibitors of several enzymes that act in digestive track such as α-glucosidase and α-amylase [35,36]. Also, it was reported that tannin derivatives could cause irritation in stomach and potential damage in liver [37]. However, the doses that shown to exert serious side effects are significantly higher and not comparable to the tested concentrations. Although, PFF-A was assumed to be possess minimal side effects further studies to evaluate its dosage and relative side-effects would be urged.

In conclusion, present study showed that PFF-A could enhance the osteogenic differentiation of pre-osteoblast and huBM-MSCs. Results demonstrated that PFF-A stimulated the osteoblastogenesis by interacting with BMP and Wnt/β-catenin pathways and consequently by increasing ALP activity and mineralization. As a result, PFF-A was suggested to be potential natural compound with therapeutic properties against osteoporosis and related bone complications that urged further studies to elucidate its detailed action mechanism and bioavailability.

4. Materials and Methods

4.1. Isolation of PFF-A from Plant Material

E. cava was collected between October 2014 and March 2015 from seashores of Jeju Island, Korea. Specimens were identified by the corresponding author and a voucher sample was stored. Fronds of *E. cava* were freeze-dried and stored at 25°C prior to experiments. Freeze-dried samples were ground to powder (4.0 kg) and extracted three times using EtOH (3 × 10 L). Solvent-partition of the crude extracts (584.3 g) was carried out with n-hexane, n-BuOH and 85% aq. MeOH to yield solvent fractions along with H_2O residue. Forty grams of the n-BuOH fraction was further separated by a silica gel column chromatography (60, 0/063–0.200 mm) (Merck, Kenilworth, NJ, USA) with the gradient mixture

(30:1 to 1:1) of CH_2Cl_2:MeOH and $CHCl_3$:MeOH, respectively. Phlorofucofuroeckol A (52.2 mg) was isolated as a light brown powder from 3:1 $CHCl_3$:MeOH sub-fraction by high performance liquid chromatography (HPLC) (C18, 40% aq. MeOH). The chemical structure of phlorofucofuroeckol A (Figure 1) was verified by spectral data as previously reported [20].

4.2. Cell Culture and Differentiation

Murine osteoblast-like MC3T3-E1 cells obtained from ATCC (CRL-2593™) and huBM-MSCs obtained from PromoCell (C-12974) were cultured in 6-well plates unless otherwise noted. MC3T3-E1 cells were fed with α-Modified minimal essential medium (αMEM) containing 10% fetal bovine serum (heat-inactivated, *v/v*), 1 mM sodium pyruvate, 100 units/L penicillin and 100 mg/L streptomycin in an atmosphere of 5% CO_2 at 37 °C. Following the confluence, osteoblast differentiation was induced with a differentiation cocktail of 50 μg/mL ascorbic acid and 10 mM β-glycerophosphate in cell culture medium. PFF-A was introduced to the cells with differentiation medium and included in every medium change (every second day).

huBM-MSCs were fed with mesenchymal stem cell growth medium (C-28009, PromoCell) and incubated in an atmosphere of 5% CO_2 at 37 °C. Following 100% confluence cells were induced by MSC osteogenic differentiation medium (C-28013, PromoCell) and incubated until the time of analysis. PFF-A was introduced with differentiation medium and included in all medium changes (every third day).

4.3. Cell Viability Assay

Effect of PFF-A on the viability of MC3T3-E1 osteoblasts, and non-induced and osteoinduced huBM-MSCs was evaluated by 3-(4,5-dimethylthiazol-2-yl)-2,5-diphenyltetrazolium bromide (MTT) assay as previously described [20]. Viability of MC3T3-E1 cells were measured at the day 5 of differentiation inducement. Viability of non-induced huBM-MSCs that fed growth medium only was measured after 2 days of incubation with or without PFF-A treatment while the viability of osteoinduced huBM-MSCs was analyzed at the day 5 of differentiation. PFF-A was introduced to the cells at the beginning of incubation or differentiation inducement and included in all medium changes until the day of analysis (every second day for MC3T3-E1 and every third day for huBM-MSC). Following the treatment period of the cells with PFF-A, cell culture medium was aspired, and wells were added 100 μL of MTT reagent (1 mg/mL) prior to 4 h incubation. Formazan salt was dissolved with the addition of 100 μL DMSO to the wells. Viability and proliferation of the cells were calculated by measuring the absorbance values of the formed formazan salts for each well at 540 nm using a Multiskan GO microplate reader (Tecan Austria GmbH, Grodig, Austria) and plotting the values as a percentage of untreated control cells.

4.4. Cellular Alkaline Phosphatase (ALP) Activity

ALP activity was evaluated in differentiated MC3T3-E1 and osteoinduced huBM-MSCs treated with or without PFF-A. Differentiated MC3T3-E1 osteoblasts were used for ALP activity assay at day 7 of differentiation. Also, a blank group which was not induced and untreated was analyzed after 7 days of incubation in growth medium. Cells were washed with phosphate buffer saline (PBS) and lysed with 0.1% Triton X-100 and 25 mM carbonate buffer. The cellular ALP activity was assessed using the supernatants of the cell lysates following centrifugation at 4 °C 12,000× *g* for 15 min. Total protein contents of the supernatants were analyzed and normalized by the Bradford protein determination method. The absorbance of the reactive solution containing the supernatants and the enzyme reaction buffer (15 mM p-nitrophenyl phosphate, 1.5 mM $MgCl_2$ and 200 mM carbonate buffer) was measured at 405 nm after 15 min. of incubation, using a Multiskan GO microplate reader (Tecan Austria GmbH, Grodig, Austria).

Osteoinduced huBM-MSCs were analyzed for ALP activity using a commercial kit following the manufacturer's instructions. Cells were prepared for ALP activity assay at day 7 of treatment and the data was obtained as absorbance values at 560 nm using a Multiskan GO microplate reader (Tecan Austria GmbH). Also, a non-induced huBM-MSC blank group was analyzed after 7 days of incubation with growth medium instead of differentiation medium.

4.5. Alizarin Red Staining

Extracellular calcium deposits of differentiated MC3T3-E1 and osteoinduced huBM-MSCs were investigated and quantified by Alizarin Red staining with or without PFF-A treatment at day 14 of differentiation. Also, a non-induced untreated blank group which was fed growth medium instead of differentiation medium was stained for calcium deposits after 14 days of incubation. Cells were fixed on 6-well plates with 30 min incubation in 70% ice-cold ethanol which was followed by removal of ethanol and washing with distilled H_2O. Staining was carried out by introduction of Alizarin Red solution followed by 10 min incubation at room temperature. After the incubation, the Alizarin Red solution was aspired from wells and cells were washed with distilled H_2O to remove to unbound Alizarin red stain. Stained calcium deposit images showing extracellular mineralization were taken by an Olympus microscope (Tokyo, Japan). Subsequently, the Alizarin red dye was eluted from wells with 10% cetylpyridinium chloride (Sigma-Aldrich, St. Louis, MO, USA) solution and mineralization was quantified by absorbance values at 560 nm using a Multiskan GO microplate reader (Tecan Austria GmbH).

4.6. Reverse Transcription-Polymerase Chain Reaction Analysis

Total RNA was obtained from differentiated MC3T3-E1 and osteoinduced huBM-MSCs with or without PFF-A treatment using Trizol reagent (Invitrogen, CA, USA) at day 12 of differentiation. Also, a non-induced untreated blank group which was fed growth medium instead of differentiation medium was analyzed after 12 days of incubation. Synthesis of cDNA was started with addition of total RNA (2 µg) to RNase-free water containing oligo (dT) and followed by denaturation at 70 °C for 5 min. Next, the mixture was reverse transcribed in a master mix (1 X RT buffer, 1 mM dNTPs, 500 ng oligo (dT), 140 U M-MLV reserve transcriptase and 40 U RNase inhibitor) using an automatic T100 Thermal Cycler (Bio-Rad, Hercules, CA, USA) with a cycle of 42 °C for 60 min and 72 °C for 5 min. Sense and antisense primers previously detailed [38] were used for the amplification of the target cDNA. The cDNA amplification was carried out using T100 Thermal Cycler (Bio-Rad) with cycle settings at 95 °C for 45 s, 60 °C for 1 min and 72 °C for 45 s for 30 cycles Final PCR products were separated by gel electrophoresis on 1.5% agarose gel for 30 min at 100 V. Bands were then observed following the staining with 1 mg/*mL* ethidium bromide under UV light using CAS-400SM Davinch-Chemi imager™ (Seoul, Korea).

4.7. Western Blotting

Protein immunoblotting was performed with standard Western blotting procedures. Differentiated MC3T3-E1 and osteoinduced huBM-MSCs with or without PFF-A treatment were lysed by pipetting with 1 mL of RIPA lysis buffer (Sigma–Aldrich) at 4 °C for 30 min at day 12 of differentiation. Also, a non-induced untreated blank group which was fed growth medium instead of differentiation medium was analyzed after 12 days of incubation. The acquired cell lysate (25 µg) was used for Western blot analysis. Proteins in cell lysate were separated by sodium dodecyl sulphate-polyacrylamide gel electrophoresis (SDS-PAGE) on 4% stacking and 12% separating gels. Proteins subjected to separation were then electrotransferred to a polyvinylidene fluoride membrane (Amersham Biosciences, Little Chalfont, England, UK), which was blocked with 5% skim milk powder in TBST buffer after transfer. The membrane was hybridized with primary antibodies diluted (1:1000) in primary antibody dilution buffer containing 1X TBST with 5% bovine serum albumin at 4 °C overnight and incubated with horseradish-peroxidase-conjugated secondary antibody at room temperature for 2 h. Immunoreactive proteins bands were visualized by a luminol-based chemiluminescence assay kit (Amersham Biosciences) according to the manufacturer's manual. Images of protein bands were captured using a Davinch-Chemi imager™ (CAS-400SM, Seoul, Korea).

4.8. Statistical Analysis

The data were presented as mean of three independent experiments ± SD. Statistically significant differences among the means of the individual test groups were determined by one-way analysis of variance (ANOVA) followed by Duncan's multiple range tests using SAS v9.1 software (SAS Institute, Cary, NC, USA), $p < 0.05$ being the defining level for the significance of differences.

Author Contributions: Conceptualization, B.-N.A. and F.K.; methodology, B.-N.A., J.-A.K. and J.H.O.; validation, B.-N.A., J.-A.K. and J.H.O.; formal analysis, F.K. and Y.S.; investigation, B.-N.A., J.-A.K., J.I.L. and J.H.O.; writing—original draft preparation, F.K.; visualization, F.K., J.I.L., Y.S. and C.-S.K.; supervision, Y.S. and C.-S.K.; project administration, C.-S.K.; funding acquisition, C.-S.K.

Funding: This work was supported by the National Research Foundation of Korea (NRF) grant funded by the Korea government (MSIP) (No. NRF-2017R1A2B4009588).

Conflicts of Interest: The authors declare no conflict of interest.

References

1. Rachner, T.D.; Khosla, S.; Hofbauer, L.C. Osteoporosis: Now and the future. *Lancet* **2011**, *377*, 1276–1287. [CrossRef]
2. Harvey, N.; Dennison, E.; Cooper, C. Osteoporosis: Impact on health and economics. *Nat. Rev. Rheumatol.* **2010**, *6*, 99–105. [CrossRef] [PubMed]
3. Eriksen, E.F.; Díez-Pérez, A.; Boonen, S. Update on long-term treatment with bisphosphonates for postmenopausal osteoporosis: A systematic review. *Bone* **2014**, *58*, 126–135. [CrossRef] [PubMed]
4. Reginster, J.Y.; Burlet, N. Osteoporosis: A still increasing prevalence. *Bone* **2006**, *38*, 4–9. [CrossRef]
5. Silverman, S.; Christiansen, C. Individualizing osteoporosis therapy. *Osteoporos. Int.* **2012**, *23*, 797–809. [CrossRef] [PubMed]
6. Djouad, F.; Guérit, D.; Marie, M.; Toupet, K.; Jorgensen, C.; Noël, D. Mesenchymal stem cells: New insights into bone regenerative applications. *J. Biomater. Tissue Eng.* **2012**, *2*, 14–28. [CrossRef]
7. Rawadi, G.; Vayssière, B.; Dunn, F.; Baron, R.; Roman-Roman, S. BMP-2 controls alkaline phosphatase expression and osteoblast mineralization by a Wnt autocrine loop. *J. Bone Miner. Res.* **2003**, *18*, 1842–1853. [CrossRef]
8. Gimble, J.M.; Nuttall, M.E. The relationship between adipose tissue and bone metabolism. *Clin. Biochem.* **2012**, *45*, 874–879. [CrossRef] [PubMed]
9. Rho, J.R.; Hwang, B.S.; Joung, S.; Byun, M.R.; Hong, J.H.; Lee, H.Y. Phorbasones A and B, sesterterpenoids isolated from the marine sponge *Phorbas* sp. and induction of osteoblast differentiation. *Org. Lett.* **2011**, *13*, 884–887. [CrossRef] [PubMed]
10. An, J.; Yang, H.; Zhang, Q.; Liu, C.; Zhao, J.; Zhang, L.; Chen, B. Natural products for treatment of osteoporosis: The effects and mechanisms on promoting osteoblast-mediated bone formation. *Life Sci.* **2016**, *147*, 46–58. [CrossRef]
11. Li, Y.X.; Li, Y.; Qian, Z.J.; Ryu, B.; Kim, S.K. Suppression of vascular endothelial growth factor (VEGF) induced angiogenic responses by fucodiphloroethol G. *Process Biochem.* **2011**, *46*, 1095–1103. [CrossRef]
12. Kim, M.-M.; Van Ta, Q.; Mendis, E.; Rajapakse, N.; Jung, W.-K.; Byun, H.-G.; Jeon, Y.-J.; Kim, S.-K. Phlorotannins in *Ecklonia cava* extract inhibit matrix metalloproteinase activity. *Life Sci.* **2006**, *79*, 1436–1443. [CrossRef] [PubMed]
13. Li, Y.; Qian, Z.J.; Ryu, B.M.; Lee, S.H.; Kim, M.M.; Kim, S.K. Chemical components and its antioxidant properties in vitro: An edible marine brown alga, *Ecklonia cava*. *Bioorganic Med. Chem.* **2009**, *17*, 1963–1973. [CrossRef] [PubMed]
14. Shibata, T.; Kawaguchi, S.; Hama, Y.; Inagaki, M.; Yamaguchi, K.; Nakamura, T. Local and chemical distribution of phlorotannins in brown algae. *J. Appl. Phycol.* **2004**, *16*, 291–296. [CrossRef]

15. Kang, H.S.; Chung, H.Y.; Kim, J.Y.; Son, B.W.; Jung, H.A.; Choi, J.S. Inhibitory phlorotannins from the edible brown alga *Ecklonia stolonifera* on total reactive oxygen species (ROS) generation. *Arch. Pharm. Res.* **2004**, *27*, 194–198. [CrossRef] [PubMed]

16. Eom, S.H.; Kim, Y.M.; Kim, S.K. Antimicrobial effect of phlorotannins from marine brown algae. *Food Chem. Toxicol.* **2012**, *50*, 3251–3255. [CrossRef] [PubMed]

17. Kong, C.S.; Kim, J.A.; Ahn, B.N.; Kim, S.K. Potential effect of phloroglucinol derivatives from *Ecklonia cava* on matrix metalloproteinase expression and the inflammatory profile in lipopolysaccharide-stimulated human THP-1 macrophages. *Fish. Sci.* **2011**, *77*, 867–873. [CrossRef]

18. Kim, S.K.; Lee, D.Y.; Jung, W.K.; Kim, J.H.; Choi, I.; Park, S.G.; Seo, S.K.; Lee, S.W.; Lee, C.M.; Yea, S.S.; et al. Effects of *Ecklonia cava* ethanolic extracts on airway hyperresponsiveness and inflammation in a murine asthma model: Role of suppressor of cytokine signaling. *Biomed. Pharmacother.* **2008**, *62*, 289–296. [CrossRef]

19. Kong, C.S.; Kim, J.A.; Yoon, N.Y.; Kim, S.K. Induction of apoptosis by phloroglucinol derivative from Ecklonia Cava in MCF-7 human breast cancer cells. *Food Chem. Toxicol.* **2009**, *47*, 1653–1658. [CrossRef]

20. Kim, H.; Kong, C.S.; Lee, J.I.; Kim, H.; Baek, S.; Seo, Y. Evaluation of inhibitory effect of phlorotannins from *Ecklonia cava* on triglyceride accumulation in adipocyte. *J. Agric. Food Chem.* **2013**, *61*, 8541–8547. [CrossRef]

21. Yaturu, S.; Humphrey, S.; Landry, C.; Jain, S.K. Decreased bone mineral density in men with metabolic syndrome alone and with type 2 diabetes. *Med. Sci. Monit.* **2009**, *15*, 5–9.

22. McFarlane, S.I.; Muniyappa, R.; Shin, J.J.; Bahtiyar, G.; Sowers, J.R. Osteoporosis and cardiovascular disease: Brittle bones and boned arteries, is there a link? *Endocrine* **2004**, *23*, 1–10. [CrossRef]

23. Fakhry, M. Molecular mechanisms of mesenchymal stem cell differentiation towards osteoblasts. *World J. Stem Cells* **2013**, *5*, 136–148. [CrossRef] [PubMed]

24. Guo, X.; Wang, X.F. Signaling cross-talk between TGF-β/BMP and other pathways. *Cell Res.* **2009**, *19*, 71–88. [CrossRef] [PubMed]

25. Van Straalen, J.P.; Sanders, E.; Prummel, M.F.; Sanders, G.T.B. Bone-alkaline phosphatase as indicator of bone formation. *Clinica Chimica Acta* **1991**, *201*, 27–33. [CrossRef]

26. Gaur, T.; Lengner, C.J.; Hovhannisyan, H.; Bhat, R.A.; Bodine, P.V.N.; Komm, B.S.; Javed, A.; Van Wijnen, A.J.; Stein, J.L.; Stein, G.S.; et al. Canonical WNT signaling promotes osteogenesis by directly stimulating Runx2 gene expression. *J. Biol. Chem.* **2005**, *280*, 33132–33140. [CrossRef]

27. Komori, T. Regulation of osteoblast differentiation by Runx2. In *Proceedings of the Advances in Experimental Medicine and Biology*; Springer: Boston, MA, USA, 2010; Volume 658, pp. 43–49.

28. Kang, Q.; Song, W.-X.; Luo, Q.; Tang, N.; Luo, J.; Luo, X.; Chen, J.; Bi, Y.; He, B.-C.; Park, J.K.; et al. A Comprehensive Analysis of the dual roles of BMPs in regulating adipogenic and osteogenic differentiation of mesenchymal progenitor cells. *Stem Cells Dev.* **2008**, *18*, 545–558. [CrossRef]

29. Wan, M.; Cao, X. BMP signaling in skeletal development. *Biochem. Biophys. Res. Commun.* **2005**, *328*, 651–657. [CrossRef]

30. Liang, W.; Lin, M.; Li, X.; Li, C.; Gao, B.; Gan, H.; Yang, Z.; Lin, X.; Liao, L.; Yang, M. Icariin promotes bone formation via the BMP-2/Smad4 signal transduction pathway in the hFOB 1.19 human osteoblastic cell line. *Int. J. Mol. Med.* **2012**, *30*, 889–895. [CrossRef]

31. Tang, D.Z.; Yang, F.; Yang, Z.; Huang, J.; Shi, Q.; Chen, D.; Wang, Y.J. Psoralen stimulates osteoblast differentiation through activation of BMP signaling. *Biochem. Biophys. Res. Commun.* **2011**, *405*, 256–261. [CrossRef]

32. Hyung, J.H.; Ahn, C.B.; Je, J.Y. Osteoblastogenic activity of ark shell protein hydrolysates with low molecular weight in mouse mesenchymal stem cells. *RSC Adv.* **2016**, *6*, 29365–29370. [CrossRef]

33. Ming, L.G.; Chen, K.M.; Xian, C.J. Functions and action mechanisms of flavonoids genistein and icariin in regulating bone remodeling. *J. Cell. Physiol.* **2013**, *228*, 513–521. [CrossRef] [PubMed]

34. Taipaleenmäki, H.; Abdallah, B.M.; AlDahmash, A.; Säämänen, A.M.; Kassem, M. Wnt signalling mediates the cross-talk between bone marrow derived pre-adipocytic and pre-osteoblastic cell populations. *Exp. Cell Res.* **2011**, *317*, 745–756. [CrossRef] [PubMed]

35. Roy, M.C.; Anguenot, R.; Fillion, C.; Beaulieu, M.; Berube, J.; Richard, D. Effect of a commercially-available algal phlorotannins extract on digestive enzymes and carbohydrate absorption in vivo. *Food Res. Int.* **2011**, *44*, 3026–3029. [CrossRef]

36. Lee, S.H.; Li, Y.; Karadeniz, F.; Kim, M.M.; Kim, S.K. α-Glucosidase and α-amylase inhibitory activities of phloroglucinal derivatives from edible marine brown alga, *Ecklonia cava. J. Sci. Food Agric.* **2009**, *89*, 1552–1558. [CrossRef]

37. Sieniawska, E.; Baj, T. Chapter 10—Tannins. In *Pharmacognosy: Fundamentals, Applications and Strategies*; Badal, S., Delgoda, R., Eds.; Academic Press: Cambridge, MA, USA, 2017; pp. 192–232.

38. Karadeniz, F.; Ahn, B.N.; Kim, J.A.; Seo, Y.; Jang, M.S.; Nam, K.H.; Kim, M.; Lee, S.H.; Kong, C.S. Phlorotannins suppress adipogenesis in pre-adipocytes while enhancing osteoblastogenesis in pre-osteoblasts. *Arch. Pharm. Res.* **2015**, *38*, 2172–2182. [CrossRef] [PubMed]

Article

Dietary Supplementation with Low-Molecular-Weight Fucoidan Enhances Innate and Adaptive Immune Responses and Protects against *Mycoplasma pneumoniae* Antigen Stimulation

Pai-An Hwang [1,*,†], Hong-Ting Victor Lin [2,†], Hsin-Yuan Lin [1] and Szu-Kuan Lo [3]

1   Department of Bioscience and Biotechnology, National Taiwan Ocean University, No. 2, Beining Road, Keelung 20246, Taiwan; 2063B003@mail.ntou.edu.tw
2   Department of Food Science, National Taiwan Ocean University, Keelung 20246, Taiwan; HL358@ntou.edu.tw
3   Gi-Kang Clinic, No. 155, Yanping Rd., Zhongli Dist., Taoyuan 32043, Taiwan; a912164@yahoo.com.tw
*   Correspondence: amperehwang@gmail.com or amperehwang@ntou.edu.tw; Tel.: +886-2-24622192 (ext. 5570); Fax: +886-2-24634732
†   These authors equally contributed to this work.

Received: 15 February 2019; Accepted: 14 March 2019; Published: 18 March 2019

Abstract: In this study, the low-molecular-weight (LMW) fucoidan, rich in fucose and sulfate, was extracted and purified from the edible brown seaweed, *Laminaria japonica*. In this study, we orally administered LMW fucoidan to mice for 6 weeks. We then examined fucoidan's effects on innate immunity, adaptive immunity, and *Mycoplasma pneumoniae* (MP)-antigen-stimulated immune responses. Our data showed that LMW fucoidan stimulated the innate immune system by increasing splenocyte proliferation, natural killer (NK) cell activity, and phagocytic activity. LMW fucoidan also increased interleukin (IL)-2, IL-4, and interferon (IFN)-γ secretion by splenocytes and immunoglobulin (Ig)-G and IgA content in serum, which help regulate adaptive immune cell functions, and decreased allergen-specific IgE. In MP-antigen-stimulated immune responses, the IgM and IgG content in the serum were significantly higher in the LMW fucoidan group after MP-antigen stimulation. Our study provides further information about the immunomodulatory effects of LMW fucoidan and highlights a potential role in preventing *M. pneumoniae* infection.

Keywords: low molecular weight fucoidan; *Mycoplasma pneumoniae*; NK cell; antigen-specific antibody; adjuvant

1. Introduction

Before the 1950s, seaweeds were used as traditional folk medicines [1]. Over the past two decades, polysaccharides of brown seaweeds, such as fucoidan, alginate, and laminarin, have been investigated for their biological activities [2,3]. The health-promoting effects of these compounds, especially fucoidan, give brown seaweed great value for developing natural dietary supplements.

Fucoidans are a class of fucose-containing sulfated polysaccharides found in brown seaweed [4]. Most fucoidans have complex chemical compositions, and their structures and biological activities vary from species to species [5]. Its non-animal origin has been related to particular pharmacological activities [6]. Fucoidan has been associated with antitumor [7–9], antiviral [10], anti-inflammatory [11,12], anticoagulant [13], and osteogenic-enhancing differentiation activities [14]. These activities are closely related to its molecular weight [15] and sulfate content [16].

Fucoidan's immunomodulatory effects have been reported in different experimental models. It has been shown to activate and promote the maturation of human monocyte-derived dendritic cells [17], and stimulate lymphocyte [18,19] and macrophage activity [18,20] in vitro. According to Do et al. [21], fucoidan is an immunomodulating nutrient which alters the sensitivity of glia and macrophages. Jin et al. [22] found that intraperitoneal (i.p.) fucoidan injection in mice upregulated CD40, CD80, and CD86 expression. It also increased production of interleukin (IL)-6, IL-12, and tumor necrosis factor-α (TNF-α) in spleen dendritic cells (DC). Furthermore, fucoidan promoted interferon (IFN)-γ generation by producing Th1 and Tc1 cells in an IL-12-dependent manner. Zhang et al. [23] also showed that i.p. administration of fucoidan in the mice promoted natural killer (NK) cell, DC, and T-cell activation. In addition, when mice are injected IP with ovalbumin (OVA) and fucoidan, mice were found to produce remarkably higher amounts of anti-OVA Immunoglobulin G (IgG) than OVA-control mice, and fucoidan was found to promote the generation of effector/memory T-cells based on the surface expression of CD44. Intravenous administration of fucoidan upregulated CD40, CD80, CD86, and the major histocompatibility complex in mouse spleen DCs [23]. Interestingly, oral administration of 300 mg fucoidan per day to elderly men and women increased their immune responses to influenza vaccination [24], but the underlying mechanisms of this are still unclear.

There are some reports on the relationship between the structure and immunomodulatory properties of fucoidan [25]; however, less is known about the relationship between their molecular weight and immunomodulatory properties. In our previous studies, we demonstrated that low-molecular-weight (LMW) fucoidan has more favorable bioactivity in vitro and in vivo than high-molecular-weight fucoidan [7–9,12], with no toxicological effects found in rats after 28 days of repeated oral administration [26]. LMW fucoidan is expected to be a safe food supplement for immunomodulation.

Mycoplasma pneumoniae is one of the most common agents of community-acquired pneumonia in previously healthy people, and infection can occur at any age [27]. *M. pneumoniae* is considered a self-limiting disease, but some patients suffer from protracted and complicated courses [28]. In addition, *M. pneumoniae* can change its cell membrane composition to mimic the host's cell membrane, thus avoiding immune system detection [27]. Currently, there is no vaccine to prevent *M. pneumoniae* infection. Macrolide is one of the first-line drugs used to treat *M. pneumoniae*, but the prevalence of macrolide-resistant *M. pneumonia* is increasing worldwide. In 2011, the macrolide-resistance rate was up to 23% in Taiwan [29]. A recent study has reported that early additional immune-modulators might can prevent *M. pneumoniae* disease progression and reduce morbidity [30]. A natural and safe food supplement that enhances the immune response to prevent *M. pneumoniae* infection could address this unmet need.

In this study, we evaluated the immunomodulatory effects of orally administered LMW fucoidan on innate and adaptive immune responses in mice. We also evaluated the effect of LMW fucoidan supplementation on immune responses to the *M. pneumonia* (MP) antigen.

2. Results and Discussion

2.1. Fractionation and Composition of Crude Fucoidan and LMW Fucoidan

A step gradient was used to selectively elute the sulfate and fucose-rich fucoidan from a crude brown seaweed extract. The yields and compositions of the three fractions, purified from *Laminaria japonica*, are shown in Figure 1 and Table 1.

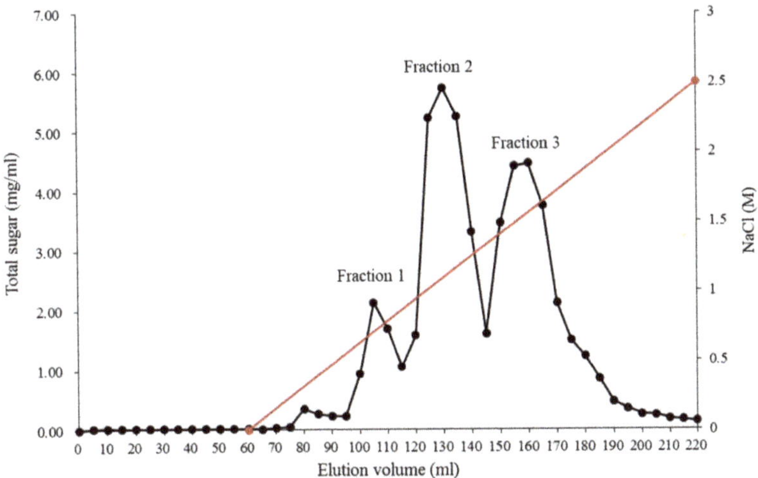

Figure 1. Fractionation of crude fucoidan isolated from the *L. japonica* on a DEAE (Diethylaminoethyl)-Sephadex A-25 column.

The fucoidan fraction 1, which was eluted at the lowest NaCl concentration, was rich in galactose (35.8 ± 0.5% mol) and glucose (20.6 ± 0.2% mol), but had low concentrations of fucose (12.9 ± 0.2% mol) and sulfate (10.6 ± 0.6%). The fucoidan fraction 3, eluted at a higher concentration of NaCl, was rich in fucose (48.2 ± 0.3% mol) and sulfate (39.5 ± 0.8%). Although the fucoidan fraction 2 had the highest yield, the fucose (34.6 ± 0.4% mol) and sulfate (25.2 ± 0.5%) contents were lower than in fraction 3. Fucoidan fractions presenting diversities in their sulfation degrees and patterns also differed in their biological activities [31,32]. Thus, we selected fraction 3 as the purified fucoidan and hydrolyzed it as our LMW fucoidan sample. The fraction 3 was hydrolyzed with glycolytic enzyme preparation to obtain the LMW fucoidan sample with a molecular weight lower than 3000 Da. The LMW fucoidan consisted of 40.5 ± 0.8% mol fucose, 5.7 ± 0.7% mol glucose, 28.3 ± 0.8% mol galactose, 5.4 ± 0.5% mol myo-inositol, 15.6 ± 0.5% mol mannose, 3.3 ± 0.6% mol xylose, 1.2 ± 0.4% mol rhamnose, and its degree of sulfation was 31.4 ± 1.6% (Table 1). Myo-inositol has been reported to be the minor sugars in brown seaweed, such as *Fucus vesiculosus* [33], *Himanthalia elongate* [34], and *Bzjizrcaria bifurcate* [35]. Tarakhovskaya et al. [33] indicated that the myo-inositols might have several biological functions, which serve as an initial precursor of different cell-wall polysaccharides.

Table 1. Structure characteristics of fucoidan fractions isolated from the *L. japonica*.

	Elution Volume (mL)	Yield (%)	Sulfate (%)	Neutral Monosaccharide (% mol)						
				Fucose	Glucose	Galactose	Myo-Inositol	Mannose	Xylose	Rhamnose
Fraction 1	No. 100–115	16.1 ± 1.5	10.6 ± 0.6	12.9 ± 0.2	20.6 ± 0.2	35.8 ± 0.5	6.2 ± 0.5	17.5 ± 0.1	3.6 ± 0.1	3.4 ± 0.1
Fraction 2	No. 120–140	40.8 ± 1.2	25.2 ± 0.5	34.6 ± 0.4	25.2 ± 0.5	11.0 ± 0.7	3.6 ± 0.3	21.7 ± 0.6	1.9 ± 0.1	2.0 ± 0.2
Fraction 3	No. 150–165	34.9 ± 2.0	39.5 ± 0.8	48.2 ± 0.3	3.8 ± 0.2	26.7 ± 0.3	1.6 ± 0.4	18.4 ± 0.6	0.4 ± 0.2	0.9 ± 0.1
LMW fucoidan	NA	NA	31.4 ± 1.6	40.5 ± 0.8	5.7 ± 0.7	28.3 ± 0.8	5.4 ± 0.5	15.6 ± 0.5	3.3 ± 0.6	1.2 ± 0.4

NA, not applicable.

2.2. Effects of LMW Fucoidan on Innate Immune Responses in Mice

Innate immunity is the first line of defense against almost any substance that threatens the body. Poor environmental factors, such as malnutrition, stress, and wake-sleep disorders may cause innate immune system deficiencies. Functional nutrient studies indicate a positive correlation between the innate immune system and health [36]. Here, we first examined the effects of oral LMW fucoidan on the innate immune response of non-immunized mice.

It was shown in our previous study that fucoidan was not toxic after intragastric administration to Sprague Dawley (SD) rats at 2000 mg/kg/day [26]. Anti-diabetes activity of LMW fucoidan was observed at the dose of 300–600 mg/kg/day in C57BLKS/J Iar-+Leprdb/+Leprdb (db/db) mice [37]. As a result, dosages of 200, 600, and 1000 mg/kg was chosen in this study for mice.

2.2.1. Spleen Weight and Proliferative Responses of Splenocytes

The spleen plays an important role in host defense, and is the site for innate and adaptive immune processes. We first examined spleen weight and the proliferative response of splenocytes in LMW fucoidan-treated mice. Mice were orally administered 200, 600, and 1000 mg/kg LMW fucoidan or distilled water (control group) daily for 6 weeks, and then sacrificed. During the 6 weeks, a steady rise in body weight (from 17.5 ± 0.2 g to 20.4 ± 0.3 g) was observed, and there were no statistically significant differences between the treated and control mice. The spleen weights were obtained post-mortem, and the spleen-to-body-weight ratio was calculated. No significant differences were observed in the spleen-to-body-weight ratio among the four groups (Figure 2).

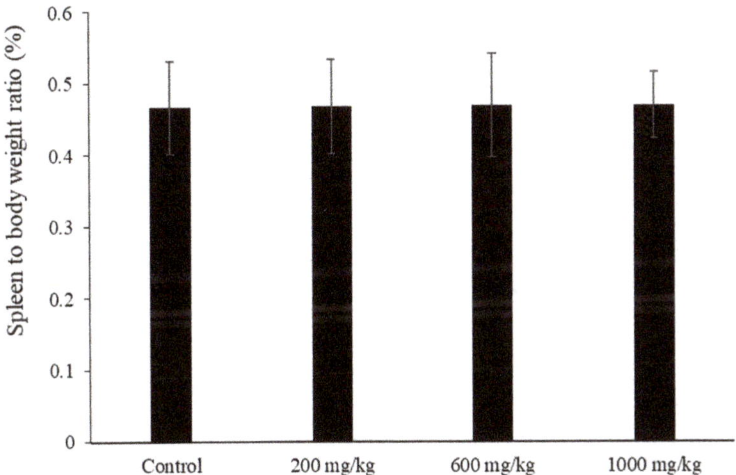

Figure 2. The spleen-to-body-weight ratios of mice treated with low-molecular-weight (LMW) fucoidan for 6 weeks. Spleen-to-body-weight ratio = [spleen weight (g)/body weight (g)] × 100%. Data were expressed as mean ± SD of ten mice.

This result is consistent with a previous study showing that feeding mice fucoidan from *Fucus vesiculosus* had no effects on the spleen-to-body-weight ratio [38]. In addition, Jang et al. [39] showed that LMW fucoidan was less toxic to spleen cells than high-molecular-weight fucoidan. Therefore, LMW fucoidan is not a sensitive indicator to cause cell stress or immunotoxicity. After harvesting spleen cells, splenocytes were prepared and stimulated with the T-cell mitogen concanavalin A (Con A) or B-cell mitogen lipopolysaccharides (LPS) for 72 h. Splenocytes from mice fed with 200, 600, and 1000 mg/kg LMW fucoidan showed significantly increased Con A and LPS-stimulated proliferation compared to the control mice (Table 2), indicating that LMW fucoidan had increased both T- and B-cell proliferation. This is consistent with the results of previous studies by Jang et al. [39], who reported that both low- and high-molecular-weight fucoidan increase spleen-cell viability. Our results confirm that LMW fucoidan has immunostimulatory effects.

Table 2. Proliferative response of splenocytes from mice treated with LMW fucoidan.

	Stimulation Index [‡]	
	Con A	**LPS**
Control	4.24 ± 0.66 [a]	1.78 ± 0.07 [a]
200 mg/kg	4.85 ± 0.74 [b]	1.95 ± 0.07 [b]
600 mg/kg	4.91 ± 0.45 [b]	1.93 ± 0.09 [b]
1000 mg/kg	4.92 ± 0.51 [b]	1.96 ± 0.10 [b]

[‡] Stimulation index was expressed as OD490 of Con A, LPS, or OVA-stimulated cells/OD490 of unstimulated cells. Data were expressed as mean ± SD of ten mice, and analyzed using one-way ANOVA followed by Duncan's multiple range test. Values with different letters in the same column were significantly different ($p < 0.05$).

2.2.2. Natural Killer (NK) Cell Activity

NK cells are a component of the innate immune system, which plays an important role in the early stages of tumor cell elimination. Previous research has shown that NK cell activity is relatively sensitive to diet and the intake of specific food components [40]. To investigate LMW fucoidan's effect on enhancing NK cell activity, we assessed whether LMW fucoidan could induce cytotoxic NK cell activity against YAC-1 cells which are sensitive to lysis by activated NK cells. Our results showed that NK cell activity was enhanced in mice receiving 200, 600, and 1000 mg/kg LMW fucoidan compared to controls (Table 3). Namkoong et al. [41] and Maruyama et al. [42] also reported that fucoidan can increase NK cell activity via oral administration. Together, these data indicate that LMW fucoidan consumption can activate NK cells and, thus, might be a good way of enhancing the immune system.

Table 3. Natural killer (NK) cell activity of splenocytes from mice treated with LMW fucoidan.

	NK Cell Activity [‡]	
	E/T Ratio [#]: 10	**E/T Ratio: 25**
Control	14.4 ± 3.8 [a]	33.1 ± 1.8 [a]
200 mg/kg	20.8 ± 7.6 [b]	42.1 ± 1.6 [b]
600 mg/kg	21.0 ± 7.1 [b]	41.4 ± 1.7 [b]
1000 mg/kg	21.1 ± 4.4 [b]	42.0 ± 1.9 [b]

[‡] NK cell activity was expressed as (test group fluorescence − spontaneous fluorescence)/(total target cell fluorescence − spontaneous fluorescence) × 100%. [#] Effector cell (splenocytes) to target cell (YAC-1 cell) ratio. Data were expressed as mean ± SD of ten mice, and analyzed using one-way ANOVA followed by Duncan's multiple range test. Values with different letters in the same column were significantly different ($p < 0.05$).

2.2.3. Phagocytic Activity of Peritoneal Cells

As part of the innate immune system, the body has developed defenses mediated by specialized cells which destroy invading microorganisms by ingestion and phagocytosis. In the present study, the phagocytic activity of peritoneal cells from mice which were fed LMW fucoidan was measured by quantifying internalized Fluorescein isothiocyanate (FITC)-*Escherichia coli* (*E. coli*). Cells isolated from mice treated with 1000 mg/kg LMW fucoidan exhibited a significant increase in phagocytosis at multiplicities of infection of 12.5 and 25 (Table 4). These results are similar to those reported by Anisimova et al. [43], who showed that fucoidan intensifies the engulfment of microorganisms by human blood, increasing both the relative number and efficiency of phagocytes.

Table 4. Phagocytic activity of peritoneal cells from mice treated with LMW fucoidan.

	Phagocytic Activity [‡] (%)	
	MOI 12.5	**MOI 25**
Control	20.1 ± 3.6 [a]	28.7 ± 6.2 [a]
200 mg/kg	22.9 ± 3.0 [ab]	32.4 ± 4.4 [ab]
600 mg/kg	23.7 ± 4.4 [ab]	36.4 ± 5.0 [b]
1000 mg/kg	24.8 ± 3.8 [bc]	37.4 ± 5.6 [b]

[‡] Phagocytic activity was expressed as FITC-positive peritoneal cells/total peritoneal cells × 100%. Data were expressed as mean ± SD of ten mice, and analyzed using one-way ANOVA followed by Duncan's multiple range test. Values with different letters in the same column were significantly different ($p < 0.05$).

In summary, we found that LMW fucoidan effectively stimulates innate immunity, not only by phagocytosis (Table 4) but also by activating extracellular killing by NK cells (Table 3) to destroy invading pathogens.

2.3. Effects of LMW Fucoidan on Adaptive Immune Responses (OVA-Specific Immunity) in Mice

The above results prompted us to further investigate the adjuvant effects of LMW fucoidan on adaptive immune responses. We orally administered LMW fucoidan daily for 6 weeks and immunized mice with OVA to examine the effect on cytokine response and specific antibody production against OVA.

2.3.1. Proliferative Response by Con A, LPS, and OVA Stimulation

Mice were orally administered distilled water (control group) or 200, 600, and 1000 mg/kg LMW fucoidan daily for 6 weeks, and OVA immunization was conducted at the third and fifth weeks. During the 6 weeks, a steady rise in body weight (from 18.6 ± 0.3 g to 21.4 ± 0.4 g) was observed, and there were no statistically significant differences between the groups of mice. The spleen weights were obtained, and the spleen-to-body-weight ratio was calculated. Interestingly, the ratios of OVA-immunized and OVA-immunized + LMW fucoidan-treated mice were higher than those of controls (Figure 3), although feeding LMW fucoidan alone for 6 weeks did not alter the ratio (Figure 2).

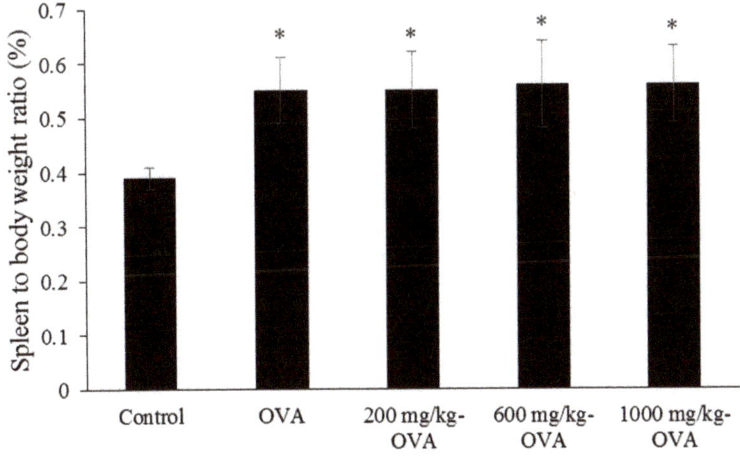

Figure 3. The spleen-to-body-weight ratios of ovalbumin (OVA)-immunized mice treated with LMW fucoidan for 6 weeks. Spleen-to-body-weight ratio = [spleen weight (g)/body weight (g)] × 100%. Data were expressed as mean ± SD of ten mice. Means with asterisks were significantly different from the control ($p < 0.05$).

A previous study reported that the spleen-to-body-weight ratio of LPS + fucoidan-treated mice was higher than LPS-treated mice when fucoidan was administered intraperitoneally [38], suggesting that different routes of fucoidan administration may lead to significantly different physiological reactions. Our findings suggest that OVA plays a major role in increasing the spleen-to-body-weight ratio, and that oral administration of LMW fucoidan does not support this effect. LMW fucoidan (200, 600, and 1000 mg/kg doses) significantly enhanced the mitogen (Con A and LPS)- and OVA-stimulated splenocyte proliferation in OVA-immunized mice compared to the OVA-control group (Table 5). These findings indicate that LMW fucoidan could significantly increase cell-mediated immune responses in OVA-immunized mice.

Table 5. Proliferative response by ConA, LPS, or OVA stimulated splenocytes from OVA-immunized mice treated with LMW fucoidan.

	Stimulation Index [‡]		
	Con A	**LPS**	**OVA**
Control	3.63 ± 0.56 [a]	1.52 ± 0.25 [a]	0.81 ± 0.03 [a]
OVA-immunized	3.86 ± 0.27 [a]	1.81 ± 0.09 [a]	1.10 ± 0.03 [b]
200 mg/kg-OVA	4.96 ± 0.28 [b]	2.25 ± 0.25 [c]	1.54 ± 0.04 [c]
600 mg/kg-OVA	5.02 ± 0.23 [b]	2.44 ± 0.18 [c]	1.54 ± 0.03 [c]
1000 mg/kg-OVA	5.05 ± 0.29 [b]	2.75 ± 0.24 [c]	1.60 ± 0.04 [c]

[‡] Stimulation index was expressed as OD490 of Con A, LPS, or OVA-stimulated cells/OD490 of unstimulated cells. Data were expressed as mean $\pm$ SD of ten mice, and analyzed using one-way ANOVA followed by Duncan's multiple range test. Values with different letters in the same column were significantly different ($p < 0.05$).

2.3.2. Cytokine Secretion by OVA-Stimulation

Cytokines serve as chemical messengers within the immune system and are primarily produced by T-helper (Th) cells. In the activation phase of adaptive immune responses, cytokines stimulate the growth and differentiation of lymphocytes. In the effector phases, they activate different effector cells to eliminate microbes and other antigens [44]. We examined the effect of LMW fucoidan on the production of cytokines that mediate and regulate adaptive immune responses. Th 1 cytokines, including IL-2 and IFN-γ, and the Th 2 cytokine, IL-4, significantly increased following LMW fucoidan administration. No significant differences in the production of the Th2 cytokine IL-5 were observed. LMW fucoidan treatment during OVA-stimulation significantly decreased TNF-α, a multifunctional, pro-inflammatory cytokine (Table 6). NK cell activity has been shown to be enhanced by IL-2 [45], IL-4 [46], and IFN-γ [47], and our data also showed that IL-2, IL-4, and IFN-γ production (Table 6) and NK cell activity (Table 3) were increased in splenocytes from mice treated with oral LMW fucoidan. Both Jin et al. [22] and Zhang et al. [23] reported that IP injection of fucoidan led to increased IFN-γ production compared to control mice immunized with OVA alone. These results indicate that fucoidan, administered either orally or via IP injection, could function as an adjuvant by promoting Th1-type immune responses.

The Th1/Th2 concept suggests that regulating the relative contribution of Th1- or Th2-type cytokines will likely modulate the development and strength of immune responses [48]. A previous study reported that ingesting brown seaweed enhanced OVA-specific Th1 and Th2 cytokine responses by draining lymph nodes in mice [49]. Similarly, our results showed that both the Th1 and Th2 responses in OVA-stimulated mice significantly increased after oral administration of LMW fucoidan. Together, these results demonstrate that oral LMW fucoidan stimulates a healthy Th1/Th2 balance.

Table 6. Cytokines secretion by OVA-stimulated splenocytes from OVA-immunized mice treated with LMW fucoidan.

	Unstimulated Basal Level	Mitogen Stimulation OVA
	IL-2 (pg/mL)	
Control	13.31 ± 2.40 [a]	193.8 ± 59.9 [a]
OVA-immunized	13.67 ± 1.44 [a]	219.8 ± 12.1 [a]
200 mg/kg-OVA	14.48 ± 3.66 [a]	299.9 ± 16.6 [b]
600 mg/kg-OVA	14.58 ± 2.70 [a]	368.0 ± 88.7 [b]
1000 mg/kg-OVA	14.51 ± 2.76 [a]	538.4 ± 122.3 [c]
	IL-4 (pg/mL)	
Control	3.36 ± 0.31 [a]	5.09 ± 0.64 [a]
OVA-immunized	3.61 ± 0.29 [b]	9.91 ± 0.42 [b]
200 mg/kg-OVA	3.43 ± 0.28 [b]	9.36 ± 0.61 [b]
600 mg/kg-OVA	3.45 ± 0.29 [b]	13.18 ± 0.59 [c]
1000 mg/kg-OVA	3.27 ± 0.25 [b]	14.23 ± 0.48 [c]
	IL-5 (pg/mL)	
Control	1.00 ± 0.50 [a]	5.07 ± 0.64 [a]
OVA-immunized	2.66 ± 0.15 [b]	9.91 ± 0.47 [b]
200 mg/kg-OVA	2.62 ± 0.13 [b]	9.36 ± 0.92 [b]
600 mg/kg-OVA	2.47 ± 0.13 [b]	9.30 ± 0.99 [b]
1000 mg/kg-OVA	2.56 ± 0.24 [b]	9.41 ± 1.07 [b]
	IFN-γ (pg/mL)	
Control	1.68 ± 0.23 [a]	1014.90 ± 38.52 [a]
OVA-immunized	51.13 ± 1.14 [b]	1628.13 ± 32.26 [b]
200 mg/kg-OVA	51.62 ± 2.86 [b]	1934.48 ± 39.83 [c]
600 mg/kg-OVA	50.91 ± 2.80 [b]	1944.67 ± 86.95 [c]
1000 mg/kg-OVA	51.66 ± 2.40 [b]	1975.95 ± 44.42 [c]
	TNF-α (pg/mL)	
Control	6.46 ± 0.96 [a]	9.08 ± 1.62 [a]
OVA-immunized	6.49 ± 0.35 [a]	21.12 ± 3.23 [c]
200 mg/kg-OVA	6.37 ± 0.64 [a]	16.46 ± 2.14 [b]
600 mg/kg-OVA	6.39 ± 0.46 [a]	15.25 ± 0.84 [b]
1000 mg/kg-OVA	6.31 ± 0.68 [a]	15.02 ± 1.45 [b]

Data were expressed as mean $\pm$ SD of ten mice, and analyzed using one-way ANOVA followed by Duncan's multiple range test. Values with different letters in the same column were significantly different ($p < 0.05$).

2.3.3. Serum Immunoglobulins Levels

Immunoglobulins, also known as antibodies, are glycoprotein molecules produced by plasma cells that protect us from pathogen infections and antigen sensitization. Immunoglobulin M (IgM) antibodies play an important role in the early stages of immune surveillance. Immunoglobulin A (IgA) is a major humoral factor of mucosal immunity, and IgE is an allergen-specific immunoglobulin that is stimulated by different allergens to produce the corresponding IgE [50]. We examined the effect of LMW fucoidan on the serum levels of anti-OVA IgG, anti-OVA IgA, and anti-OVA IgE that mediate adaptive immune responses. As expected, immunization with OVA significantly increased the serum levels of anti-OVA IgG, anti-OVA IgA, and anti-OVA IgE compared to control mice. Anti-OVA IgG and anti-OVA IgA were significantly increased in the serum from mice administered with LMW fucoidan compared with OVA-immunized mice (Table 7). Similar results were found in the production of IL-2 and IL-4, which stimulate B-cell activation and differentiation [51]. These data suggested that LMW fucoidan functions as an adjuvant to enhance antigen-specific immune responses. Furthermore, in these mice, anti-OVA IgE was significantly decreased (Table 7), so LMW fucoidan might be useful in counteracting allergic responses. Collectively, LMW fucoidan had a significant effect on cytokine expression by enhancing the production of IL-2, IL-4, and IFN-γ (Table 6). Thus, it is possible that LMW fucoidan favors Th1 differentiation and decreases IgE [52].

Table 7. Serum immunoglobulins levels of OVA-immunized mice treated with LMW fucoidan.

	Serum Immunoglobulin (EU [‡])		
	Anti-OVA IgG	Anti-OVA IgA	Anti-OVA IgE
Control	0.04 ± 0.01 [a]	0.06 ± 0.01 [a]	0.02 ± 0.01 [a]
OVA-immunized	2.12 ± 0.78 [b]	0.96 ± 0.03 [b]	0.85 ± 0.29 [c]
200 mg/kg-OVA	2.39 ± 0.46 [bc]	1.05 ± 0.07 [b]	0.84 ± 0.11 [c]
600 mg/kg-OVA	2.61 ± 0.24 [c]	1.93 ± 0.04 [c]	0.57 ± 0.07 [b]
1000 mg/kg-OVA	2.77 ± 0.42 [c]	2.21 ± 0.08 [c]	0.37 ± 0.12 [b]

[‡] EU was expressed as (OD450 of sample − OD450 of blank)/(OD450 of positive- OD450 of blank). Data were expressed as mean $\pm$ SD of ten mice, and analyzed using one-way ANOVA followed by Duncan's multiple range test. Values with different letters in the same column were significantly different ($p < 0.05$).

2.4. Effects of LMW Fucoidan on MP-Antigen-Stimulated Immune Responses in Mice

According to the above data, LMW fucoidan showed strong immunomodulatory effects in innate and adaptive immunity. As shown in Table 2; Table 3, the mice treated with LMW fucoidan showed better proliferative responses and NK cell activity, and there were no significant differences between the 200, 600, and 1000 mg/kg groups. As shown in Table 4, the mice treated with LMW fucoidan showed better phagocytic activity, and the 600 and 1000 mg/kg groups exhibited the best activities (Table 4). As shown in Tables 5 and 6, the mice treated with LMW fucoidan showed better proliferative responses and cytokines secretion, and there were no significant differences between the 200, 600, and 1000 mg/kg groups (except IL-2 in Table 6). As a result, we chose the 600 mg/kg dose concentration to investigate the effects of orally administered LMW fucoidan on MP antigen immunogenicity.

This evaluation was done to determine whether LMW fucoidan could function as an early additional immunomodulator, preventing *M. pneumoniae* disease infection and reducing disease morbidity. In this experiment, MP antigen inoculation was carried out twice. The first MP antigen inoculation was considered the first infection by *M. pneumoniae*, and the second MP antigen inoculation was considered the re-infection. As expected, after the MP antigen inoculation, IgM, IgG, and IgA contents in the serum were higher in the LMW fucoidan group than in the control group. IgM and IgG significantly increased ($p < 0.05$) in the serum from 600 mg/kg LMW fucoidan + MP-antigen-inoculated group mice compared to the MP-antigen-inoculated group mice after both the first and second inoculations. There was no significant difference in IgA with or without LMW fucoidan after MP antigen inoculation (Figure 4). Our data show a possible adjunctive role of LMW fucoidan by stimulating antibody production upon repeat infection of *M. pneumoniae*. Fucoidan has also been reported to increase immune responses to seasonal influenza vaccinations [24] and improve *M. pneumoniae* vaccine efficacy [40]. Based on our previous research and these results, we suggest that LMW fucoidan might help prevent the spread of *M. pneumoniae*.

In this study, we investigated the immunomodulatory effects of orally administrated LMW fucoidan on innate immunity, adaptive immunity, and MP-antigen-stimulated immune responses. Our results indicate that LMW fucoidan can increase splenocyte proliferation, NK cell activity, and phagocytic activity, as well as IL-2, IL-4, and IFN-γ production. In addition, LMW fucoidan enhanced antigen-specific antibody production in OVA- and MP-antigen-stimulated mice. LMW fucoidan, in the form of a natural food supplement, could enhance immune responses and attenuate the effects of *M. pneumoniae* infection.

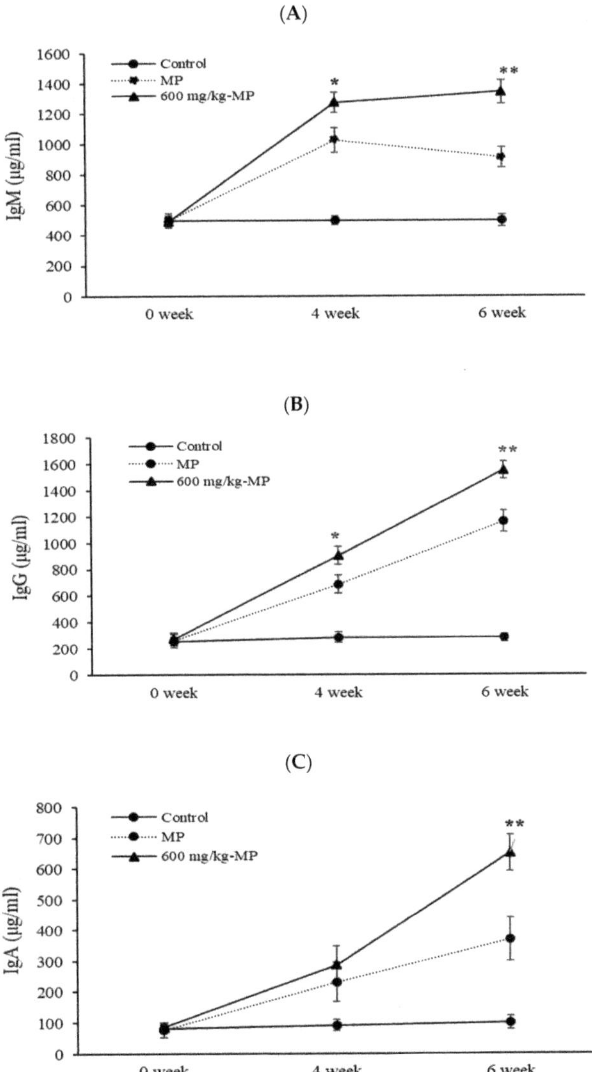

Figure 4. The IgM (**A**), IgG (**B**), and IgA (**C**) productions of *Mycoplasma pneumoniae* (MP) antigen-inoculated mice treated with 600 mg/kg LMW fucoidan for 6 weeks. Data were expressed as mean ± SD of ten mice. Values are expressed as mean ± SD. Means with asterisk were significantly different from the other groups ($p < 0.05$).

3. Materials and Methods

3.1. Animal

The female BALB/c mice (Albino mice maintained by Halsey Bagg) were obtained from BioLASCO Taiwan CO., Ltd (Taipei, Taiwan). All mice were weighted and given provisional numbers upon arrival. The animals were subjected to acclimation in a housing facility for at least eight days before starting the experiment. Only healthy mice were selected for use in the study. The mice were eight weeks of age at the start of the experiment, and were housed in a normal, environmentally controlled animal room with free access to pathogen-free feed and water ad libitum. These mice were randomly divided, with ten mice in each group.

3.2. LMW Fucoidan Preparation

LMW fucoidan extracted from *Laminaria japonica* were collaboratively manufactured by Hi-Q Marine Biotech International Ltd. (New Taipei City, Taiwan). Seaweed samples were ground to flour with a miniblender, and then dried with a dryer at 50 °C. The 100 g dried seaweed was treated with 5 L of distilled water and boiled at 100 °C for 30 min, and the extract was centrifuged at $10,000 \times g$ for 20 min. The supernatant was added with 4 M $CaCl_2$ incubated for 1 h to separate alginic acid, and re-centrifuged at $10,000 \times g$ for 20 min. All polysaccharides were dialyzed (cut off 10,000 Da) using deionized water for 48 h, and precipitated by the addition of ethanol at the ratio of 1:3 (V/V) and left overnight to give the crude fucoidan. The fucoidan so obtained was fractionated by anion-exchange chromatography using DEAE-Sephadex A-25 (Cl^- form, Pharmacia, Uppsala, Sweden) using gradually increasing concentrations of sodium chloride (0–2.5 M) at a flow rate of 1 mL/min, and eluents (5 mL/tube) were separately collected. Each fraction was analyzed for sulfate content (Section 3.3) and monosaccharide composition (Section 3.4). The fraction of rich sulfate and fucose content (fraction 3) was extracted and hydrolyzed with a mixture of crude glycolytic enzymes, isolated from *Bacillus subtilis* by our lab (Section 3.5). The reaction mixture was passed through a filter (3000 Da cut off) by using Amicon Stirred Cells (Merck Germany) to collect the LMW fucoidan with a molecular weight lower than 3000 Da. The hydrolysis reaction of the fucoidan by the crude glycolytic enzymes was performed at pH 6.5, 37 °C for 24 h and heated at 95 °C for 20 min for inactivation of the enzyme.

3.3. Analysis of Degree of Sulfation

Sulfate content was determined according to the gelatinbarium method [53] using sodium sulfate (1 mg/mL) as a standard after acid hydrolysis of the polysaccharide extract sample in 6 N HCl at 100 °C for 6 h.

3.4. Monosaccharide Composition Analysis

The monosaccharides compositions of the eluted fractions from anion-exchange chromatography were separated by a High-performance anion-exchange chromatography (HPAEC) (Dionex BioLC system) using an anion-exchange column (DionexTM Carbopac PA-10, 4 × 250 mm, Dionex, Thermo Fisher Scientific, Waltham, MA, USA). The analysis of monosaccharides was carried out at an isocratic NaOH concentration of 18 mM at ambient temperature [54].

3.5. Preparation of the Crude Enzyme Mixtures from B. subtilis Cultures

One colony of B. subtilis, isolated from natto in Keelung local market, was cultivated in tryptic soy broth (TSB, Becton, Dickinson and Company, Sparks, MD, USA) at 30 °C, 150 rpm for 24 h. The cell cultures were centrifuged to remove bacterial cells to obtain the mixtures of extracellular glycosidic enzymes.

3.6. Animal Experiments

The LMW fucoidan was a light brownish-white powder, and well-soluble below the highest concentration. In this study, three independent experiments were undertaken to evaluate the innate immune, adaptive immune, and MP antigen-inoculated immune response, respectively. The first experiment was performed to assay innate immune, mice were assigned to four groups (n = 10); concentrations of 200, 600, and 1000 mg/kg LMW fucoidan were dissolved in distilled water and then administered by oral gavage in 10 mL/kg on a daily basis for 6 weeks. The control group (0 mg/kg LMW fucoidan) mice were treated identically with equal volumes of distilled water also via oral gavage throughout the study. Mice were weighted and sacrificed, and blood samples were collected at the end of the experiment for serum immunoglobulins assay. Spleens were collected and weighted, and the splenocytes were analyzed for cell proliferation, the cytokines secretion assay, and natural killer cell activity assay. Peritoneal cells were collected form mice abdominal cavity for the phagocytic activity assay. Each mouse was weighted once a week during the study period.

The second experiment was performed byadaptive immune mice assay, and five groups of mice (control, OVA-immunized, 200, 60,0 and 1000 mg/kg LMW fucoidan, then OVA-immunized) were orally administered distilled water or LMW fucoidan for 6 weeks. The first OVA IP immunization was conducted after 3 weeks of administration. 6.25 μg OVA was emulsified in the complete Freund's adjuvant (CFA) and intraperitoneally injected into the mice. Two weeks after the first OVA immunization, mice were given a second injection of 12.5 μg OVA emulsified with incomplete Freund's adjuvant (IFA) in order to enhance the OVA-specific immune responses. Blood samples were collected before the OVA inducement at the end of the experiment for the serum immunoglobulins assay. Mice were then sacrificed by isoflurane inhalation. Spleens were collected and weighted, and the splenocytes were analyzed for cell proliferation and the cytokines secretion assay.. Each mouse was weighted once a week during the study period.

The third experiment was performed to assay MP antigen-inoculated immune responds, three groups of mice (control, MP antigen-stimulation and 600 mg/kg LMW fucoidan then MP antigen stimulation) were orally administered distilled water or LMW fucoidan for 6 weeks. First, MP antigen (10 μg/mouse, MP901, Native Antigen, UK) IP stimulation was conducted after 3 weeks of administration. Two weeks after the first stimulation, mice were given a second injection of MP antigen (10 μg/mouse) [40]. Blood samples were collected before the first MP antigen stimulation, after one week of stimulation, and at the end of the experiment for the serum immunoglobulins assay. Each mouse was weighted once a week during the study period.

3.7. Splenocyte Proliferation Assay

Splenocytes were prepared from treated mice and resuspended in Roswell Park Memorial Institute (RPMI) 1640 medium supplemented with 10% fetal bovine serum (FBS), 50 U/mL penicillin, 50 mg/mL streptomycin, and 0.5 mg/mL amphotericin B. 2×10^5 cells/well of splenocytes were treated with Con A or LPS for 72 h in an innate immune experiment, and treated with Con A, LPS, or OVA for 72 h in an adaptive immune experiment. Cell proliferation was measured by OD490 using the CellTiter 96 Aqueous One Solution Cell Proliferation Assay (Promega). Results were expressed as the stimulation index, and the formula for calculating stimulation index is shown below:

Stimulation index = OD490 of Con A, LPS, or OVA-stimulated cells/OD490 of un-stimulated cells.

3.8. NK Cell Activity Aassay

NK cell activity was measured by the release of the fluorescent marker from YAC-1 target cells (A lymphoma which was induced by inoculation of the Moloney leukemia virus (MLV) into a newborn A/Sn mouse) [American Type Culture Collection (ATCC) TIB-160, Food Industry Research and Development Institute, Hsinchu, Taiwan]. Splenocytes (effector cells, including NK cells) were cultured in the supplemented RPMI 1640 medium. YAC-1 cells were incubated with 2′-7′-bis (carboxyethyl)-5(6)-carboxyfluorescein (BCECF) (ThermoFisher) for 30 min at 37 °C. After being washed with culture medium, cells were re-suspended in the medium and cultured for 2 h at 37 °C. 2′,7′ –bis-(Carboxyethyl)-5-(and-6)-carboxyfluorescein (BCECF)-labeled YAC-1 cells were mixed with splenocytes as effector cells to make a 10 and 25 ratio of effector cells to target cells (E/T). After incubation for 4 h, the cultured supernatant was harvested and measured fluorescence with excitation and emission wavelengths of 485 nm and 538 nm. The total target cell fluorescence was determined by lysis of cells with 0.25% (*v/v*) Triton X-100. Spontaneous fluorescence was measured from target cells incubated without effector cells. Test group fluorescence was measured from the effector cell mixing with target cells. Results were expressed as NK cell activity, and the formula for calculating NK cell activity is shown below:

NK cell activity = (test group fluorescence-spontaneous fluorescence)/(total target cell fluorescence- spontaneous fluorescence) × 100%.

3.9. Phagocytic Activity Assay by Peritoneal Cells

Mice were scarified and peritoneal cells were harvested by injecting 5 mL of Hank's balanced salt solution (HBSS) into the peritoneal cavity and gently shaking it. Then, peritoneal fluid was collected and centrifuged at 450× *g*, 4 °C for 10 min. The cell pellet was resuspended at 1×10^6 cells/mL in supplemented RPMI 1640 medium. The phagocytosis was performed by monitoring the engulfment of FITC-labeled *E. coli* by the peritoneal exudate cells. Cells were mixed with FITC-*E. coli* at Multiplicity of infection (MOI) of 12.5 and 25 for 30 min. Fluorescence from non-internalized FITC-*E. coli* was quenched by the addition of acidified trypan blue, and the cells were then washed and analyzed by flow cytometry (BD FACSAria, Becton Dickinson Biosciences, San Jose, CA, US). Results were expressed as phagocytic activity, and the formula for calculating phagocytic activity is shown below:

Phagocytic activity = FITC positive peritoneal cells/total peritoneal cells × 100%.

3.10. Cytokines Secretion Assay

Cytokines including IL-2, IL-4, IL-5, TNF-α, and IFN-γ were measured by using enzyme-linked immunosorbent assay (ELISA) kits (R&D Systems Inc., Minneapolis, MN, USA) according to the manufacturer's instructions. Splenocytes were prepared from treated mice and resuspended in supplemented RPMI 1640 medium. Cells were incubated at 1×10^6 cells/well with 25 μg/mL OVA for 72 h in the second experiment for adaptive immune assay. After 72 h incubation, the cell-free supernatants were collected, centrifuged 450× *g*, 4°C for 10 min, and cytokines were measured by using the ELISA kits.

3.11. Serum Immunoglobulins Assay

In the second experiment for the adaptive immune assay, blood was centrifuged at 450× *g*, 4 °C for 10 min, and serum samples were collected for OVA-specific anti-IgG, OVA-specific anti-IgA, and OVA-specific anti-IgE. OVA-specific antibodies were detected by an indirect ELISA (Bethyl Laboratories). Results were expressed as ELISA units (EU), and the formula for calculating EU is shown below: EU = (OD450 of sample- OD_{450}nm of blank)/(OD450 of positive- OD450 of blank). A positive OD450 indicated a serum with a high titer of OVA-specific antibodies [52]. In the third experiment, serum samples were collected for serum immunoglobulins assay by using a mouse immunoglobin quantitation set (Bethyl Laboratories) to measure the concentrations of IgM, IgG, and IgA, respectively.

3.12. Statistical Analysis

Numerical data are presented as means ± standard deviation. The data was analyzed by a one-way analysis of variance (ANOVA) and Student's *t*-test.

4. Conclusions

In this study, we investigated the immune modulatory effects of orally administered LMW fucoidan on the innate immune, adaptive immune, and MP antigen-stimulated immune response. Our results provided evidence that LMW fucoidan could increase splenocyte proliferation, NK cell activity, and phagocytic activity, as well as IL-2, IL-4, and IFN-γ secretion. In addition, LMW fucoidan enhanced antigen-specific antibody production in OVA- and MP antigen-stimulated mice. Our hope is that LMW fucoidan, a natural food supplement, could enhance immune responses, be needed for immunopotentiation, and attenuate the *M. pneumoniae* infectious disease.

Author Contributions: Conceptualization, P.-A.H.; Methodology, P.-A.H., H.-T.V.L.; Validation, P.-A.H.; Investigation, H.-T.V.L. and H.-Y.L.; Resources, P.-A.H.; Data Curation, P.-A.H., H.-T.V.L. and S.-K.L.; Writing—Original Draft Preparation, P.-A.H.; Writing—Review & Editing, P.-A.H., H.-T.V.L.

Funding: This research was funded by Ministry of Science and Technology, MOST 107-2320-B-019-001-MY2, and funded by Center of Excellence for the Oceans, National Taiwan Ocean University, NTOU-RD-AA-2016-1-02012.

Conflicts of Interest: The authors declare no conflict of interest.

References

1. Lincoln, R.A.; Strupinski, K.; Walker, J.M. Bioactive compounds from algae. *Life Chem. Rep.* **1991**, *8*, 183.
2. Okolie, C.L.; Rajendran, S.R.C.K.; Udenigwe, C.C.; Aryee, A.N.A.; Mason, B. Prospects of brown seaweed polysaccharides (BSP) as prebiotics and potential immunomodulators. *J. Food Biochem.* **2017**, *41*, e12392. [CrossRef]
3. Kraan, S. Algal polysaccharides, novel applications and outlook. In *Carbohydrates: Comprehensive Studies on Glycobiology and Glycotechnology*; Chang, C., Ed.; InTech: Maastricht, The Netherlands, 2012; pp. 489–532.
4. Ale, M.T.; Meyer, A.S. Fucoidans from brown seaweeds: an update on structures, extraction techniques and use of enzymes as tools for structural elucidation. *Rsc Adv.* **2013**, *3*, 8131–8141. [CrossRef]
5. Fitton, J.H. Therapies from fucoidan: Multifunctional marine polymers. *Mar Drugs* **2011**, *9*, 1731–1760. [CrossRef]
6. Senni, K.; Gueniche, F.; Foucault-Bertaud, A.; Igondjo-Tchen, S.; Fioretti, F.; Colliec-Jouault, S.; Durand, P.; Guezennec, J.; Godeau, G.; Letourneur, D. Fucoidan a sulfated polysaccharide from brown algae is a potent modulator of connective tissue proteolysis. *Arch. Biochem. Biophys.* **2006**, *445*, 56–64. [CrossRef] [PubMed]
7. Hsu, H.Y.; Lin, T.Y.; Hwang, P.A.; Tseng, L.M.; Chen, R.H.; Tsao, S.M.; Hsu, J. Fucoidan induces changes in the epithelial to mesenchymal transition and decreases metastasis by enhancing ubiquitin-dependent TGF beta receptor degradation in breast cancer. *Carcinogenesis* **2013**, *34*, 874–884. [CrossRef] [PubMed]
8. Hsu, H.Y.; Lin, T.Y.; Wu, Y.C.; Tsao, S.M.; Hwang, P.A.; Shih, Y.W.; Hsu, J. Fucoidan inhibition of lung cancer in vivo and in vitro: Role of the Smurf2-dependent ubiquitin proteasome pathway in TGF beta receptor degradation. *Oncotarget* **2014**, *5*, 7870–7885. [CrossRef]
9. Chen, M.C.; Hsu, W.L.; Hwang, P.A.; Chou, T.C. Low molecular weight fucoidan inhibits tumor angiogenesis through downregulation of HIF-1/VEGF signaling under hypoxia. *Mar. Drugs* **2015**, *13*, 4436–4451. [CrossRef] [PubMed]
10. Tengdelius, M.; Lee, C.J.; Grenegard, M.; Griffith, M.; Pahlsson, P.; Konradsson, P. Synthesis and biological evaluation of fucoidan-mimetic glycopolymers through cyanoxyl-mediated free-radical polymerization. *Biomacromolecules* **2014**, *15*, 2359–2368. [CrossRef]
11. Hwang, K.A.; Yi, B.R.; Choi, K.C. Molecular mechanisms and in vivo mouse models of skin aging associated with dermal matrix alterations. *Lab. Anim. Res.* **2011**, *27*, 1–8. [CrossRef]
12. Hwang, P.A.; Hung, Y.L.; Chien, S.Y. Inhibitory activity of *Sargassum hemiphyllum* sulfated polysaccharide in arachidonic acid-induced animal models of inflammation. *J. Food Drug Anal.* **2015**, *23*, 49–56. [CrossRef] [PubMed]
13. Irhimeh, M.R.; Fitton, J.H.; Lowenthal, R.M. Pilot clinical study to evaluate the anticoagulant activity of fucoidan. *Blood Coagul. Fibrin.* **2009**, *20*, 607–610. [CrossRef]
14. Hwang, P.A.; Hung, Y.L.; Phan, N.N.; Hieu, B.T.N.; Chang, P.M.; Li, K.L.; Lin, Y.C. The in vitro and in vivo effects of the low molecular weight fucoidan on the bone osteogenic differentiation properties. *Cytotechnology* **2016**, *68*, 1349–1359. [CrossRef] [PubMed]
15. Yang, C.; Chung, D.; Shina, I.S.; Lee, H.; Kim, J.; Lee, Y.; You, S. Effects of molecular weight and hydrolysis conditions on anticancer activity of fucoidans from sporophyll of *Undaria pinnatifida*. *Int. J. Biol. Macromol.* **2008**, *43*, 433–437. [CrossRef] [PubMed]
16. Soeda, S.; Sakaguchi, S.; Shimeno, H.; Nagamatsu, A. Fibrinolytic and anticoagulant activities of highly sulfated fucoidan. *Biochem. Pharmacol.* **1992**, *43*, 1853–1858. [CrossRef]
17. Yang, M.; Ma, C.H.; Sun, J.T.; Shao, Q.Q.; Gao, W.J.; Zhang, Y.; Li, Z.W.; Xie, Q.; Dong, Z.G.; Qu, X. Fucoidan stimulation induces a functional maturation of human monocyte-derived dendritic cells. *Int. Immunopharmacol.* **2008**, *8*, 1754–1760. [CrossRef]

18. Choi, E.M.; Kim, A.J.; Kim, Y.O.; Hwang, J.K. Immunomodulating activity of arabinogalactan and fucoidan in vitro. *J. Med. Food* **2005**, *8*, 446–453. [CrossRef]

19. Takai, M.; Miyazaki, Y.; Tachibana, H.; Yamada, K. The enhancing effect of fucoidan derived from *Undaria pinnatifida* on immunoglobulin production by mouse spleen lymphocytes. *Biosci. Biotechnol. Biochem.* **2014**, *78*, 1743–1747. [CrossRef] [PubMed]

20. Teruya, T.; Takeda, S.; Tamaki, Y.; Tako, M. Fucoidan isolated from *Laminaria angustata* var. longissima induced macrophage activation. *Biosci. Biotechnol. Biochem.* **2010**, *74*, 1960–1962. [CrossRef]

21. Do, H.; Kang, N.S.; Pyo, S.; Billiar, T.R.; Sohn, E.H. Differential regulation by fucoidan of IFN-γ-induced NO production in glial cells and macrophages. *J. Cell Biochem.* **2010**, *111*, 1337–1345. [CrossRef] [PubMed]

22. Jin, J.-O.; Zhang, W.; Du, J.-Y.; Wong, K.-W.; Oda, T.; Yu, Q.J.P.O. Fucoidan can function as an adjuvant in vivo to enhance dendritic cell maturation and function and promote antigen-specific T cell immune responses. *PLoS ONE* **2014**, *9*, e99396. [CrossRef] [PubMed]

23. Zhang, W.; Oda, T.; Yu, Q.; Jin, J.O. Fucoidan from *Macrocystis pyrifera* has powerful immune-modulatory effects compared to three other fucoidans. *Mar. Drugs* **2015**, *13*, 1084–1104. [CrossRef] [PubMed]

24. Negishi, H.; Mori, M.; Mori, H.; Yamori, Y. Supplementation of Elderly Japanese Men and Women with Fucoidan from Seaweed Increases Immune Responses to Seasonal Influenza Vaccination. *J. Nutr.* **2013**, *143*, 1794–1798. [CrossRef]

25. Li, B.; Lu, F.; Wei, X.; Zhao, R. Fucoidan: Structure and bioactivity. *Molecules* **2008**, *13*, 1671–1695. [CrossRef] [PubMed]

26. Hwang, P.A.; Yan, M.D.; Lin, H.T.V.; Li, K.L.; Lin, Y.C. Toxicological evaluation of low molecular weight fucoidan in vitro and in vivo. *Mar. Drugs* **2016**, *14*, 121. [CrossRef]

27. Waites, K.B.; Talkington, D.F. *Mycoplasma pneumoniae* and its role as a human pathogen. *Clin. Microbiol. Rev.* **2004**, *17*, 697–728. [CrossRef] [PubMed]

28. Lee, K.Y.; Lee, H.S.; Hong, J.H.; Lee, M.H.; Lee, J.S.; Burgner, D.; Lee, B.C. Role of prednisolone treatment in severe *Mycoplasma pneumoniae* pneumonia in children. *Pediatr. Pulmonol.* **2006**, *41*, 263–268. [CrossRef] [PubMed]

29. Wu, P.S.; Chang, L.Y.; Lin, H.C.; Chi, H.; Hsieh, Y.C.; Huang, Y.C.; Liu, C.C.; Huang, Y.C.; Huang, L.M. Epidemiology and clinical manifestations of children with macrolide-resistant *Mycoplasma pneumoniae* pneumonia in Taiwan. *Pediatr. Pulmonol.* **2013**, *48*, 904–911. [CrossRef]

30. Youn, Y.S.; Lee, S.C.; Rhim, J.W.; Shin, M.S.; Kang, J.H.; Lee, K.Y. Early additional immune-modulators for *Mycoplasma pneumoniae* pneumonia in children: An observation study. *Infect. Chemother.* **2014**, *46*, 239–247. [CrossRef]

31. Boisson-Vidal, C.; Chaubet, F.; Chevolot, L.; Sinquin, C.; Theveniaux, J.; Millet, J.; Sternberg, C.; Mulloy, B.; Fischer, A.M. Relationship between antithrombotic activities of fucans and their structure. *Drug Dev. Res.* **2000**, *51*, 216–224. [CrossRef]

32. Fonseca, R.J.C.; Oliveira, S.N.M.C.G.; Melo, F.R.; Pereira, M.G.; Benevides, N.M.B.; Mourao, P.A.S. Slight differences in sulfation of algal galactans account for differences in their anticoagulant and venous antithrombotic activities. *Thromb Haemost.* **2008**, *99*, 539–545. [CrossRef]

33. Tarakhovskaya, E.; Lemesheva, V.; Bilova, T.; Birkemeyer, C. Early embryogenesis of brown alga *Fucus vesiculosus* L. is characterized by significant changes in carbon and energy metabolism. *Molecules* **2017**, *22*, 1509. [CrossRef]

34. Santoyo, S.; Plaza, M.; Jaime, L.; Ibanez, E.; Reglero, G.; Senorans, J. Pressurized liquids as an alternative green process to extract antiviral agents from the edible seaweed *Himanthalia elongata*. *J. Appl. Phycol.* **2011**, *23*, 909–917. [CrossRef]

35. Mian, A.J.; Percival, E. Carbohydrates of the brown seaweeds *himanthalia lorea, bifurcaria bifurcata*, and *Padina pavonia*: Part I. extraction and fractionation. *Carbohydr. Res.* **1973**, *26*, 133–146. [CrossRef]

36. Karacabey, K.; Ozdemir, N. The Effect of Nutritional Elements on the Immune System. *J. Obes. Wt Loss Ther.* **2012**, *2*, 152. [CrossRef]

37. Lin, H.T.V.; Tsou, Y.C.; Chen, Y.T.; Lu, W.J.; Hwang, P.A. Effects of low-molecular-weight fucoidan and high stability fucoxanthin on glucose homeostasis, lipid metabolism, and liver function in a mouse model of type II diabetes. *Mar. Drugs* **2017**, *15*, 113. [CrossRef]

38. Ko, E.-J.; Joo, H.-G. Fucoidan enhances the survival and sustains the number of splenic dendritic cells in mouse endotoxemia. *Korean J. Physiol. Pharmacol.* **2011**, *15*, 89–94. [CrossRef]

39. Jang, J.Y.; Moon, S.Y.; Joo, H.G. Differential effects of fucoidans with low and high molecular weight on the viability and function of spleen cells. *Food Chem. Toxicol.* **2014**, *68*, 234–238. [CrossRef]

40. Kim, S.-Y.; Joo, H.-G. Evaluation of adjuvant effects of fucoidan for improving vaccine efficacy. *J. Vet. Sci.* **2015**, *16*, 145–150. [CrossRef]

41. Namkoong, S.; Kim, Y.J.; Kim, T.S.; Sohn, E.H. Immunomodulatory Effects of Fucoidan on NK Cells in Ovariectomized Rats. *Korean J. Plant Res.* **2012**, *25*, 317–322. [CrossRef]

42. Maruyama, H.; Tamauchi, H.; Iizuka, M.; Nakano, T. The role of NK cells in antitumor activity of dietary fucoidan from. *Planta Med.* **2006**, *72*, 1415–1417. [CrossRef] [PubMed]

43. Anisimova, N.Y.; Ustyuzhanina, N.E.; Donenko, F.V.; Bilan, M.I.; Ushakova, N.A.; Usov, A.I.; Nifantiev, N.E.; Kiselevskiy, M.V. Influence of fucoidans and their derivatives on antitumor and phagocytic activity of human blood leucocytes. *Biochemistry (Mosc)* **2015**, *80*, 925–933. [CrossRef] [PubMed]

44. Fresno, M.; Kopf, M.; Rivas, L. Cytokines and infectious diseases. *Immunol. Today* **1997**, *18*, 56–58. [CrossRef]

45. Saxena, R.K.; Saxena, Q.B.; Adler, W.H. Interleukin-2-induced activation of natural killer activity in spleen cells from old and young mice. *Immunology* **1984**, *51*, 719.

46. Kiniwa, T.; Enomoto, Y.; Terazawa, N.; Omi, A.; Miyata, N.; Ishiwata, K.; Miyajima, A. NK cells activated by Interleukin-4 in cooperation with Interleukin-15 exhibit distinctive characteristics. *Proc. Natl. Acad. Sci. USA* **2016**, *113*, 10139–10144. [CrossRef]

47. Kubota, A.; Lian, R.H.; Lohwasser, S.; Salcedo, M.; Takei, F. IFN-γ production and cytotoxicity of IL-2-activated murine NK cells are differentially regulated by MHC class I molecules. *J. Immunol.* **1999**, *163*, 6488–6493.

48. D'elios, M.; Del Prete, G. Th1/Th2 balance in human disease. *Transplant. Proc.* **1998**, *30*, 2373–2377. [CrossRef]

49. Yan, H.; Kakuta, S.; Nishihara, M.; Sugi, M.; Adachi, Y.; Ohno, N.; Iwakura, Y.; Tsuji, N.M. *Kjellmaniella crassifolia* Miyabe (Gagome) extract modulates intestinal and systemic immune responses. *Biosci. Biotechnol. Biochem.* **2011**, *75*, 2178–2183. [CrossRef]

50. Coico, R.; Sunshine, G. *Immunology: A Short Course*; John Wiley & Sons: New York, NY, USA, 2015.

51. Galanaud, P.; Karray, S.; Llorente, L. Regulatory effects of IL-4 on human B-cell response to IL-2. *Eur. Cytokine Netw.* **1990**, *1*, 57–64.

52. Kuo, C.L.; Chen, T.S.; Liou, S.Y.; Hsieh, C.C. Immunomodulatory effects of EGCG fraction of green tea extract in innate and adaptive immunity via T regulatory cells in murine model. *Immunopharmacol. Immunotoxicol.* **2014**, *36*, 364–370. [CrossRef]

53. Dodgson, K.; Price, R. A note on the determination of the ester sulphate content of sulphated polysaccharides. *Biochem. J.* **1962**, *84*, 106. [CrossRef] [PubMed]

54. Hwang, P.A.; Chien, S.Y.; Chan, Y.L.; Lu, M.K.; Wu, C.H.; Kong, Z.L.; Wu, C.J. Inhibition of lipopolysaccharide (LPS)-induced inflammatory responses by *Sargassum hemiphyllum* sulfated polysaccharide extract in RAW 264.7 macrophage cells. *J. Agric. Food Chem.* **2011**, *59*, 2062–2068. [CrossRef] [PubMed]

 *marine drugs*

Article

Eckol Inhibits Particulate Matter 2.5-Induced Skin Keratinocyte Damage via MAPK Signaling Pathway

Ao Xuan Zhen [1,†], Yu Jae Hyun [1,†], Mei Jing Piao [1], Pincha Devage Sameera Madushan Fernando [1], Kyoung Ah Kang [1], Mee Jung Ahn [2], Joo Mi Yi [3], Hee Kyoung Kang [1], Young Sang Koh [1], Nam Ho Lee [4] and Jin Won Hyun [1,*]

1 School of Medicine, Jeju National University, Jeju 63243, Korea
2 Laboratory of Veterinary Anatomy, College of Veterinary Medicine, Jeju National University, Jeju 63243, Korea
3 Department of Microbiology and Immunology, College of Medicine, Inje University, Busan 47392, Korea
4 Department of Chemistry and Cosmetics, College of Natural Sciences, Jeju National University, Jeju 63243, Korea
* Correspondence: jinwonh@jejunu.ac.kr; Tel.: +82-64-754-3838; Fax: +82-64-702-2687
† These two authors contributed equally to this study.

Received: 1 July 2019; Accepted: 25 July 2019; Published: 27 July 2019

Abstract: Toxicity of particulate matter (PM) towards the epidermis has been well established in many epidemiological studies. It is manifested in cancer, aging, and skin damage. In this study, we aimed to show the mechanism underlying the protective effects of eckol, a phlorotannin isolated from brown seaweed, on human HaCaT keratinocytes against $PM_{2.5}$-induced cell damage. First, to elucidate the underlying mechanism of toxicity of $PM_{2.5}$, we checked the reactive oxygen species (ROS) level, which contributed significantly to cell damage. Experimental data indicate that excessive ROS caused damage to lipids, proteins, and DNA and induced mitochondrial dysfunction. Furthermore, eckol (30 μM) decreased ROS generation, ensuring the stability of molecules, and maintaining a steady mitochondrial state. The western blot analysis showed that $PM_{2.5}$ promoted apoptosis-related protein levels and activated MAPK signaling pathway, whereas eckol protected cells from apoptosis by inhibiting MAPK signaling pathway. This was further reinforced by detailed investigations using MAPK inhibitors. Thus, our results demonstrated that inhibition of $PM_{2.5}$-induced cell apoptosis by eckol was through MAPK signaling pathway. In conclusion, eckol could protect skin HaCaT cells from $PM_{2.5}$-induced apoptosis via inhibiting ROS generation.

Keywords: phlorotannin; particulate matter; reactive oxygen species; keratinocytes

---

1. Introduction

Natural compounds can be effective candidates for various skin diseases. Particularly, phlorotannins extracted from seaweeds have interesting properties that make them useful for cosmeceutical applications. They can whiten the skin by inhibiting melanin synthesis [1], and delay skin wrinkles by inhibiting matrix metalloproteinase [2]. Moreover, phlorotannins show antioxidant [3], anti-inflammatory [4], and hair-growth promotion activities [5]. Studies have shown that eckol, which is a kind of phlorotannin present in brown seaweeds (Phaeophyceae), decreases ultraviolet B (UVB)-induced oxidative stress in human keratinocytes at a dose of 27 μM [6], and inhibits cancer in SKH-1 mice via inhibiting UVB-induced inflammation [7], and declines matrix metalloproteinase-1 level in human dermal fibroblasts implying anti-aging effects at a dose of 10 μM [8]. Our earlier studies have proved that eckol could clear excess reactive oxygen species (ROS) and protect skin keratinocytes from apoptosis [6].

Air pollution by continuous emission of various pollutants into the atmosphere has led to the rapid decline of public health and accelerated climate change [9]. In addition, more than 90% population in the world breathed unhealthy air in 2017, which suggests that particulate pollution is a global challenge [10]. According to previous studies, particulate matter (PM) increases the public health risks for various diseases, such as respiratory disease [11], cardiovascular disease [12], and lung inflammation [13]. Fine particulate matter with a diameter less than 2.5 μm, denoted as $PM_{2.5}$, is present in the air for over several hours to weeks [14]. Significantly, $PM_{2.5}$ could deeply penetrate the skin and the respiratory tract [15]. Skin damage caused by $PM_{2.5}$ is manifested as inflammatory skin diseases, such as atopic dermatitis, acne, and psoriasis, aging, and cancer via multiple signaling pathways [16].

A recent study presents a comprehensive summary of the characteristics of eckol relevant for its therapeutic potential, including antioxidant activity following exposure to H_2O_2, radiation, and PM [17]. $PM_{2.5}$ is a fine particulate matter that causes skin apoptosis related to the ROS generation [16], and it would be interesting to investigate whether eckol protected skin cells from ROS-induced damage. Moreover, the mode of action of eckol on $PM_{2.5}$ is not well-established. Here, we have investigated the potential benefits of eckol on keratinocytes by studying its inhibitory effect on molecular damage, mitochondrial dysfunction, apoptosis-related factors, and MAPK signaling related proteins. In this study, our aim was also to gain insights into the mechanism underlying the protective action of eckol on $PM_{2.5}$-induced skin cell apoptosis.

2. Results

2.1. Eckol Showed Anti-oxidative Effects to Protect Cells from $PM_{2.5}$-Induced Apoptotic Cell Death

Previous studies have shown that eckol exhibited no cytotoxicity to HaCaT cells at a concentration of 30 μM [18] but showed antioxidant activity [6]. Therefore, in this study, we used 30 μM of eckol (Figure 1a) as the optimal concentration. From Figure 1b,c, it is evident that while $PM_{2.5}$ increased the levels of ROS as indicated by 2′,7′-dichlorofluorescein diacetate (DCF-DA) staining, eckol inhibited intracellular ROS generation. The results demonstrated that $PM_{2.5}$-induced ROS could accelerate cell apoptosis and death. To confirm that eckol could help cells escape from this damage, we checked nuclei integrity, and cell viability. According to results, $PM_{2.5}$ treatment led to sub-G_1 cell population after 24 h (Figure 1d), fragmented nuclei (Figure 1e), and cell death (Figure 1f). However, it was noted that following treatment with eckol, the apoptotic cell death was decreased, and the cell viability was also enhanced.

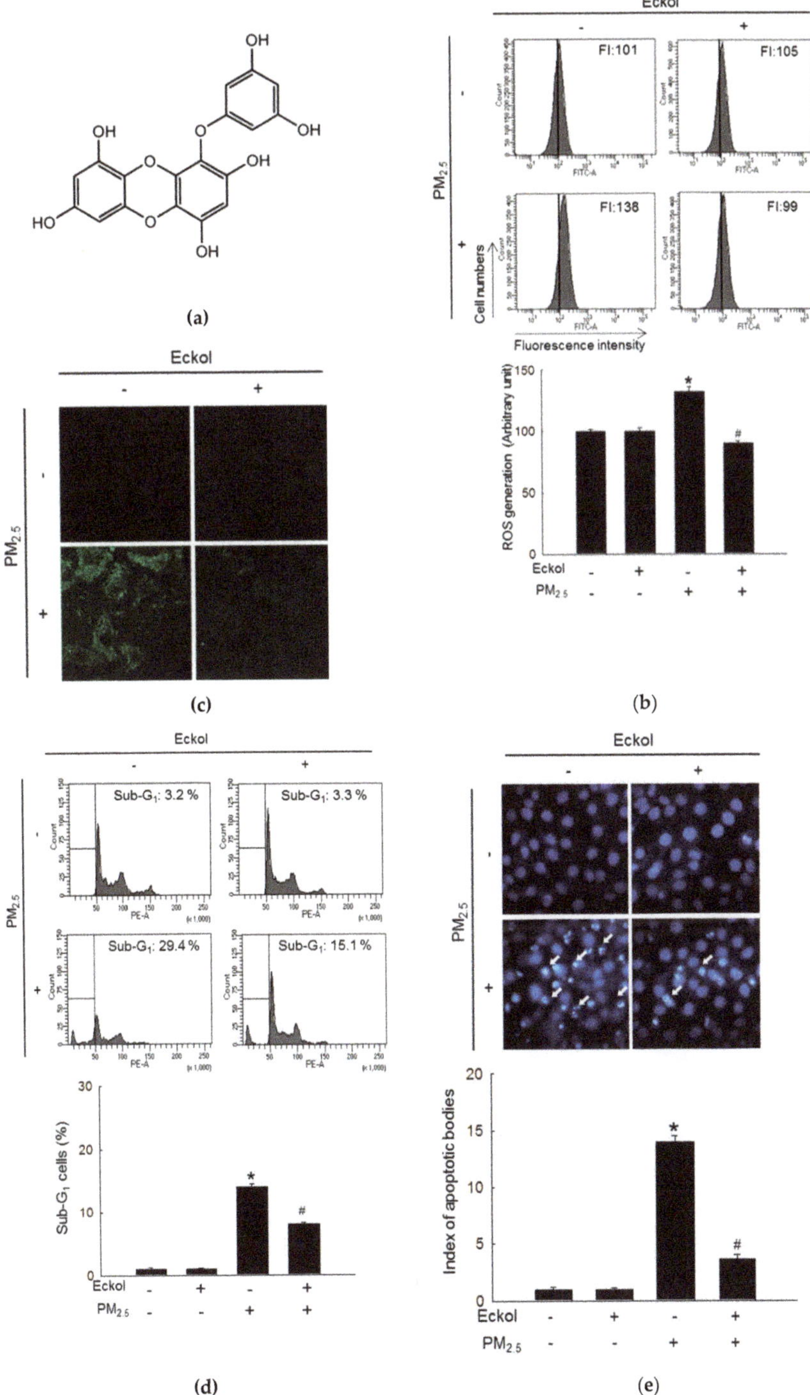

Figure 1. *Cont.*

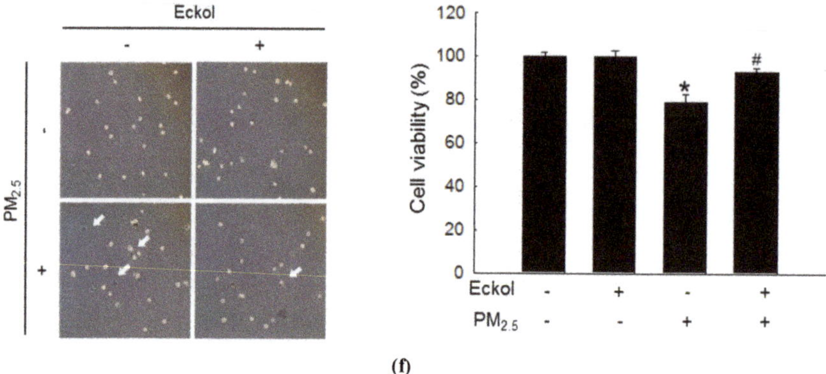

(f)

Figure 1. Eckol (30 µM) decreased cell apoptotic bodies by inhibiting PM$_{2.5}$-induced ROS level. (**a**) Chemical structure of eckol. Intracellular ROS level (DCF-DA staining) induced by PM$_{2.5}$ (50 µg/mL) was inhibited via treatment with eckol as observed by (**b**) flow cytometry and (**c**) confocal. (**d**) Sub-G$_1$ cell population induced by PM$_{2.5}$ was blocked by treatment with eckol, as determined by propidium iodide staining. (**e**) Apoptosis induced by PM$_{2.5}$ was reduced by treatment with eckol, observed by Hoechst 33342 staining. The arrow indicated the apoptotic bodies. (**f**) Cell deaths induced by PM$_{2.5}$ were reduced via treatment with eckol, determined by trypan blue assay. The arrow indicated the dead cell (stained cells by trypan blue). All experiments were performed after treatment with PM$_{2.5}$ for 24 h, and $n = 3$ for every group. * $p < 0.05$ and # $p < 0.05$ compared to control cells and PM$_{2.5}$-exposed cells, respectively.

2.2. Eckol Protected Cells against PM$_{2.5}$-Induced Intracellular Molecular Damage

Previous studies have shown that increment in ROS disrupted intracellular molecules involved in apoptosis [19,20]. Thus, we detected lipid peroxidation, protein carbonylation, and DNA damage. The confocal images show that PM$_{2.5}$ caused generation of phosphine oxide, which is a marker of lipid peroxidation. However, this was reversed by treatment with eckol (Figure 2a). Moreover, PM$_{2.5}$ aggravated protein carbonylation level, which was decreased by eckol treatment (Figure 2b). DNA lesions and strand breaks were studied by staining the cells with avidin-tetramethylrhodamine isothiocyanate (TRITC) conjugate (Figure 2c) and comet assay (Figure 2d). The data show that eckol guarded DNA against PM$_{2.5}$.

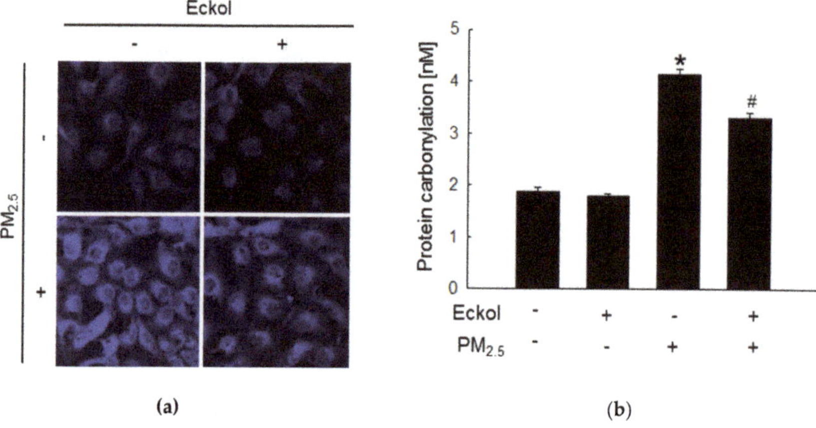

(a)   (b)

Figure 2. *Cont.*

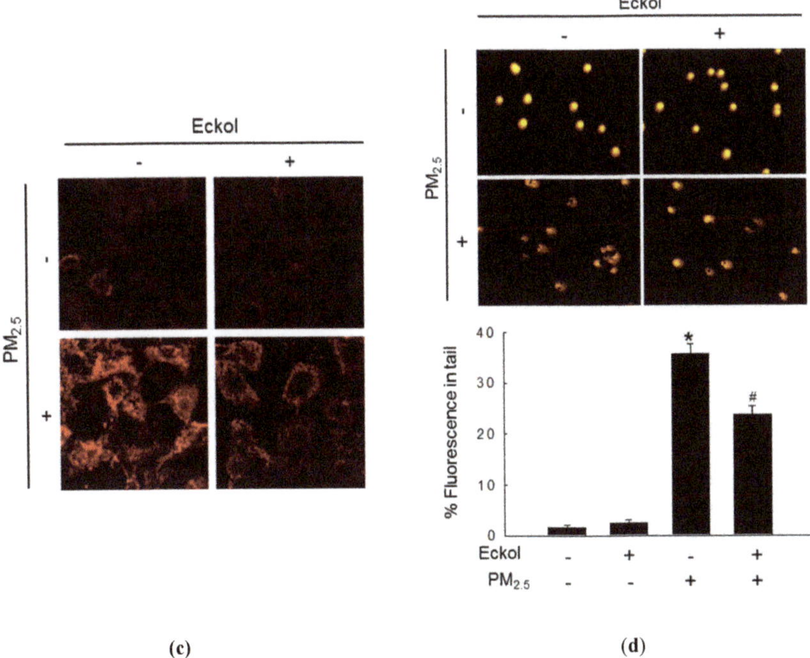

(c)   (d)

Figure 2. Eckol (30 μM) protected intracellular molecules from $PM_{2.5}$-induced damage. (**a**) Lipid oxidation induced by $PM_{2.5}$ was mitigated via treatment with eckol through diphenylpyrenylphosphine (DPPP) staining. (**b**) Protein carbonylation induced by $PM_{2.5}$ was declined via treatment with eckol as observed by a protein carbonylation assay. DNA damage induced by $PM_{2.5}$ was inhibited via treatment with eckol, as confirmed through (**c**) avidin-TRITC staining and (**d**) comet assay. All experiments were performed after treatment with $PM_{2.5}$ for 24 h, and n = 3 for every group. * $p < 0.05$ and # $p < 0.05$ compared to control cells and $PM_{2.5}$-exposed cells, respectively.

2.3. Eckol Prevented PM2.5-Induced Mitochondrial Dysfunction

Mitochondria play an important role in cellular energy production, and their biogenesis is related to synthesis of molecules, such as lipids and proteins, DNA transcription, and even cell apoptosis [21]. Next, we examined mitochondrial functions. Dihydrorhodamine 123 (DHR123) staining images show that mitochondrial ROS was accumulated in $PM_{2.5}$-treated group. Whereas, ROS level was decreased by pretreatment with eckol (Figure 3a). Both flow cytometry (Figure 3b) and confocal microscopy (Figure 3c) data demonstrate that $PM_{2.5}$ caused mitochondrial depolarization, which was arrested by treatment with eckol. Furthermore, the flux of mitochondrial calcium was increased in the $PM_{2.5}$-treatment group, and it was decreased in eckol-treatment group, which was monitored using the calcium indicator, Rhod-2 acetoxymethyl ester (Rhod-2 AM), by confocal microscopy (Figure 3d) and flow cytometry (Figure 3e).

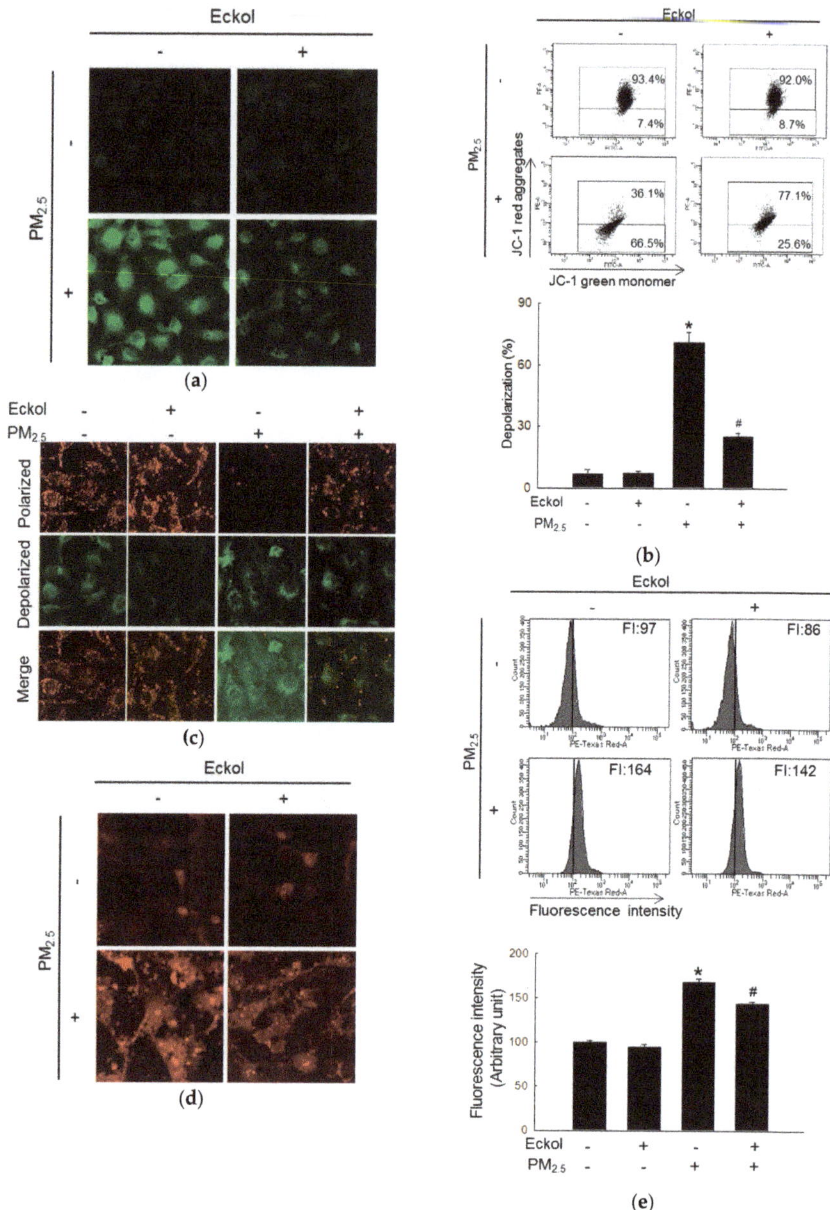

Figure 3. Eckol (30 μM) prevented PM$_{2.5}$-induced mitochondrial dysfunction by balancing mitochondrial membrane potential and calcium level. (**a**) Mitochondrial ROS induced by PM$_{2.5}$ was decreased via treatment with eckol through DHR123 staining. Depolarization of mitochondrial membrane potential (JC-1 staining) induced by PM$_{2.5}$ was repolarized via treatment with eckol through (**b**) flow cytometry and (**c**) confocal microscopy. Extra-mitochondrial Ca^{2+} (Rhod-2 AM staining) induced by PM$_{2.5}$ was blocked by treatment with eckol was monitored using (**d**) confocal microscopy and (**e**) flow cytometry. All experiments were performed after treatment with PM$_{2.5}$ for 24 h, and *n* = 3 for every group. * $p < 0.05$ and # $p < 0.05$ compared to control cells and PM$_{2.5}$-exposed cells, respectively.

2.4. Eckol Modulated PM$_{2.5}$-Induced Apoptotic Factors

It has been reported that urban particulate pollution penetrates the skin barrier and causes apoptosis in keratinocytes by activating caspase-3 [22]. Therefore, we evaluated the levels of the proapoptotic protein-Bax, antiapoptotic protein-Bcl-2, and cleaved caspase-3 (Figure 4a). The protein levels of Bax and activated caspase-3 were increased by PM$_{2.5}$, but expression of Bcl-2 was decreased by treatment with PM$_{2.5}$. However, these were reversed by eckol treatment. To investigate whether PM$_{2.5}$ could induce apoptosis, we counted apoptotic bodies via Hoechst 33342 dye staining (Figure 4b). The number of apoptotic cells in PM$_{2.5}$ group surged four times compared to that in the control group. However, both eckol and Z-VAD-FMK (the caspase inhibitor) halted the apoptotic bodies induced by PM$_{2.5}$.

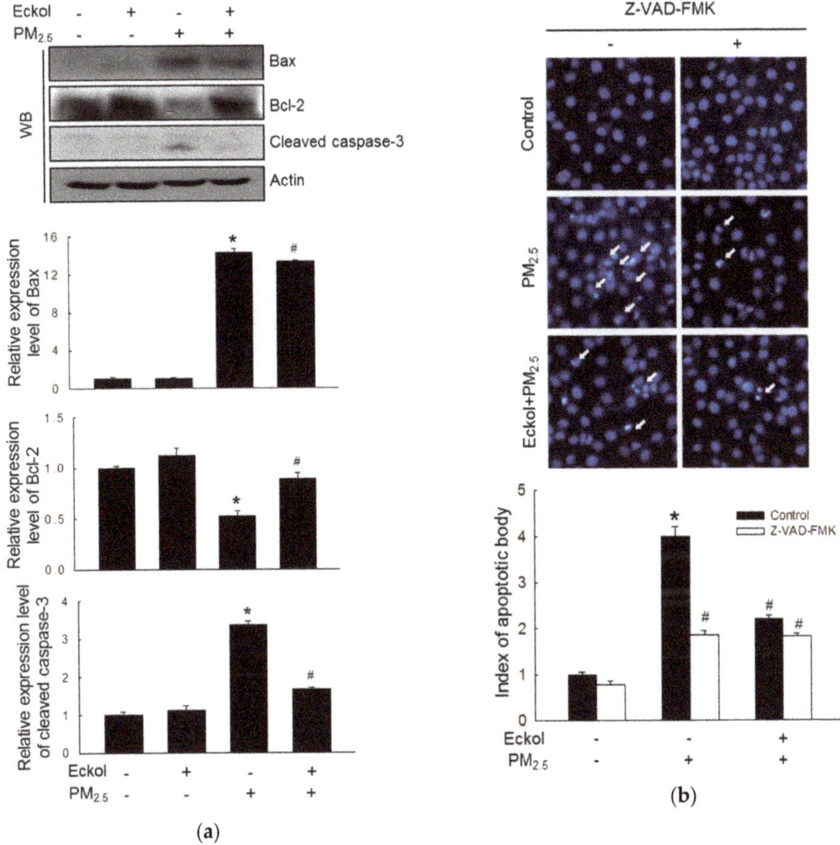

Figure 4. Eckol (30 μM) regulated apoptosis-related proteins induced by PM$_{2.5}$. (**a**) Increase of Bax and cleaved caspase-3 and decrease of Bcl-2 by PM$_{2.5}$ were reversed by treatment with eckol, as observed by western blotting (WB). (**b**) Apoptosis induced by PM$_{2.5}$ was reduced by treatment with eckol or caspase inhibitor (Z-VAD-FMK), as seen by Hoechst 33342 staining. All experiments were performed after treatment with PM$_{2.5}$ for 24 h, and $n = 3$ for every group. * $p < 0.05$ and # $p < 0.05$ compared to control cells and PM$_{2.5}$-exposed cells, respectively.

2.5. Eckol Reduced MAPK Signaling Pathway Activated by PM$_{2.5}$

In a review, Sun et al. have pointed out that many anti-cancer therapeutics induced apoptosis by modulating the MAPK/ERK signaling pathway [23]. Thus, we checked the expression levels of MAPK-related proteins, ERK, p38, and JNK, and the results showed that $PM_{2.5}$ could stimulate ERK, p38, and JNK (Figure 5a). However, eckol inhibited the activation of ERK, p38, and JNK. Next, we examined $PM_{2.5}$-induced apoptotic bodies by treatment with MAPK pathway inhibitors, U0126, SB203580, and SP600125 (inhibitors of ERK, p38, and JNK, respectively), and the results showed that all these three inhibitors could reduce the number of apoptotic bodies (Figure 5b). In addition, eckol enhanced the anti-apoptotic effect of MAPK-related inhibitors.

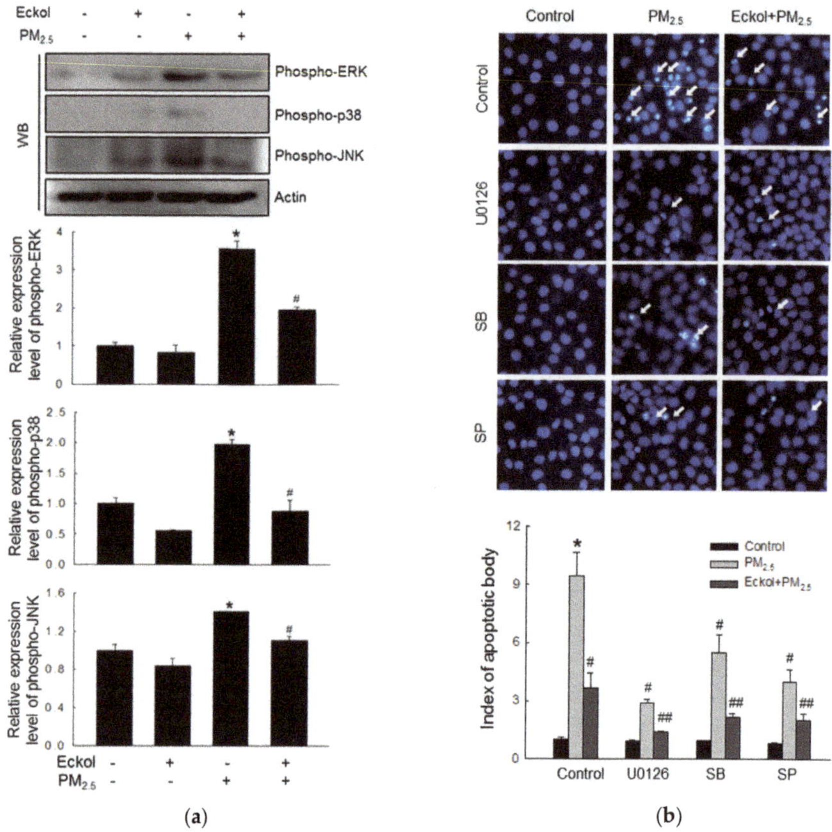

Figure 5. Eckol (30 µM) reduced $PM_{2.5}$-induced MAPK signaling pathway. (a) Western blot showing that activation of ERK, p38, and JNK induced by $PM_{2.5}$ was reversed via treatment with eckol. (b) Apoptosis induced by $PM_{2.5}$ was reduced by treatment with eckol or ERK, p38, and JNK inhibitors (U0126, SB203580 (SB), and SP600125 (SP), respectively), as observed through Hoechst 33342 staining. All experiments were performed after treatment with $PM_{2.5}$ for 24 h, and $n = 3$ for every group. * $p < 0.05$, # $p < 0.05$ and ## $p < 0.05$ compared to control cells, $PM_{2.5}$-exposed cells, and both inhibitor and $PM_{2.5}$-exposed cells respectively.

3. Discussion

There have been several investigations into the bioactivities of eckol, since it was first isolated from *Ecklonia cava* [3]. Eckol has multi-protective effects towards several cell lines, including lung fibroblast cells [24], human dermal fibroblasts [8], Chang liver cells [25], and human keratinocytes [6]. Furthermore, eckol is a compound with therapeutic potential in many areas, such as anti-oxidative stress [24], radioprotective action [26], antithrombotic and profibrinolytic activities [27], and anticancer activity [28]. Piao et al. studied $PM_{2.5}$-induced ROS generation at different concentrations (25, 50, 75, and 100 μg/mL) for 24 h in HaCaT keratinocytes, and found that $PM_{2.5}$ 50 μg/mL caused excessive ROS and skin dysfunction [29]. Usually, oxidative stress is caused by excessive accumulation of ROS or lack of the ability to eliminate them. $PM_{2.5}$ produces large amounts of ROS beyond the clearance ability of cells [30]. In our study, eckol showed its ability to protect cells against $PM_{2.5}$-induced ROS, cell cycle arrest, and apoptosis, and improved cell viability.

To explore the mechanism in detail, we checked the state of intracellular molecules such as lipids, protein, and DNA, which play various important roles in the cells [31]. Furthermore, intracellular macromolecular damage can be recognized as oxidative stress [32]. Our results demonstrated that $PM_{2.5}$ indeed induced oxidation of molecules, whereas eckol relieved intracellular molecular damage. The review also points out that mitochondria-dependent ROS generation subsequently caused cell cycle arrest and apoptosis, which is ROS-mediated apoptosis via mitochondrial mechanism [33]. In addition, our previous studies showed that calcium level and mitochondrial membrane potential affect the function of mitochondria [30,34]. The data in Figure 3 show that $PM_{2.5}$ increased the calcium level and depolarized mitochondrial membrane potential as compared to the control cells, whereas eckol regulated the mitochondria and maintained a stable state. The mechanism of mitochondrial damage is related to Bcl-2 proteins, which maintains mitochondrial membrane integrity [35]. The interaction between Bcl-2 and Bax also influences antiapoptosis [36]. Bcl-2 plays an anti-apoptotic role, whereas Bax is proapoptotic [37]. There is a complex crosslink between Bcl-2 family proteins and caspase proteins in cell apoptosis, in which Bcl-2 indirectly activates the caspase cascade [38]. The caspase-3 results in apoptosis induced by both extrinsic and intrinsic stimulus [39]. The results elucidated that except for the decrease of Bcl-2, Bax and cleaved caspase-3 (activated caspase-3) were increased by $PM_{2.5}$. However, eckol reversed these effects. Then, we treated cells with caspase inhibitor (Z-VAD-FMK) and found that upon pretreatment with caspase inhibitor, the apoptotic bodies were decreased significantly. These data prove that caspase proteins contributed to cell apoptosis induced by $PM_{2.5}$. MAPK signaling pathway plays a role in many systems of cell proliferation, migration, and apoptosis [23]. Many drugs are used to modulate MAPK signaling pathway to induce cell apoptosis in cells, such as lung cancer [40], human colorectal cancer [41], and cervical cancer HeLa cells [42]. Finally, we checked MAPK signaling pathway-related proteins, ERK, p38, and JNK. The results show that $PM_{2.5}$ activated all three proteins, but eckol exhibited the ability to inactivate them. When we used inhibitors of ERK, p38, and JNK to treat $PM_{2.5}$-damaged cells, the numbers of apoptotic bodies were decreased, similar to eckol. These data further prove that MAPK signaling pathway plays a vital role in the inhibition of $PM_{2.5}$-induced apoptosis by eckol.

Although the protective effects of eckol on human keratinocytes from $PM_{2.5}$-induced skin damage has been shown, there are limitations to this study. These results from in vitro experiments need to be validated by animal studies and clinical trials. Moreover, the concentration of air pollutants in the natural environment is different from the $PM_{2.5}$ purchased from the company, which provide certain ingredients for reference. In the future, there should be in vivo animal trials on skin protection to elucidate the protective effects and side effects of eckol under the complicated living environments.

4. Materials and Methods

4.1. Eckol and $PM_{2.5}$

Eckol was provided by Professor Nam Ho Lee of Jeju National University (Jeju, Korea), which belonged to Phaeophyceae. Preparation of the extract and its purification was following the reported protocol [43]. The dried brown alga *Ecklonia cava* was extracted with 80% methanol and the crude extract was purified by HPLC. After purification, 20 mg of pure eckol was obtained from 1 kg dry weight of the brown algae. A stock solution of eckol was prepared by dimethyl sulfoxide (DMSO). The NIST particulate matter SRM 1650b ($PM_{2.5}$) was bought from Sigma-Aldrich (St. Louis, MO, USA) and a stock solution in DMSO was prepared to obtain a concentration of $PM_{2.5}$ at 25 mg/mL. DMSO was as the control.

4.2. Cell Culture

The human HaCaT keratinocytes were purchased from Cell Lines Service (Heidelberg, Germany) and were grown in Dulbecco's modified Eagle's medium (Life Technologies Co., Grand Island, NY, USA) with 10% heat-inactivated fetal calf serum, streptomycin (100 µg/mL), and penicillin (100 units/mL). The cells were cultured at 37 °C in an incubator in an atmosphere containing 5% CO_2 [6].

4.3. ROS Detection

To examine the intracellular ROS scavenging ability of eckol, DCF-DA (Sigma-Aldrich) staining assay was performed. Cells (1.5×10^5 cells/mL) were treated with eckol (30 µM) for 30 min, $PM_{2.5}$ (50 µg/mL) for another 24 h, and DCF-DA (25 µM) sequentially. Then, the stained cells were detected by using a flow cytometer (Becton Dickinson, Franklin Lakes, NJ, USA) and confocal microscope (Carl Zeiss, Oberkochen, Germany). Similarly, mitochondrial ROS levels were detected by DHR123 (Molecular Probes) staining (10 µM) [30].

4.4. Sub-G_1 Cell Detection

Cells were treated with eckol (30 µM) and $PM_{2.5}$ (50 µg/mL) sequentially. After 24 h, cells were harvested and dyed with PI and RNase A (1:1000) for 30 min. Finally, the fluorescence emission was detected with a FACSCalibur flow cytometer (Becton Dickinson) [30].

4.5. Hoechst 33342 Staining

The apoptotic bodies were examined with Hoechst 33342 (Sigma-Aldrich), which is a nucleus-specific dye. All cells were stained with Hoechst 33342, and the images were captured under a Cool SNAP-Pro color digital camera (Media Cybernetics, Silver Spring, MD, USA) in a fluorescence microscope [44].

4.6. Cell Viability

Cells were cultured with eckol and/or $PM_{2.5}$ for 24 h, and the dead cells were stained with 0.1% trypan blue solution. Then, unstained bodies (live cells) and stained bodies (dead cells) were counted separately. Cell viability (%) was determined as: Live cells/ (live cells + dead cells) $\times$ 100% [45].

4.7. Lipid Peroxidation Assay

Cells (1.5×10^5 cells/mL) were cultured with eckol and/or $PM_{2.5}$ for 24 h in the chamber slides. The lipid hydroperoxides in cells were reacted with diphenylpyrenylphosphine (DPPP, Molecular Probes) and the lipid adducts of DPPP oxide were detected by a confocal microscope [30].

4.8. Protein Carbonylation Assay

Cells (1.5×10^5 cells/mL) were cultured with eckol and/or $PM_{2.5}$ for 24 h in the culture dish. The Oxiselect^{TM} protein carbonyl ELISA kit (Cell Biolabs, San Diego, CA, USA) was used to detect protein oxidation following the manufacturer's instructions [34].

4.9. Detection of 8-Oxoguanine (8-oxoG)

Cells (1.5×10^5 cells/mL) were cultured with eckol and/or $PM_{2.5}$ for 24 h in the chamber slides. The avidin-TRITC conjugate was used to detect 8-oxoG, an indicator of oxidative DNA damage. The stained cells were visualized under a confocal microscope [27].

4.10. Single Cell Gel Electrophoresis (Comet Assay)

Cells (1.5×10^5 cells/mL) were cultured with eckol and/or $PM_{2.5}$ for 30 min in the microtubes, and the harvested cells were fixed on a microscopic slide with low-melting agarose (1%). The slides with cells were permeated into lysis buffer (pH 10) for 1 h, which contained NaCl (2.5 M), Na-EDTA (100 mM), Tris (10 mM), Triton X-100 (1%), and DMSO (10%). Finally, the slides were subjected to electrophoresis, and the samples were dyed with ethidium bromide. The fluorescent images of the tails were captured with a fluorescence microscope and the tail lengths (50 cells per slide) were analyzed by the image analysis software (Kinetic Imaging, Komet 5.5, Liverpool, UK) [29].

4.11. Quantification of Ca^{2+} Level

Cells (1.5×10^5 cells/mL) were cultured with eckol and/or $PM_{2.5}$ for 24 h in the chamber slides. Cells were stained with Rhod-2 AM to check mitochondrial calcium levels, and the images were captured by confocal microscopy and flow cytometry [30].

4.12. Mitochondrial Membrane Potential ($\Delta\Psi m$) Analysis

Cells (1.5×10^5 cells/mL) were cultured with eckol and/or $PM_{2.5}$ for 24 h in the chamber slides. Cells were stained with 5,5′,6,6′-tetrachloro-1,1′,3,3′-tetraethylbenzimidazolylcarbocyanine iodide (JC-1, Invitrogen, Carlsbad, CA, USA) to observe changes in cell membrane potential. The fluorescence emission was analyzed by confocal microscopy and flow cytometry [6].

4.13. Western Blotting

The protein lysates from harvested cells were subjected to SDS-PAGE and transferred into membranes in sequence. The membranes were incubated with primary and secondary antibody (Pierce, Rockford, IL, USA) for 1 h separately. Protein bands were visualized via X-ray film (AGFA, Belgium). The following primary antibodies were used: caspase-3, phospho-p38, phospho-ERK, and phospho-JNK (Cell Signaling Technology, Beverly, MA, USA), Bax and Bcl-2 (Santa Cruz Biotechnology, Santa Cruz, CA, USA), and actin (Sigma-Aldrich) [44].

4.14. Statistical Analysis

All data were collected from three experiments and presented as mean ± standard error, which were analyzed by variance (ANOVA) with Tukey's test using Sigma Stat (v12) software (SPSS, Chicago, IL, USA). *P*-values < 0.05 were considered statistically significant.

5. Conclusions

$PM_{2.5}$ causes skin cell damage by generating ROS, excess of which oxidizes intracellular molecules and causes mitochondrial dysfunction, and activates MAPK signaling pathways. The skin is the first protection from various pollutants in the air, and it is important to identify effective biological compounds to prevent skin damage. Eckol has been known to inhibit UVB-induced ROS generation in keratinocytes [6]. In this study, we show that eckol could protect keratinocytes from $PM_{2.5}$-induced apoptosis by halting ROS generation, suggesting that eckol is a suitable candidate for skin protection and can be useful in the cosmetics industry as well as in medicine.

Author Contributions: A.X.Z., Y.J.H. and J.W.H. designed the experiments and wrote the paper; A.X.Z., M.J.P., K.A.K. and P.D.; S.M.F. performed the experiments; M.J.A., J.M.Y., H.K.K., Y.S.K. and N.H.L. contributed to the analytical tools.

Funding: This work was supported by a grant from the Basic Research Laboratory Program (NRF-2017R1A4A1014512) by the National Research Foundation of Korea (NRF) grant funded by the Korea government (MSIP).

Conflicts of Interest: The authors declare no conflict of interest.

References

1. Heo, S.J.; Ko, S.C.; Cha, S.H.; Kang, D.H.; Park, H.S.; Choi, Y.U.; Kim, D.; Jung, W.K.; Jeon, Y.J. Effect of phlorotannins isolated from *Ecklonia cava* on melanogenesis and their protective effect against photo-oxidative stress induced by UV-B radiation. *Toxicol. In Vitro* **2009**, *23*, 1123–1130. [CrossRef] [PubMed]

2. Kong, C.S.; Kim, J.A.; Ahn, B.N.; Kim, S.K. Potential effect of phloroglucinol derivatives from *Ecklonia cava* on matrix metalloproteinase expression and the inflammatory profile in lipopolysaccharide-stimulated human THP-1 macrophages. *Fish. Sci.* **2011**, *77*, 867–873. [CrossRef]

3. Ko, S.C.; Cha, S.H.; Heo, S.J.; Lee, S.H.; Kang, S.M.; Jeon, Y.J. Protective effect of *Ecklonia cava* on UVB-induced oxidative stress: In Vitro and In Vivo zebrafish model. *J. Appl. Phycol.* **2011**, *23*, 697–708. [CrossRef]

4. Sanjeewa, K.K.A.; Kim, E.A.; Son, K.T.; Jeon, Y.J. Bioactive properties and potential cosmeceutical applications of phlorotannins isolated from brown seaweeds: A review. *J. Photochem. Photobiol. B* **2016**, *162*, 100–105. [CrossRef] [PubMed]

5. Muhammad, K.; Mohamed, S. Ethanolic extract of *Eucheuma cottonii* promotes in vivo hair growth and wound healing. *J. Anim. Vet. Adv.* **2011**, *10*, 601–605.

6. Piao, M.J.; Lee, N.H.; Chae, S.; Hyun, J.W. Eckol inhibits ultraviolet B-induced cell damage in human keratinocytes via a decrease in oxidative stress. *Biol. Pharm. Bull.* **2012**, *35*, 873–880. [CrossRef] [PubMed]

7. Hwang, H.; Chen, T.; Nines, R.G.; Shin, H.C.; Stoner, G.D. Photochemoprevention of UVB-induced skin carcinogenesis in SKH-1 mice by brown algae polyphenols. *Int. J. Cancer* **2006**, *119*, 2742–2749. [CrossRef] [PubMed]

8. Joe, M.J.; Kim, S.N.; Choi, H.Y.; Shin, W.S.; Park, G.M.; Kang, D.W.; Kim, Y.K. The inhibitory effects of eckol and dieckol from *Ecklonia stolonifera* on the expression of matrix metalloproteinase-1 in human dermal fibroblasts. *Biol. Pharm. Bull.* **2006**, *29*, 1735–1739. [CrossRef] [PubMed]

9. Kim, K.H.; Jahan, S.A.; Kabir, E. A review on human health perspective of air pollution with respect to allergies and asthma. *Environ. Int.* **2013**, *59*, 41–52. [CrossRef]

10. Ngoc, L.; Park, D.; Lee, Y.; Lee, Y.C. Systematic review and meta-analysis of human skin diseases due to particulate matter. *Int. J. Environ. Res. Public Health* **2017**, *14*, 1458. [CrossRef]

11. Guaita, R.; Pichiule, M.; Mate, T.; Linares, C.; Diaz, J. Short-term impact of particulate matter (PM2.5) on respiratory mortality in Madrid. *Int. J. Environ. Health Res.* **2011**, *21*, 260–274. [CrossRef] [PubMed]

12. Halonen, J.I.; Lanki, T.; Tuomi, T.Y.; Tiittanen, P.; Kulmala, M.; Pekkanen, J. Particulate air pollution and acute cardiorespiratory hospital admissions and mortality among the elderly. *Epidemiology* **2009**, *20*, 143–153. [CrossRef] [PubMed]

13. Perez, L.; Tobías, A.; Querol, X.; Pey, J.; Alastuey, A. Saharan dust, particulate matter and cause-specific mortality: A case-crossover study in Barcelona (Spain). *Environ. Int.* **2012**, *48*, 150–155. [CrossRef] [PubMed]

14. Atkinson, R.W.; Fuller, G.W.; Anderson, H.R.; Harrison, R.M.; Armstrong, B. Urban ambient particle metrics and health: A time series analysis. *Epidemiology* **2010**, *21*, 501–511. [CrossRef] [PubMed]

15. Kim, K.H.; Kabir, E.; Kabir, S. A review on the human health impact of airborne particulate matter. *Environ. Int.* **2015**, *74*, 136–143. [CrossRef] [PubMed]

16. Kim, K.E.; Cho, D.; Park, H.J. Air pollution and skin diseases: Adverse effects of airborne particulate matter on various skin diseases. *Life Sci.* **2016**, *152*, 126–134. [CrossRef] [PubMed]

17. Manandhar, B.; Paudel, P.; Seong, S.H.; Jung, H.A.; Choi, J.S. Characterizing eckol as a therapeutic aid: A systematic review. *Mar. Drugs* **2019**, *17*, 361. [CrossRef]

18. Kang, N.J.; Koo, D.H.; Kang, G.J.; Han, S.C.; Lee, B.W.; Koh, Y.S.; Hyun, J.W.; Lee, N.H.; Ko, M.H.; Kang, H.K.; et al. Dieckol, a component of *Ecklonia cava*, suppresses the production of MDC/CCL22 via down-regulating STAT1 pathway in interferon-γ stimulated HaCaT human keratinocytes. *Biomol. Ther. (Seoul)* **2015**, *23*, 238–244.

19. Hyun, Y.J.; Piao, M.J.; Kang, K.A.; Zhen, A.X.; Madushan Fernando, P.D.S.; Kang, H.K.; Ahn, Y.S.; Hyun, J.W. Effect of fermented fish oil on fine particulate matter-induced skin aging. *Mar. Drugs* **2019**, *17*, 61. [CrossRef]

20. Ghosh, D.; LeVault, K.R.; Barnett, A.J.; Brewer, G.J. A reversible early oxidized redox state that precedes macromolecular ROS damage in aging nontransgenic and 3xTg-AD mouse neurons. *J. Neurosci.* **2012**, *32*, 5821–5832. [CrossRef]

21. Shin, E.J.; Tran, H.Q.; Nguyen, P.T.; Jeong, J.H.; Nah, S.Y.; Jang, C.G.; Nabeshima, T.; Kim, H.C. Role of mitochondria in methamphetamine-induced dopaminergic neurotoxicity: Involvement in oxidative stress, Neuroinflammation, and pro-apoptosis-a review. *Neurochem. Res.* **2017**, *43*, 57–69. [CrossRef] [PubMed]

22. Pan, T.L.; Wang, P.W.; Aljuffali, I.A.; Huang, C.T.; Lee, C.W.; Fang, J.Y. The impact of urban particulate pollution on skin barrier function and the subsequent drug absorption. *J. Dermatol. Sci.* **2015**, *78*, 51–60. [CrossRef] [PubMed]

23. Sun, Y.; Liu, W.Z.; Liu, T.; Feng, X.; Yang, N.; Zhou, H.F. Signaling pathway of MAPK/ERK in cell proliferation, differentiation, migration, senescence and apoptosis. *J. Recept. Signal Transduct. Res.* **2015**, *35*, 600–604. [CrossRef] [PubMed]

24. Kang, K.A.; Lee, K.H.; Chae, S.; Zhang, R.; Jung, M.S.; Lee, Y.; Kim, S.Y.; Kim, H.S.; Joo, H.G.; Park, J.W.; et al. Eckol isolated from *Ecklonia cava* attenuates oxidative stress induced cell damage in lung fibroblast cells. *FEBS Lett.* **2005**, *579*, 6295–6304. [CrossRef] [PubMed]

25. Kim, A.D.; Kang, K.A.; Piao, M.J.; Kim, K.C.; Zheng, J.; Yao, C.W.; Cha, J.W.; Hyun, C.L.; Kang, H.K.; Lee, N.H.; et al. Cytoprotective effect of eckol against oxidative stress-induced mitochondrial dysfunction: Involvement of the FoxO3a/AMPK pathway. *J. Cell. Biochem.* **2014**, *115*, 1403–1411. [CrossRef] [PubMed]

26. Park, E.; Ahn, G.N.; Lee, N.H.; Kim, J.M.; Yun, J.S.; Hyun, J.W.; Jeon, Y.J.; Wie, M.B.; Lee, Y.J.; Park, J.W.; et al. Radioprotective properties of eckol against ionizing radiation in mice. *FEBS Lett.* **2008**, *582*, 925–930. [CrossRef] [PubMed]

27. Kim, T.H.; Ku, S.K.; Bae, J.S. Antithrombotic and profibrinolytic activities of eckol and dieckol. *J. Cell. Biochem.* **2012**, *113*, 2877–2883. [CrossRef] [PubMed]

28. Hyun, K.H.; Yoon, C.H.; Kim, R.K.; Lim, E.J.; An, S.; Park, M.J.; Hyun, J.W.; Suh, Y.; Kim, M.J.; Lee, S.J. Eckol suppresses maintenance of stemness and malignancies in glioma stem-like cells. *Toxicol. Appl. Pharmacol.* **2011**, *254*, 32–40. [CrossRef] [PubMed]

29. Piao, M.J.; Ahn, M.J.; Kang, K.A.; Ryu, Y.S.; Hyun, Y.; Shilnikova, K.; Zhen, A.X.; Jeong, J.W.; Choi, Y.H.; Kang, H.K.; et al. Particulate matter 2.5 damages skin cells by inducing oxidative stress, subcellular organelle dysfunction, and apoptosis. *Arch. Toxicol.* **2018**, *92*, 2077–2091. [CrossRef] [PubMed]

30. Zhen, A.X.; Piao, M.J.; Hyun, Y.J.; Kang, K.A.; Madushan Fernando, P.D.S.; Cho, S.J.; Ahn, M.J.; Hyun, J.W. Diphlorethohydroxycarmalol attenuates fine particulate matter-induced subcellular skin dysfunction. *Mar. Drugs* **2019**, *17*, 95. [CrossRef] [PubMed]

31. Jorge, A.T.; Arroteia, K.F.; Lago, J.C.; de Sa-Rocha, V.M.; Gesztesi, J.; Moreira, P.L. A new potent natural antioxidant mixture provides global protection against oxidative skin cell damage. *Int. J. Cosmet. Sci.* **2011**, *33*, 113–119. [CrossRef] [PubMed]

32. Trachana, V.; Petrakis, S.; Fotiadis, Z.; Siska, E.K.; Balis, V.; Gonos, E.S.; Kaloyianni, M.; Koliakos, G. Human mesenchymal stem cells with enhanced telomerase activity acquire resistance against oxidative stress-induced genomic damage. *Cytotherapy* **2017**, *19*, 808–820. [CrossRef] [PubMed]

33. Kiang, J.G.; Olabisi, A.O. Radiation: A poly-traumatic hit leading to multi-organ injury. *Cell Biosci.* **2019**, *9*, 25. [CrossRef] [PubMed]

34. Zhen, A.X.; Piao, M.J.; Hyun, Y.J.; Kang, K.A.; Ryu, Y.S.; Cho, S.J.; Kang, H.K.; Koh, Y.S.; Ahn, M.J.; Kim, T.H.; et al. Purpurogallin protects keratinocytes from damage and apoptosis induced by Ultraviolet B radiation and particulate matter 2.5. *Biomol. Ther.* **2019**, *27*, 395–403. [CrossRef] [PubMed]

35. Zheng, J.H.; Viacava Follis, A.; Kriwacki, R.W.; Moldoveanu, T. Discoveries and controversies in BCL-2 protein-mediated apoptosis. *FEBS J.* **2016**, *283*, 2690–2700. [CrossRef] [PubMed]

36. Moldoveanu, T.; Follis, A.V.; Kriwacki, R.W.; Green, D.R. Many players in BCL-2 family affairs. *Trends Biochem. Sci.* **2014**, *39*, 101–111. [CrossRef] [PubMed]

37. Czabotar, P.E.; Lessene, G.; Strasser, A.; Adams, J.M. Control of apoptosis by the BCL-2 protein family: Implications for physiology and therapy. *Nat. Rev. Mol. Cell Biol.* **2014**, *15*, 49–63. [CrossRef]

38. Elmore, S. Apoptosis: A review of programmed cell death. *Toxicol. Pathol.* **2007**, *35*, 495–516. [CrossRef]

39. Asweto, C.O.; Wu, J.; Alzain, M.A.; Hu, H.; Andrea, S.; Feng, L.; Yang, X.; Duan, J.; Sun, Z. Cellular pathways involved in silica nanoparticles induced apoptosis: A systematic review of In Vitro studies. *Environ. Toxicol. Pharmacol.* **2017**, *56*, 191–197. [CrossRef]

40. Jeanson, A.; Boyer, A.; Greillier, L.; Tomasini, P.; Barlesi, F. Therapeutic potential of trametinib to inhibit the mutagenesis by inactivating the protein kinase pathway in non-small cell lung cancer. *Expert Rev. Anticancer Ther.* **2018**, *4*, 1–7. [CrossRef]

41. Pan, H.; Wang, Y.; Na, K.; Wang, Y.; Wang, L.; Li, Z.; Guo, C.; Guo, D.; Wang, X. Autophagic flux disruption contributes to Ganoderma lucidum polysaccharide-induced apoptosis in human colorectal cancer cells via MAPK/ERK activation. *Cell Death Dis.* **2019**, *10*, 456. [CrossRef] [PubMed]

42. Yao, W.; Lin, Z.; Wang, G.; Li, S.; Chen, B.; Sui, Y.; Huang, J.; Liu, Q.; Shi, P.; Lin, X.; et al. Delicaflavone induces apoptosis via mitochondrial pathway accompanying G2/M cycle arrest and inhibition of MAPK signaling cascades in cervical cancer HeLa cells. *Phytomedicine* **2019**, *62*, 152973. [CrossRef] [PubMed]

43. Moon, C.; Kim, S.H.; Kim, J.C.; Hyun, J.W.; Lee, N.H.; Park, J.W.; Shin, T. Protective effect of phlorotannin components phloroglucinol and eckol on radiation-induced intestinal injury in mice. *Phytother. Res.* **2008**, *22*, 238–242. [CrossRef] [PubMed]

44. Han, X.; Kang, K.A.; Piao, M.J.; Zhen, A.X.; Hyun, Y.J.; Kim, H.M.; Ryu, Y.S.; Hyun, J.W. Shikonin exerts cytotoxic effects in human colon cancers by inducing apoptotic cell death via the endoplasmic reticulum and mitochondria-mediated pathways. *Biomol. Ther.* **2019**, *27*, 41–47. [CrossRef] [PubMed]

45. Kim, D.Y.; Kim, J.H.; Lee, J.C.; Won, M.H.; Yang, S.R.; Kim, H.C.; Wie, M.B. Zinc oxide nanoparticles exhibit both cyclooxygenase- and lipoxygenase-mediated apoptosis in human bone marrow-derived mesenchymal stem cells. *Toxicol. Res.* **2019**, *35*, 83–91. [CrossRef] [PubMed]

Article

Anticoagulant Activity of Sulfated Ulvan Isolated from the Green Macroalga *Ulva rigida*

Amandine Adrien [1,2], Antoine Bonnet [1], Delphine Dufour [2], Stanislas Baudouin [2], Thierry Maugard [1] and Nicolas Bridiau [1,*]

[1] Equipe BCBS (Biotechnologies et Chimie des Bioressources pour la Santé), La Rochelle Université, UMR CNRS 7266 LIENSs, Avenue Michel Crépeau, 17042 La Rochelle, France; amandine.adrien@seprosys.com (A.A.); antoine.bonnet@univ-lr.fr (A.B.); thierry.maugard@univ-lr.fr (T.M.)

[2] SEPROSYS, Séparations, Procédés, Systèmes, 12 Rue Marie-Aline Dusseau, 17000 La Rochelle, France; delphine.dufour@seprosys.com (D.D.); stanislas.baudouin@seprosys.com (S.B.)

* Correspondence: nicolas.bridiau@univ-lr.fr

Received: 2 April 2019; Accepted: 7 May 2019; Published: 14 May 2019

Abstract: (1) Background: Brown and red algal sulfated polysaccharides have been widely described as anticoagulant agents. However, data on green algae, especially on the *Ulva* genus, are limited. This study aimed at isolating ulvan from the green macroalga *Ulva rigida* using an acid- and solvent-free procedure, and investigating the effect of sulfate content on the anticoagulant activity of this polysaccharide. (2) Methods: The obtained ulvan fraction was chemically sulfated, leading to a doubling of the polysaccharide sulfate content in a second ulvan fraction. The potential anticoagulant activity of both ulvan fractions was then assessed using different assays, targeting the intrinsic and/or common (activated partial thromboplastin time), extrinsic (prothrombin time), and common (thrombin time) pathways, and the specific antithrombin-dependent pathway (anti-Xa and anti-IIa), of the coagulation cascade. Furthermore, their anticoagulant properties were compared to those of commercial anticoagulants: heparin and Lovenox®. (3) Results: The anticoagulant activity of the chemically-sulfated ulvan fraction was stronger than that of Lovenox® against both the intrinsic and extrinsic coagulation pathways. (4) Conclusion: The chemically-sulfated ulvan fraction could be a very interesting alternative to heparins, with different targets and a high anticoagulant activity.

Keywords: *Ulva rigida*; ulvan; chemical sulfation; anticoagulant activity

---

1. Introduction

Marine macroalgae are used in several industrial applications and have represented a sharply increasing annual market over the last decades, going from about USD 6 billion in 2003 to currently USD 10.6 billion, and exhibiting an annual growth rate close to 10%, due to the high expansion of the aquaculture sector, which represents today 97% of the global seaweed production worldwide [1–3]. Seaweed and seaweed products are mainly used for human consumption (85%) and in the hydrocolloid industry. The three main groups of marine algae (Phaeophycae, Rhodophyta, and Chlorophyta) are rich sources of various compounds of interest such as proteins, pigments, and polysaccharides, whose potential market is estimated at several USD billion [3–5], and have been studied for their potential pharmaceutical, cosmetic, or nutraceutical applications. Polysaccharides from brown and red marine macroalgae are widely used in the industry for their gelling properties, alginates, agar, and carrageenans in particular [1,3]. It has also been shown that sulfated polysaccharides from marine macroalgae have numerous biological activities, including immunoinflammatory [6–8], antioxidant [9,10], antitumor [11–13], antiviral [7,14–16], and anticoagulant [17–20] properties. Conversely, although they have been consumed for centuries, green macroalgae are still relatively unexploited. Nevertheless, the discovery of ulvans, the sulfated polysaccharides from green algae of the *Ulva* genus (including species from the formerly genus *Enteromorpha*), has increased the interest for these seaweeds. Ulvans are mainly composed of rhamnose and uronic acids (glucuronic or iduronic acid) [21,22]. The main repeated disaccharide units of ulvans are [→4)-β-D-Glc$p$-(1→4)-α-l-Rha$p$3S-(1→]$_n$ (type A) and [→4)-α-l-Ido$p$-(1→4)-α-l-Rha$p$3S-(1→]$_n$ (type B) [7].

Heparin, a polysaccharide belonging to the glycosaminoglycan family, is mainly composed of L-iduronic-2-O-sulfate acid and D-glucosamine-N-sulfate, 6-O-sulfate. Unfractionated heparin (UFH) has an average molecular weight (MW) of 15 kDa [23]. Its anticoagulant activity is primarily due to its specific antithrombin-binding pentasaccharide sequence, where the central glucosamine is not only 2- and 6-O-sulfated but also 3-O-sulfated [24]. Despite of its major anticoagulant activity, UFH may cause serious adverse events (AE), such as heparin-induced thrombocytopenia [25] or hemorrhage [26]. Moreover, its low bioavailability [27] makes such a treatment really expensive. To reduce the risks of AE, heparin may be depolymerized to produce smaller molecules, known as low-molecular-weight heparins (LMWH, MW$_{avg}$ ≥ ~6 kDa) [28]. Although LMWH significantly reduce the risks of AE associated with UFH, their recommended use as first-line treatment for cancer-associated venous thromboembolism (VTE), for at least 3 to 6 months, still exposes patients to allergic reactions, heparin-induced thrombocytopenia, recurrent VTE, and major bleeding events, leading to high rates of treatment discontinuation [29,30]. Given the risks and high costs of these treatments, there is a compelling need for further investigations to identify new sources of anticoagulants.

The first studies assessing the anticoagulant activity of sulfated polysaccharides from macroalgae have shown that polysaccharides from brown algae could be alternative sources of new anticoagulant compounds. Sulfated fucoidans from brown algae and sulfated galactans from red seaweeds (also referred to as carrageenans) seem to have a strong anticoagulant activity. Data on the anticoagulant activity of polysaccharides from green macroalgae are limited compared to brown and red algae, but a few studies have demonstrated their potential. Regarding the anticoagulant activity, the most studied genera are *Codium* [31–36], which contains sulfated arabinans and arabinogalactans, and then *Monostroma*, which contains sulfated rhamnans [37–42]. Only two studies have shown the potential anticoagulant properties of ulvans extracted from *Ulva conglobata* and *Ulva reticulata* [43,44].

The relationship between the polysaccharide chemical structure and the anticoagulant activity is complex but previous studies have shown that several factors, such as MW, osidic composition, and sulfate content and substitution pattern, may significantly affect the anticoagulant activity [17,45–47]. Polysaccharide sulfate content appears to have a major impact on its anticoagulant potential. Thus, Cianca et al. have demonstrated that sulfated galactans extracted from *Codium fragile* and *Codium vermilara* with a high MW and high sulfate content have a higher anticoagulant activity than LMW and low sulfate content polysaccharides [36]. Furthermore, the only published studies that have highlighted a strong activity of sulfated rhamnans from the *Ulva* genus were based on highly sulfated ulvans (sulfate content of 26–35%) [43,44].

In this study, a solvent- and acid-free process to extract and purify ulvan from *Ulva rigida* was developed. To study the effect of sulfate content, a chemical sulfation procedure based on the sulfur-trioxide pyridine complex method was performed on the ulvan fraction obtained. The anticoagulant activity of both the native and chemically-sulfated ulvan fractions was then assessed against the intrinsic and/or common (activated partial thromboplastin time: APTT), extrinsic (prothrombin time: PT), and common (thrombin time: TT) pathways, and the specific antithrombin-dependent pathway (anti-Xa and anti-IIa), of the coagulation cascade.

2. Results and Discussion

2.1. Extraction, Purification, and Sulfation of Ulvan from U. rigida

The extraction of ulvan from the green macroalga U. rigida was carried out in hot water and was followed by a multistep purification procedure previously developed [48]: ultrafiltration,

ion exchange, and precipitation of remaining proteins. After neutralization with 1 M NaOH, the sulfated polysaccharides were recovered in a sodium salt form in the ULVAN-01 fraction, with an extraction yield of about 5% of the non-desalinated dry biomass (initially containing about 50% of salt).

The sulfation procedure was finally carried out starting from 500 mg of ULVAN-01 fraction, to give 200 mg of ULVAN-02 fraction.

2.2. Chemical Characterization of the ULVAN-01 and ULVAN-02 Fractions

The efficacy of the extraction and purification procedure of ulvan from *U. rigida* in terms of sulfated polysaccharide production was high with less than 4% of remaining contaminant proteins in the ULVAN-01 fraction. The purity of the polysaccharide was even higher after chemical sulfation with a percentage of remaining proteins less than 1% in the ULVAN-02 fraction (Table 1). This better purification was very likely due to the last step of the sulfation procedure, i.e., ethanol precipitation of sulfated polysaccharides.

Table 1. Chemical composition of the ULVAN-01 and ULVAN-02 fractions.

		ULVAN-01	ULVAN-02
Neutral sugars		41	33
Uronic acids		34	31
Sulfates on the polysaccharide backbone (sulfate to uronic acid molar ratio)	% (w/w$_{dry\ fraction}$)	11 (0.65)	20 (1.30)
Proteins		4	<1
Polyphenols		0	0
Lipids		0	0
Ashes		23	nd [d]
	Glc/Gal	12.2	nd [d]
	Xyl	8.0	nd [d]
Monosaccharide composition (% molar ratio)	Rha	42.6	nd [d]
	GlcN	6.9	nd [d]
	GlcA/IdoA	30.3	nd [d]
M_n [a] (kDa)		31.3	39.0
M_w [b] (kDa)		56.7	55.3
I [c]		1.8	1.4

[a] number-averaged molecular weight; [b] weight-averaged molecular weight; [c] polydispersity index; [d] not determined.

Both ulvan fractions contained about 30% of uronic acids, a value that was in the higher range of what could be observed in ulvans from *Ulva* species [49]. Conversely, the sulfate content of about 11%, corresponding to a sulfate to uronic acid molar ratio of 0.65, was in the lower range of ulvan sulfate content [22,49]: this could be explained by the harsh conditions of extraction, ultrafiltration and protein precipitation used during the isolation process (80 °C, 5 bars), which may have led to partial desulfation. ULVAN-01 was mainly composed of rhamnose and glucuronic and/or iduronic acid, and to a lesser extent of glucose and/or galactose, xylose, and glucosamine. It is noteworthy, however, that chemical sulfation allowed to almost double the initial sulfate content of ULVAN-01: indeed ULVAN-02 exhibited 20% of sulfate groups on the polysaccharide backbone, for a sulfate to uronic acid molar ratio of 1.30. Moreover, a very interesting feature of the extraction procedure developed by SEPROSYS was the high purity of the resulting ulvan in the ULVAN-01 fraction, which was estimated at 86% by adding the contents of neutral sugars, uronic acids, and sulfate esters, taking into account that sulfates are not burned at 550 °C and are also part of the ash content, together with sodium cations.

2.3. Structural Characterization of the ULVAN-01 and ULVAN-02 Fractions

A M_w of about 57 kDa was estimated by SEC (Figure 1 and Table 1) for ULVAN-01, based on a calibration curve of dextran standards. After chemical sulfation of ULVAN-01 using sulfur trioxide pyridine complex in pyridine and dimethylformamide (DMF), the resulting ulvan in ULVAN-02 had an estimated M_w of about 55 kDa, slightly lower than ULVAN-01 M_w. SEC chromatograms of ULVAN-01 and ULVAN-02 were very similar, except that the major peak eluted from 30 to 40 min seemed narrower for ULVAN-02 and eluted slightly sooner than that of ULVAN-01. On the other hand, a peak eluted at about 44 min, corresponding to oligosaccharides, was only found in ULVAN-01, while a peak eluted at about 47 min, likely corresponding to low-molecular-weight oligosaccharides, was only found in ULVAN-02. Besides, the polydispersity of ULVAN-02 (I = 1.4) was lower than that of ULVAN-01 (I = 1.8). The fact that the sulfate content in ULVAN-02 was double to the respective measured in ULVAN-01 should highly affect M_W, considering the hydrodynamic size of the polysaccharide that is strongly affected by sulfate groups, due to anionic charges. Therefore, the weak difference between ULVAN-01 and ULVAN-02 M_w may appear very surprising. However, these observations could be explained by a slight cleavage of the glycosidic linkages in the polysaccharides that likely occurred during chemical sulfation, as proposed by Nishino and Nagumo [50]. This would lead to the transformation of bigger polysaccharides into polysaccharides of intermediary MW, and therefore to the increase in M_n observed in ULVAN-02, concomitant with the decrease in polydispersity index, resulting in the narrower shape of the major peak eluted from 30 to 40 min. This would also cause the depolymerization of oligosaccharides to smaller oligosaccharides, which might explain the appearance of the peak eluted at 47 min for ULVAN-02 as the peak eluted at 44 min found in ULVAN-01 disappeared. Furthermore, this would be in accordance with the slightly lower value of ULVAN-02 M_w as small oligosaccharides strongly affect the M_w value.

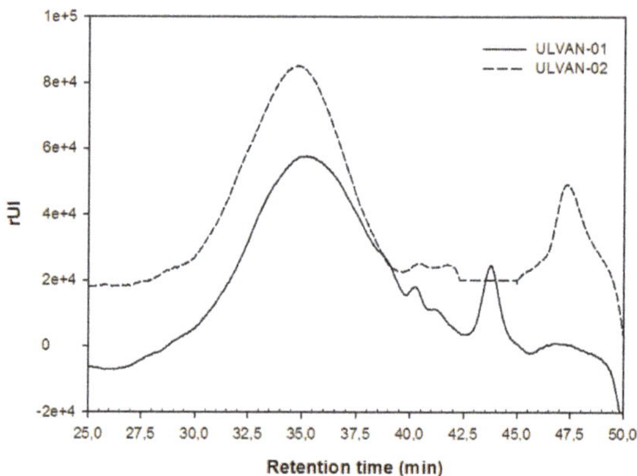

Figure 1. Size exclusion chromatograms of the ULVAN-01 and ULVAN-02 fractions.

ULVAN-01 was also characterized in a previous work by ^{13}C NMR [48]. The chemical shifts were attributed on the basis of references reporting assignments of ulvans and oligosaccharides [51]. As previously reported, the assignment signals corresponding to the carbons of both type A (β-D-Glc*p*-(1→4)-α-l-Rha*p*3S) and type B (α-l-Ido*p*-(1→4)-α-l-Rha*p*3S) ulvanobiouronic acid 3-sulfate were detected, confirming the presence of this sequence. On the other hand, the ultra high pressure liquid chromatography-high resolution mass spectrometry (UHPLC-HRMS) analyses of the 9kDa ulvan fraction obtained after controlled depolymerization of ULVAN-01 revealed characteristic ions that could be attributed after selective pseudo-MS^3 fragmentation to oligosaccharide sequences with a polymerization degree up to 12 (Table 2).

Table 2. Oligosaccharide sequences identified by ultra high pressure liquid chromatography-mass spectrometry untargeted in-source fragmentation (UHPLC-MS^2) from ULVAN-01 (sequences 1–4 [48]) and ULVAN-02 (sequences 5–8).

Oligosaccharide Sequence	Ion (m/z)			Ionization	Molecular Formula	Symbolic Representation of Oligosaccharide Fragments [a] (CFG Nomenclature [b])
	Experimental Value	Predicted Value	Resolution (ppm)			◁ Rha, ◇ GlcA (or IdoA) [c] ● Glc (or Gal) [c], ☆ Xyl S = Sulfate Group
1	1561.3492	1561.3516	1.54	$[M - H]^-$	$C_{54}H_{81}O_{50}S$	
2	1561.3492	1561.3516	1.54	$[M - H]^-$	$C_{54}H_{81}O_{50}S$	
3	1723.4083	1723.4045	2.20	$[M - H]^-$	$C_{60}H_{91}O_{55}S$	
4	1839.4554	1839.4518	1.95	$[M - H]^-$	$C_{65}H_{99}O_{58}S$	
5	401.0385	401.0390	1.25	$[M - H - H_2O]^-$	$C_{12}H_{17}O_{13}S$	
6	502.9772	502.9777	0.99	$[M - 2H - H_2O + Na]^-$	$C_{12}H_{16}O_{16}S_2Na$	
7	604.9165	604.9165	0.00	$[M - 3H - H_2O + 2Na]^-$	$C_{12}H_{15}O_{19}S_3Na_2$	
8	706.8561	706.8553	1.13	$[M - 4H - H_2O + 3Na]^-$	$C_{12}H_{14}O_{22}S_4Na_3$	

[a] Identified ions correspond to oligosaccharide fragments derived from untargeted fragmentation in the ionization source of the mass spectrometer; [b] Consortium for Functional Glycomics. Symbolic representations were obtained using Glycoworkbench software [52]; [c] GlcA and IdoA residues, as well as Glc and Gal, are isomers and therefore cannot be distinguished by MS analysis: symbols used for GlcA et Glc can thus correspond to IdoA and Gal, respectively, two monosaccharides that can be found in ulvans but in smaller proportions.

Sequences 1 to 4 were determined by pseudo-MS^3 analysis and shown to exhibit a repeated disaccharide sequence, sulfated or not, composed of a rhamnose moiety and an uronic (glucuronic or iduronic) acid, i.e., ulvanobiouronic acid 3-sulfate [48]. However, this typical structure of ulvans was found associated to 3 variable moieties, despite being quite uncommon: glucose (or possibly a galactose since these epimers cannot be distinguished by MS/MS analysis) (sequence 3), xylose (sequence 4), or uronic acid (sequence 1). Together, ^{13}C NMR and MS analyses were thus in line with the biochemical and monosaccharide compositions of ULVAN-01, which put forward a high proportion of neutral sugars (mainly rhamnose) and uronic acids. Glucosamine was not found in these products of ulvan-01 depolymerization.

The MS^2 analyses also revealed four disaccharide sequences in ULVAN-02 that were characterized as Z$_2$ ions (*m/z*) obtained after fragmentation of the polysaccharidic chain in the spectrometer source, in line with the work by Saad and Leary [53], who observed the same dissociation mechanism when analyzing heparin disaccharides by electrospray ionization MS^2 analysis. The dissociation mechanism of ULVAN-02 giving access to the Z$_2$ ion (*m/z*) 401.0385 (sequence 5 in Table 2) is shown as an example on Scheme 1. The regio- and stereo-specificities of the osidic linkages could not be determined from these analyses so that the reported structures were designed according to the established structure of ulvans [22].

Scheme 1. Dissociation mechanism resulting in Z$_2$ ion formation (example of the ion (*m/z*) 401.0385: sequence 5 in Table 2 and Figure 2A) (adapted from [53]).

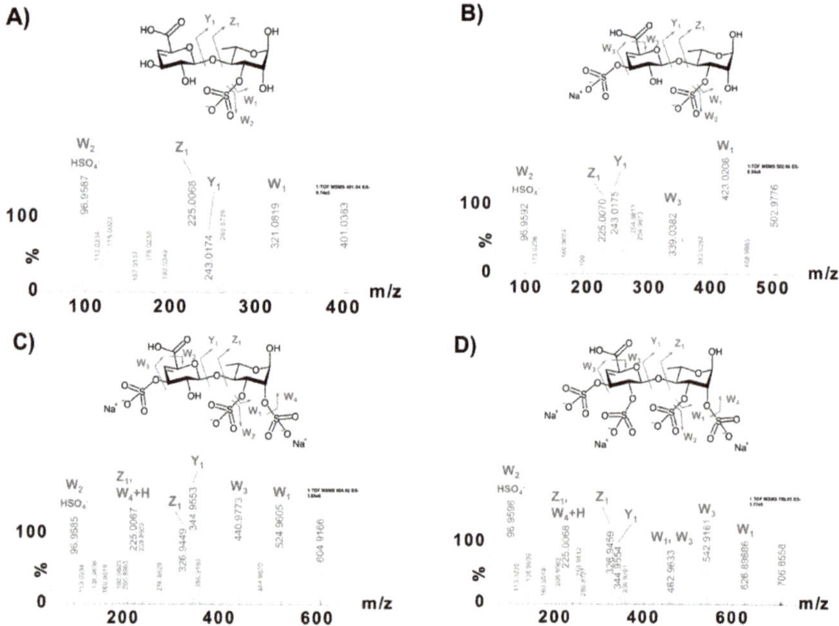

Figure 2. Negative-ion mode electrospray ionization-mass spectrometry targeted fragmentation spectra (ESI$^-$-pseudo-MS^3) of the ions (*m/z*) 401.0383 (**A**, sequence 5 in Table 2), 502.9776 (**B**, sequence 6 in Table 2), 604.9166 (**C**, sequence 7 in Table 2), and 706.8558 (**D**, sequence 8 in Table 2).

The fragmentation profiles of the four disaccharide sequences obtained after pseudo-MS^3 analysis proved that these sequences exhibited sulfate groups on both rhamnose and uronic acid moieties, with one to two sulfate groups on each residue (sequences 5–8 in Table 2 and Figure 2). This result demonstrated that the chemical sulfation method used was completely random and could sulfate up to all four free alcohols of the ulvanobiouronic acid disaccharide: two of rhamnose and two of uronic acid (GlcA or IdoA). Besides, given both the sulfate content of ULVAN-02 of about 20%, indicating an average sulfation of two sulfate groups every three residues, and the observation of disaccharide sequences with up to four sulfate groups, it was concluded that the chemical sulfation of the ULVAN-01 polysaccharide led to a high heterogeneous sulfation pattern, involving both hyper-sulfated and scarcely sulfated zones within ULVAN-02 polysaccharidic structure.

2.4. Anticoagulant Activity

The potential anticoagulant activity of ULVAN-01 and ULVAN-02 was investigated using different clotting and enzymatic assays, targeting the intrinsic and/or common (activated partial thromboplastin time, APTT), extrinsic (prothrombin time, PT) or common (thrombin time, TT) pathways of the coagulation cascade, as well as the specific antithrombin-dependent pathway (anti-Xa and anti-IIa). Furthermore, the anticoagulant properties of ulvan fractions were compared to those of commercial anticoagulants, one UFH and one LMWH, the so-called Lovenox®.

The APTT test mainly assesses the anticoagulant activity on the intrinsic pathway, and, to a lesser extent, on the common pathway. ULVAN-01 activity on the intrinsic pathway was very low (Figure 3A). Indeed, at a concentration of 1000 µg/mL, the clotting time of the plasma was less than 100 s, i.e., 100- and 1000-fold higher than that of Lovenox® and UFH, respectively. Conversely, ULVAN-02 showed a very interesting anticoagulant activity that was higher than that of Lovenox® and just slightly lower than that of UFH. While 500 µg/mL of ULVAN-01 were needed to double the clotting time, only 2.4 µg/mL of ULVAN-02 gave the same result: this concentration was two-fold lower than that of Lovenox® (4.7 µg/mL) and only three-fold higher than that of UFH (Table 3). ULVAN-02 appeared thus to be more effective than Lovenox® on the intrinsic pathway.

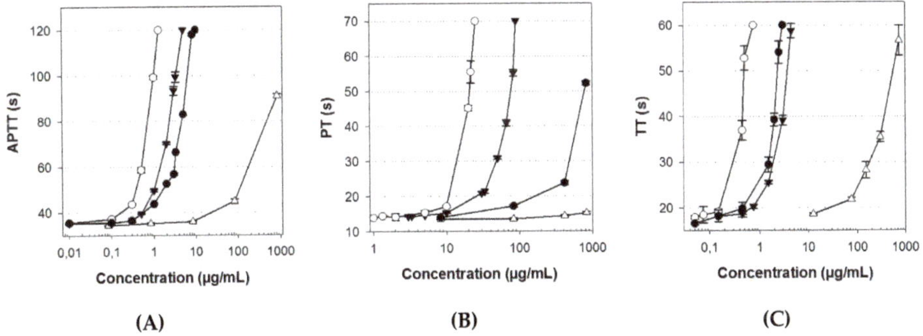

(A)   (B)   (C)

Figure 3. (**A**) Activated partial thromboplastin time (APTT), (**B**) prothrombin time (PT), and (**C**) thrombin time (TT) assays of ULVAN-01 (Δ), ULVAN-02 (▼), unfractionated heparin (UFH) (○), and Lovenox® (●). The clotting times of the plasma in the absence of fractions (negative control, 0.9% NaCl) were (**A**) 38.2 s, (**B**) 13.1 s, and (**C**) 12.9 s. The maximum clotting times measured by the coagulometer were (**A**) 120 s, (**B**) 70 s, and (**C**) 60 s (no coagulation within this time range). Data shown as the mean +/−SD, n = 6.

Therefore, the chemical sulfation procedure significantly increased the activity of ULVAN-01 on the intrinsic and/or common coagulation pathways. Several sulfated polysaccharides from green macroalgae species have been studied for their anticoagulant activity, including high arabinose-containing sulfated polysaccharides, arabinogalactans, or galactans found in *Codium* [32,34,36] or *Ulva* (formerly *Enteromorpha*) [54,55]; polysaccharides rich in sulfated galactose from *Caulerpa* [56]; and high-rhamnose-containing sulfated polysaccharides from *Monostroma* [38,40,41,57]. It has mainly been shown that these polysaccharides could extend the APTT. Only one study has assessed, by APTT assay, a significant anticoagulant activity of sulfated ulvans extracted from an *Ulva* species (*Ulva conglobata*) [43]. These polysaccharides were naturally highly sulfated (between 23.0 and 35.2%), with a rhamnose content between 63 and 72%. The most effective ulvan contained 35.2% of sulfates and prolonged the clotting time to 120 s at a concentration of 2.5 µg/mL, versus 4 µg/mL and 5 µg/mL for two other ulvans with 23 and 28% of sulfate content, respectively. Interestingly, the concentration of 4 µg/mL in ULVAN-02 (sulfated at 20%) resulted in the same clotting time, thus making its activity very similar to that of *Ulva conglobata* ulvans.

Table 3. Overall analysis of the anticoagulant activity of ULVAN-01 and ULVAN-02.

	Concentration (µg/mL)			
	Unfractionated Heparin (UFH)	Lovenox®	ULVAN-01	ULVAN-02
Activated partial thromboplastin time (APTT) [a]	0.75	4.7	500	2.4
Prothrombin time (PT) [a]	14.25	480	inactive	45
Thrombin time (TT) [a]	0.45	1.82	117	2.62
Anti-Xa [b]	0.2	0.125	inactive	17.5
Anti-IIa [b]	0.25	1.75	inactive	18

[a] Sample concentration needed to double the clotting time compared to the negative control (0.9% NaCl); [b] sample concentration for which the enzyme maintained 50% of residual activity.

The PT test, assessing the ulvan activity on the extrinsic coagulation pathway, also showed a major effect of the sulfation procedure on the polysaccharide activity. Indeed, ULVAN-01 was inactive on this coagulation pathway whereas the activity of the sulfated fraction ULVAN-02 was high and only three-fold lower than that of UFH (Figure 3B). ULVAN-02 activity was more than ten-fold higher than that of Lovenox®, with respective concentrations needed to double the clotting time of 45 and 480 µg/mL (Table 3).

Furthermore, a significantly increased anticoagulant activity was associated with ULVAN-01 sulfation, based on the TT assay, which evaluates the anticoagulant activity on the common pathway of the coagulation process (Figure 3C). Indeed, the concentration of ULVAN-01 needed to double the coagulation time of the negative control (117 µg/mL) was more than 40-fold higher than that of ULVAN-02 (2.6 µg/mL) (Table 3). However, ULVAN-02 activity on this coagulation pathway was slightly lower than that of Lovenox®, which was not the case on the other coagulation pathways. Thus, for the common pathway, the chemical sulfation procedure led to a significantly increased anticoagulant activity of *U. rigida* ulvan.

All these clotting time assays tended to show that the anticoagulant activity of ULVAN-02 was mostly oriented towards the early steps of the coagulation cascade, both on the intrinsic and extrinsic pathways, since its activity on the common pathway, which is the final coagulation with fibrin complex formation (red thrombus), was low compared to both Lovenox® and UFH.

Finally, the anticoagulant activity of the ulvan fractions ULVAN-01 and ULVAN-02 was assessed on two central enzymes of the coagulation process, factors Xa and IIa (also called thrombin). The anticoagulant activity was measured in the presence of antithrombin, a coagulation inhibitor, since UFH and Lovenox® are known for their antithrombin-mediated anticoagulant activity.

ULVAN-01 was devoid of any antithrombin-mediated anti-Xa activity (no activity at 1000 µg/mL, data not shown) while ULVAN-02 showed a maximal activity at a concentration of 100 µg/mL (Figure 4A). However, despite a much higher antithrombin-mediated anti-Xa activity of ULVAN-02 compared to that of ULVAN-01, this activity was 100-fold lower than that of UFH and Lovenox®. Indeed, factor Xa residual activity was of 50% with UFH, Lovenox®, and ULVAN-02 concentrations of 0.2 µg/mL, 0.125 µg/mL, and 17.5 µg/mL, respectively (Table 3). The lack of antithrombin-mediated anti-Xa activity of the native ulvan fraction ULVAN-01 is in line with the study by Mao et al. [39], which has concluded that rhamnan sulfates from *Monostroma nitidum* have a very low antithrombin-mediated anti-Xa activity (concentration higher than 1000-fold than that of heparin for the same activity). Two sulfated rhamnan extracts were compared by these authors, with MW of 70 kDa and 870 kDa, and sulfate contents of 34.4 and 28.2%, respectively. Although their sulfate content was much higher than that of ULVAN-01 or ULVAN-02 ulvans (11 and 20%, respectively), these sulfated polysaccharides were not more active. Moreover, despite its sulfate content being about 65% lower than that of these sulfated rhamnans described by Mao et al., the concentration of ULVAN-02 needed to achieve a similar antithrombin-mediated activity was 10-fold lower. The work by Majdoub et al. on sulfated rhamnans (20% of sulfates) from the cyanobacteria *Arthrospira platensis* (formerly *Spirulina platensis*) also showed a lack of anti-Xa activity of this type of sulfated polysaccharide [58]. Conversely, Matsubara et al. highlighted the anticoagulant anti-Xa activity of sulfated galactans extracted from the green seaweed *Codium cylindricum*, containing 89% of galactose and only 13.1% of sulfates, and found an activity very similar to that of ULVAN-02 [33]. In conclusion, the degree of sulfation seems to play a key role in the antithrombin-mediated inhibition of factors Xa and IIa by ulvans, but it does not seem to be the only structural feature that drives this reactivity when it comes to polysaccharides from other seaweeds.

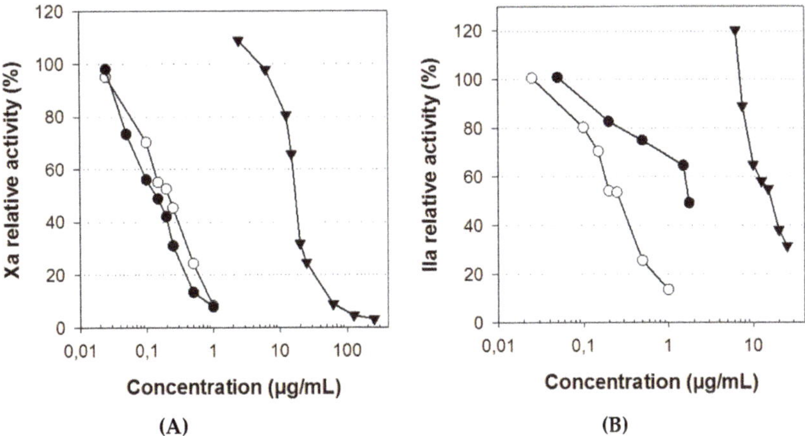

Figure 4. Dose-response curves of antithrombin-mediated (**A**) anti-Xa and (**B**) anti-IIa activities of ULVAN-02 (▼), unfractionated heparin (UFH) (○), and Lovenox® (●). Data shown as the mean, n = 6 (error bars are not indicated for better readability).

In addition, ULVAN-02 was also shown to exhibit an antithrombin-mediated anti-IIa activity (Figure 4B), with a concentration of 18 μg/mL needed to reduce the enzymatic activity of factor IIa by 50%, while ULVAN-01 had no activity (data not shown). Nevertheless, this activity was quite low, 70-fold and 10-fold lower than those of UFH and Lovenox®, respectively (Table 3). Lovenox® activity was thus much lower than that of UFH, which is consistent with the known anticoagulant effect of Lovenox® since it mainly targets factor Xa. These results are also in accordance with those obtained with other sulfated rhamnans from green macroalgae, which are able to inhibit thrombin in the presence of antithrombin, but to a lesser extent than heparin. Rhamnose-rich polysaccharides from *Monostroma nitidum* [39] or *Arthrospira platensis* (formerly *Spirulina platensis*) [55], and ulvans from *Ulva* species [43], are, for instance, 100-fold and 10-fold less active than UFH, respectively.

All together, these results confirm previous findings showing that the anticoagulant activity of macroalgal sulfated polysaccharides is affected by their degree of sulfation [46,47]. Indeed, almost all the tests showed a highly significant increase in the anticoagulant activity of *U. rigida* ulvan after chemical sulfation. The results obtained with ULVAN-02 on the PT test are of particular interest as they are not in line with previous studies on sulfated polysaccharides from green macroalgae that have concluded that sulfated polysaccharides have a positive effect on the intrinsic and/or common coagulation pathways, based on the APTT test, while they do not exhibit any activity on the extrinsic coagulation pathway, based on the PT test. For instance, an ulvan from *Capsosiphon fulvescens* (15.4% sulfate content) [59], sulfated rhamnans from *Monostroma latissima* (21–26% sulfate content) [38,40,41] or from *Monostroma nitidum* (28.2–34.4% sulfate content) [39], did not show any activity on the extrinsic pathway, whereas their sulfate content was very close to that of ULVAN-02. Thus, the degree of sulfation does not seem to be the only parameter affecting the anticoagulant activity of ulvans. Other polyanionic polysaccharides from various origins have been shown to have an anticoagulant activity strongly related to their degree of sulfation ([43,44,46,47]: sulfated fucans, in particular, have been extensively studied and their activity has been proved to be mainly mediated by thrombin inhibition by either antithrombin or heparin cofactor II, at different extents. However, numerous studies on sulfated fucans and galactans [60–62], and more recently on carrageenans [63], also established that anticoagulant activity, particularly in terms of efficiency, is not merely a function of charge density and depends critically on other structural features [46,47]: monosaccharide composition, glycosidic bounds, MW [50], branching residues, and position of sulfate groups on the sugar backbone [18,60]. Among them, the most important seem to be the sulfation pattern and monosaccharide composition [18,46,47,60]. According to Melo et al. [60], the anticoagulant activity of sulfated galactans is achieved mainly through potentiation of plasma cofactors, including antithrombin, which are the natural inhibitors of coagulation proteases but the structural basis of the interactions involved is very complex, due to the heterogeneity of these polysaccharides. Their study on interactions with thrombin and its cofactors, in particular, notably showed that sulfated galactans can link with antithrombin but require significantly longer chains than heparin to interact with the antithrombin/thrombin complex, and bind to a different site. In addition, Becker et al. highlighted, by molecular modeling techniques, that similarities obtained in the glycosidic linkages and predominant $^1C_4$ chair form of sulfated fucans and galactans could fully explain their specific interaction with antithrombin and the differences between their anticoagulant activity and that of heparin [62]. All these data indicate that the action mechanism of heparin mimetics including sulfated polysaccharides from marine seaweeds differs from that of heparin, which is well in accordance with the overall results obtained in the present study. It was indeed shown that ULVAN-01 was very slightly active on the intrinsic and common pathways, and inactive on the extrinsic pathway of the coagulation cascade, based on the PT test. It was also inactive on the coagulation factor Xa, which is the upstream common point for intrinsic and extrinsic coagulation pathways. On the contrary, ULVAN-02 was active on all tested pathways, showing that chemical sulfation strongly enhanced the anticoagulant activity of ULVAN-01 and could even enable it to inhibit the antithrombin/Xa or antithrombin/IIa complexes. It was also demonstrated by UHPLC-HRMS analyses that the chemical sulfation method used to obtain ULVAN-02 led to an ulvan with a very heterogenous sulfation pattern, involving both hyper-sulfated and scarcely sulfated zones within the polysaccharidic structure, which is very different from what is usually found in the native form of ulvans, i.e., a sulfation pattern most exclusively involving the position 3 of rhamnose. The most likely hypothesis to explain the anticoagulant activity of ULVAN-02 lies in the probability that a geometrical match between the polysaccharide and its binding site on coagulation factors is increased by sulfation, the sugar backbone providing the geometric constraints to accommodate the binding site, and the increased negative charge due to sulfation pattern modifications in the hyper-sulfated zones providing higher physical interaction properties, further enhancing the binding capacity of the polysaccharide.

2.5. Effect of ULVAN-01 and ULVAN-02 on Cell Viability

The in vitro cytotoxicity of ULVAN-01 and ULVAN-02 was assessed using the MTT assay on human fibroblastic cells, and compared to those of UFH and Lovenox® (Figure 5).

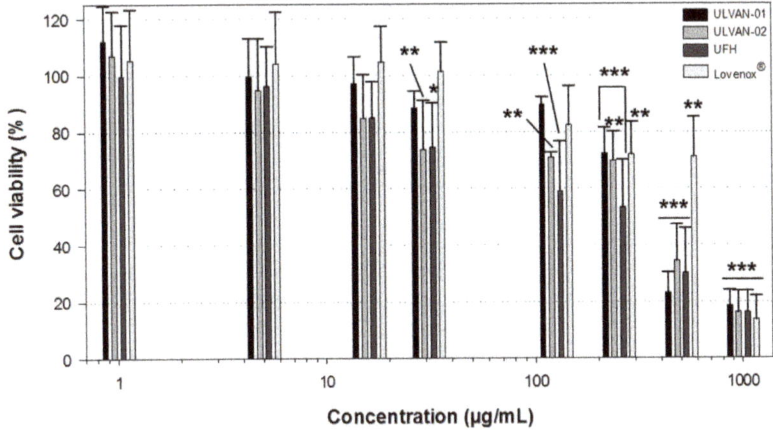

Figure 5. Viability of human fibroblasts treated for 72 h with ULVAN-01, ULVAN-02, unfractionated heparin (UFH), and Lovenox®. Results are expressed as the relative percentage of viability compared to the negative control. Data shown as the mean +/−SD, n = 5. Significant differences between samples and negative control are indicated by * ($p < 0.05$), ** ($p < 0.01$), and *** ($p < 0.001$).

Results showed that ULVAN-01, ULVAN-02, and UFH had a very similar dose-response effect on cell viability. Up to 250 µg/mL, they had a very limited impact on cell viability: about 80% viable cells were indeed enumerated, compared to the negative control. Lovenox® even seemed to have slightly less impact on cell viability. More importantly, ULVAN-02 was absolutely not cytotoxic to human fibroblasts at concentrations within the range of 1–15 µg/mL, corresponding to the concentration range where it was shown to be bioactive, regardless of the anticoagulant activity tested, which is an essential prerequisite to consider any further development of this sulfated ulvan fraction extracted from *U. rigida* as part of a therapeutic anticoagulation application.

3. Materials and Methods

3.1. Materials

Green macroalga *U. rigida* was cultivated in a pond in the "Ferme du Douhet," a marine farm in Ile d'Oléron (France). It was collected in 2012 and dried.

UFH was purchased from Interchim (Montluçon, France) (heparin sodium salt 12865E, batch 201274) and Lovenox® was kindly provided by the the Saint-Louis Hospital (La Rochelle, France).

Unless otherwise stated in the text, all chemical reagents were purchased from Merck (Darmstadt, Germany).

3.2. Methods

3.2.1. Extraction and Purification Process

First, 500 g of *U. rigida* thallus were washed in 15 L of distilled water (1/30 (*w/w*)) for 10 min at room temperature (RT) and wrung with a fabric cone to remove as much water as possible [48]. The washed algae were then ground at 80 °C in a blender, in 5 L of deionized water, until obtaining 2-mm particles. The 5 L of minced algae in water were transferred into a thermostated tank at 80 °C containing 2.5 L of distilled water. Extraction was processed under constant agitation with a bladed stirrer at a rotation speed of 10 spins/min for 2 h. The pulps were then removed from the tank and filtered with the fabric cone to collect the aqueous extract. The aqueous extract was then recovered and filtered with an ultrafiltration unit equipped with a 15 kDa Kerasep KBW membrane (Novasep Process, Pompey, France). Filtration was carried out at 80 °C at a pressure of 5 bars and a circulation flow of

450 L/h (circulation speed of 5 m/s), until obtaining a retentate around 4°Bx. Next, 1.5 L of retentate was demineralized by passage on a column containing 100 mL of Amberlite FPA 98, a strong anionic resin in the OH^- form, in series with a column containing 200 mL of Amberlite IR 120 Na, a strong cationic resin in the H^+ form. Circulation was processed with a peristaltic pump at a flow rate of 2 BV/h. The deionized product was finally decanted in a water bath at 80 °C for 2 h. After centrifugation at 5000× *g* for 15 min at RT, the fraction, referred to as ULVAN-01, was neutralized to pH 7 with 1 M NaOH and lyophilized.

3.2.2. Sulfation Procedure

The sulfation procedure was carried out on the ULVAN-01 fraction using the sulfur-trioxide pyridine complex (SO_3–pyridine) method, in dimethylformamide (DMF) and pyridine [64].

Briefly, 500 mg of ULVAN-01 were dissolved in a mixture containing 10 mL of DMF and 1.4 mL of pyridine. The mixture was heated to 60 °C and sulfation was then processed by progressive addition of 3 g of SO_3–pyridine complex for 2 h at 60 °C under constant stirring. The mixture was left for two additional hours at 60 °C under stirring. After cooling of the mixture at RT, it was centrifuged at 5000× *g* for 10 min at 4 °C. The supernatant was then removed and the pellet was dissolved in 5 mL of 2.5 M NaOH, and 45 mL of absolute ethanol were added (final concentration: 90% (*v/v*)). After 12 h at 4 °C under stirring, the precipitate was isolated by centrifugation (10,000× *g*, 10 min, 4 °C) and dissolved in 60 mL of deionized water. The solution was finally dialyzed for seven days against deionized water (cut-off: 1 kDa) and lyophilized. The resulting sulfated fraction was referred to as ULVAN-02.

3.2.3. Chemical Composition

The neutral sugar content was determined according to the phenol-sulfuric method [65], using glucose as a standard. The uronic acid content was measured as described by Bitter and Muir [66]. The protein content was determined using the Bradford protein assay [67]. The sulfate content was obtained using 3-amino-7-(dimethylamino)phenothiazin-5-ium chloride (Azure A), which binds to the sulfated groups in a polysaccharide chain [68,69]. The quantitation method of polyphenols was adapted from the original one [70], using gallic acid as a standard: briefly, 50 μL of Folin–Ciocalteu reagent and 200 μL of 20% sodium carbonate were successively added to 100 μL of sample and the mixture was incubated in the dark for 45 min at RT, prior to absorbance reading at 730 nm. The ash content of ULVAN-01 was quantified after 15 h at 550 °C. The ash content of ULVAN-02 could not be determined due to a lack of material. All analyses were performed at least in triplicate, except the determination of ULVAN-01 ash content (n = 2), and data are thus shown as the mean.

3.2.4. Analysis of the Fractions by High Performance Size Exclusion Chromatography

The structural and quantitative analyses of ulvan fractions were performed using a HPLC 1100 LC/RID system (Agilent technologies, Santa Clara, CA, USA), with two successive exclusion chromatography columns of 30 cm in size: TSK gel 5000 PW and TSK gel 4000 PW (Tosoh Bioscience, Tokyo, Japan). These analyses were carried out at 30 °C, after injection of 20 μL of 1 mg/mL fraction or sample, at a flow rate of 0.5 mL/min of elution buffer (0.1 M ammonium acetate). Products were detected and quantified by differential refractometry using the HP Chemstation software in off-line mode for processing. The standard curve was made using dextran standards from 1000 to 50,000 Da. The number-averaged molecular weight (M_n), weight-averaged molecular weight (M_w) and polydispersity index (I) were calculated as follows [71]:

$$M_n = (\Sigma\ N_i \times M_i)/\Sigma\ N_i$$
$$M_w = (\Sigma\ N_i \times M_i^2)/(\Sigma\ N_i \times M_i)$$
$$I = M_w/M_n$$

where N_i is the number of moles of polymer species and M_i the molecular weight of polymer species.

3.2.5. Nuclear Magnetic Resonance (NMR) and Ultra High Performance Liquid Chromatography-High Resolution Mass Spectrometry (UHPLC-HRMS)

Carbon (^{13}C) NMR analysis of the ULVAN-01 fraction was performed using a JEOL JNMLA400 spectrometer (400 MHz, JEOL, Peabody, MA, USA) in D_2O solution at a concentration of 50 mg/mL [48].

The monosaccharide composition and oligosaccharide sequences were determined by UHPLC-HRMS. Analyses were carried out using an UHPLC system, "Acquity UPLC H-class," coupled to a HRMS system, "XEVO G2-S Q-TOF," equipped with an electrospray ionization source (Waters, Milford, MA, USA). The UHPLC system was formed by a quaternary pump (Quaternary Solvent Manager, Waters) and an automatic injector (Sample Manager-FTN, Waters) equipped with a 10 µL injection loop. Analyses were performed according to the UHPLC and MS parameters given in Table 4.

Table 4. Ultra high pressure liquid chromatography (UHPLC) and mass spectrometry (MS) parameters used to determine the monosaccharide composition and oligosaccharide sequences.

<table>
<tr><th></th><th>Monosaccharide Composition</th><th>Oligosaccharide Sequences</th></tr>
<tr><td colspan="3">UHPLC Parameters</td></tr>
<tr><td>Column</td><td colspan="2" style="text-align:center">"Acquity UPLC BEH Amide" (Waters) (2.1 × 500 mm, 1.7 µm), maintained at 30 °C</td></tr>
<tr><td>Flow rate (µL/min)</td><td>0–15 min, 130–200; 15–18 min, 200; 18–30 min, 200–130; 30–37 min, 200</td><td>200</td></tr>
<tr><td>Gradient: water (A)/acetonitrile (B) + 0.015% (v/v) ammoniac</td><td>0–15 min, 90–55% B; 15–17 min, 55% B; 17–18 min, 55–90% B; 18–37 min, 90% B</td><td>0–3 min, 100% B; 3–3.1 min 100–70% B; 3.1–10 min, 70–50% B; 10–10.5 min, 50–45% B; 10.5–18 min, 45–100% B</td></tr>
<tr><td>Injection</td><td colspan="2" style="text-align:center">5 µL (7 °C)</td></tr>
<tr><td colspan="3">MS Parameters</td></tr>
<tr><td>Mode</td><td>ESI⁺ (centroid)</td><td>ESI[−] (centroid)</td></tr>
<tr><td>Source temperature</td><td colspan="2" style="text-align:center">120 °C</td></tr>
<tr><td>Desolvation temperature</td><td>250 °C</td><td>500 °C</td></tr>
<tr><td>Gas flow rate of the cone</td><td colspan="2" style="text-align:center">50 L/h</td></tr>
<tr><td>Desolvation gas flow rate</td><td colspan="2" style="text-align:center">800 L/h</td></tr>
<tr><td>Capillary voltage</td><td colspan="2" style="text-align:center">2.5 kV</td></tr>
<tr><td>Sampling cone voltage</td><td>35 V</td><td>130 V</td></tr>
<tr><td>Source compensation</td><td colspan="2" style="text-align:center">80 V</td></tr>
<tr><td>Acquisition mass range</td><td>50–1200 m/z</td><td>50–2100 m/z (down-regulated according to the fragmented ion with a scan time of 0.15 s)</td></tr>
<tr><td>Lock-mass</td><td colspan="2" style="text-align:center">Leucine Enkephalin (MW = 555.62 Da, 1 ng/µL)</td></tr>
</table>

The analyses of monosaccharide composition were performed after total acid hydrolysis of ULVAN-01. Briefly, 1 mL of 10 mg/mL ULVAN-01 solution in HCL 3 M was prepared and heated at 100 °C for 5 h (total hydrolysis was monitored over-time using reducing sugar assessment by the dinitrosalicylic acid method [72]). After cooling at room temperature, the sample was centrifuged at 10,000× g for 10 min and filtrated through 0.22 µm filter, prior to UHPLC-HRMS analysis.

The analyses of oligosaccharide sequences were carried out on a 9 kDa ulvan fraction filtrated through 0.22 µm filter, obtained after solid supported depolymerization of ULVAN-01. Briefly, 100 mL of ULVAN-01 solution at 25 mg/mL were prepared and depolymerized, using a circuit consisting of the ULVAN-01 solution, a peristaltic pump set at a flow rate of 12 mL/min and a column containing 10 mL of AMBERLITE™ FPC23 H resin in the H+ form. The polysaccharide solution and the column were maintained at 80 °C for 19 h. After cooling at room temperature, the samples were filtrated through 0.22 µm filter, prior to UHPLC-HRMS/MS analysis.

3.2.6. Clotting Time Assays

The anticoagulant activity of the different fractions was determined by measuring the APTT, PT, or TT. All the assays were carried out using a Start4 coagulometer and assay kits from Stago (Asnières-sur-Seine, France), according to the manufacturer's instructions. Briefly, 90 µL of normal human plasma were mixed with 10 µL of 0.9% NaCl solution containing various sample concentrations for each assay. For the APTT assay, 100 µL of APTT assay reagent were added to the mixture prior to incubation for 3 min at 37 °C and addition of 100 µL of 0.025 M $CaCl_2$, and the clotting time (APTT) was recorded. For the TT and PT assays, the mixture was first incubated for 2 min (PT) or 1 min (TT) at 37 °C, before adding 200 µL of PT assay reagent (PT) or 100 µL of TT assay reagent (TT), and recording the clotting time. The negative control used in all assays was 0.9% NaCl solution.

3.2.7. Assays of Antithrombin-Mediated Inhibition of Factors Xa and IIa

Stachrom ATIII and Stachrom Heparin kits (Stago, Asnières-sur-Seine, France) were used to assess the antithrombin-mediated inhibition of factors Xa and IIa (thrombin), respectively, according to the manufacturer's instructions. Antithrombin was diluted 1:2 in 0.1X kit buffer. Then, 25 µL of sample (aqueous solution of ulvan fraction, UFH or Lovenox® at various concentrations) were incubated with 25 µL of antithrombin (0.626 µg/µL) at 37 °C for 30 s. Thereafter, 25 µL of factor Xa or factor IIa (thrombin) (11.25 nKat/mL) were added. After 30 s of incubation, 25 µL of a 3.25 nM solution of factor Xa chromogenic substrate (CBS 31.39; CH_2-SO_2-D-Leu-Gly-Arg-pNA, AcOH) or 25 µL of a 1.4 nM solution of factor IIa chromogenic substrate (CBS 61.50; EtM-SPro-ARG-pNA, AcOH) were added. Factor Xa or IIa activities were immediately measured at 405 nm every 6 s for 5 min. The initial velocity was calculated as the slope of the linear segment of the kinetic curve. All assays were performed in 96-well NUNC microplates (Thermo Fisher Scientific, Waltham, MA, USA), using a Fluostar Omega microplate reader (BMG LABTECH, Ortenberg, Germany).

3.2.8. Evaluation of Cell Viability

Normal human dermal fibroblasts (NHDF) were obtained from ATCC (Manassas, VA, USA): product code CCD-1059Sk; ATCC® CRL-2072™, lot number 62062292, from the skin of a 20-year-old woman according to the supplier's information. Cells were cultured in Dulbecco's Modified Eagle Medium (DMEM, PAN Biotech, Aidenbach, Germany) supplemented with 10% (*v/v*) fetal bovine serum (PAN Biotech, Dutscher, France) and 1% (*v/v*) antibiotic solution (10,000 U/mL penicillin, 10 mg/mL streptomycin) (PAN Biotech, Aidenbach, Germany), referred to as complete culture medium thereafter. The cells were cultured at 37 °C with 5% CO_2, in a temperature-controlled, humidified incubator. The cells were grown in 75 cm^2 surface ventilated Falcon culture flasks (BD Biosciences, Franklin Lakes, NJ, USA) and subcultured by trypsinization (0.05% (*w/v*) trypsin, PAN Biotech, Aidenbach, Germany). The culture medium was changed every 2 or 3 days. Cells were used between the 3rd and 8th passages for the experiments.

The MTT test was used to evaluate cell viability, according to the method described by Mosmann [73]. This colorimetric assay allows assessing cell proliferation, based on the reduction of a tetrazolium salt (3-(4,5-dimethylthiazol-2-yl)-2,5-diphenyltetrazolium bromide, MTT), by living cells. MTT is a yellow salt. In the presence of mitochondrial succinate dehydrogenase produced by active living cells, the tetrazolium ring they contain is reduced and forms a violet product: formazan. The yellow solution becomes purple and the intensity of the purple coloration is proportional to the number of living cells.

Briefly, 100 µL of complete culture medium containing 5×10^4 cells/mL were seeded in 96-well Falcon microplates (BD Biosciences, Franklin Lakes, NJ, USA) and incubated for 24 h. The medium was then removed and 100 µL of samples prepared in DMEM supplemented with 1% (*v/v*) antibiotic solution, or negative control (DMEM supplemented with 1% (*v/v*) antibiotic solution alone) were added into the wells. After 48 h, 25 µL of MTT (5% (*w/v*) in PBS) were added in each well and the plates were incubated for 4 h at 37 °C. The medium was then removed and 200 µL of dimethyl sulfoxide were added into each well. The plates were then incubated for 10 min and absorbance was read at 550 nm using a Fluostar Omega microplate reader (BMG LABTECH, Ortenberg, Germany).

3.2.9. Statistical Analysis

The one-way Anova was applied to determine significant differences between samples and negative control, using Sigma Plot 12.5 (Systat Software Inc., San Jose, CA, USA).

4. Conclusions

A solvent- and acid-free procedure was developed to extract and purify ulvan from *U. rigida*, providing an ulvan fraction of high purity (ULVAN-01), with less than 4% of contaminant proteins remaining at the end of the procedure. The sulfate content of this ulvan was 11%, which is in the lower sulfation range of ulvans extracted from algae of the *Ulva* genus. Data on the anticoagulant activity of ulvans are limited but previous studies focusing on other types of sulfated polysaccharides have shown that sulfation is essential for their anticoagulant activity, although the osidic composition, MW, and sulfate groups position are also important features. To investigate the importance of the degree of sulfation of *U. rigida* ulvan, a chemical sulfation procedure was applied to ULVAN-01, which allowed almost doubling the sulfate content of the native polysaccharide, reaching 20%. The resulting chemically-sulfated ulvan fraction, ULVAN-02, thus exhibited a strongly enhanced anticoagulant activity, regardless of the coagulation pathway tested. Its most significant activity was shown to target the intrinsic and extrinsic pathways, with an activity higher than that of Lovenox® and close to that of UFH. It was indeed less active on the common pathway, although its activity was only slightly lower than that of Lovenox®. This last result was confirmed by the low antithrombin-mediated inhibition activity exhibited by ULVAN-02 on two important coagulation factors of the common pathway, Xa and IIa (thrombin), which was also significantly increased compared to ULVAN-01 but remained quite low compared to UFH and Lovenox®. Hence, the ULVAN-02 fraction could be a very interesting alternative to heparins, with different targets and a high anticoagulant activity. Moreover, ULVAN-02 did not appear to be cytotoxic on human model cells but further studies are needed to confirm these findings.

Author Contributions: Conceptualization and Methodology, A.A., D.D., S.B., T.M. and N.B.; software, validation and formal analysis, A.A., A.B. and N.B.; investigation, A.A. and A.B.; resources and data curation, A.A., A.B., D.D., S.B., T.M. and N.B.; writing – original draft preparation, A.A. and N.B.; writing – review and editing, A.A., T.M. and N.B. (T.M. did the proof reading and improved the manuscript by providing valuable suggestions; all the authors read and approved the final manuscript); funding acquisition, project administration and supervision, S.B., T.M. and N.B.

Funding: This work was supported by the "Association nationale de la recherche et de la technologie", by providing a part of a PhD funding (Grant number 2012/1356).

Acknowledgments: We are very grateful to the Saint-Louis Hospital (La Rochelle, France), which kindly provided us Lovenox®, and to the "Biomolecule High Resolution Analysis Platform" from the LIENSs Laboratory of La Rochelle University for the NMR and UHPLC-HRMS analyses.

Conflicts of Interest: The authors declare no conflict of interest.

References

1. McHugh, D.J. *A Guide to the Seaweed Industry*; FAO Fisheries Technical Paper; Food and Agriculture Organization of the United Nations: Rome, Italy, 2003; ISBN 978-92-5-104958-7.
2. Calumpong, H.P.; West, J.; Martin, G. Seaweeds. In *The First Global Integrated Marine Assessment: World Ocean Assessment I*; United Nations, Ed.; Cambridge University Press: Cambridge, UK, 2017; ISBN 978-1-316-51001-8.
3. Ferdouse, F.; Lovestad Holdt, S.; Smith, R.; Murua, P.; Yang, Z. *The Global Status of Seaweed Production, Trade and Utilization*; FAO Globefish Research Programme; Food and Agriculture Organization of the United Nations: Rome, Italy, 2018; ISBN 978-92-5-130870-7.
4. Pomponi, S.A. The bioprocess–technological potential of the sea. *J. Biotechnol.* **1999**, *70*, 5–13. [CrossRef]
5. Sasi, N.; Kriston, B. Current status of global cultivated seaweed production and markets. *World Aquac.* **2014**, *45*, 32–37.
6. Granert, C.; Raud, J.; Xie, X.; Lindquist, L.; Lindbom, L. Inhibition of leukocyte rolling with polysaccharide fucoidin prevents pleocytosis in experimental meningitis in the rabbit. *J. Clin. Investig.* **1994**, *93*, 929–936. [CrossRef] [PubMed]
7. Jiao, G.; Yu, G.; Zhang, J.; Ewart, H. Chemical Structures and Bioactivities of Sulfated Polysaccharides from Marine Algae. *Mar. Drugs* **2011**, *9*, 196–223. [CrossRef]
8. Mohamed, S.; Hashim, S.N.; Rahman, H.A. Seaweeds: A sustainable functional food for complementary and alternative therapy. *Trends Food Sci. Technol.* **2012**, *23*, 83–96. [CrossRef]

9. Rupérez, P.; Ahrazem, O.; Leal, J.A. Potential antioxidant capacity of sulfated polysaccharides from the edible marine brown seaweed *Fucus vesiculosus*. *J. Agric. Food Chem.* **2002**, *50*, 840–845. [CrossRef] [PubMed]

10. Rocha de Souza, M.C.; Marques, C.T.; Guerra Dore, C.M.; Ferreira da Silva, F.R.; Oliveira Rocha, H.A.; Leite, E.L. Antioxidant activities of sulfated polysaccharides from brown and red seaweeds. *J. Appl. Phycol.* **2007**, *19*, 153–160. [CrossRef]

11. Noda, H.; Amano, H.; Arashima, K.; Nisizawa, K. Antitumor activity of marine algae. *Hydrobiologia* **1990**, *204*, 577–584. [CrossRef]

12. Bourgougnon, N.; Roussakis, C.; Kornprobst, J.-M.; Lahaye, M. Effects in vitro of sulfated polysaccharide from *Schizymenia dubyi* (Rhodophyta, Gigartinales) on a non-small-cell bronchopulmonary carcinoma line (NSCLC-N6). *Cancer Lett.* **1994**, *85*, 87–92. [CrossRef]

13. Zorofchian Moghadamtousi, S.; Karimian, H.; Khanabdali, R.; Razavi, M.; Firoozinia, M.; Zandi, K.; Abdul Kadir, H. Anticancer and Antitumor Potential of Fucoidan and Fucoxanthin, Two Main Metabolites Isolated from Brown Algae. *Sci. World J.* **2014**, *2014*, 1–10. [CrossRef]

14. Damonte, E.B.; Matulewicz, M.C.; Cerezo, A.S. Sulfated seaweed polysaccharides as antiviral agents. *Curr. Med. Chem.* **2004**, *11*, 2399–2419. [CrossRef] [PubMed]

15. Pujol, C.A.; Ray, S.; Ray, B.; Damonte, E.B. Antiviral activity against dengue virus of diverse classes of algal sulfated polysaccharides. *Int. J. Biol. Macromol.* **2012**, *51*, 412–416. [CrossRef] [PubMed]

16. Witvrouw, M.; De Clercq, E. Sulfated polysaccharides extracted from sea algae as potential antiviral drugs. *Gen. Pharmacol.* **1997**, *29*, 497–511. [CrossRef]

17. Jin, W.; Zhang, Q.; Wang, J.; Zhang, W. A comparative study of the anticoagulant activities of eleven fucoidans. *Carbohydr. Polym.* **2013**, *91*, 1–6. [CrossRef] [PubMed]

18. Pereira, M.S.; Mulloy, B.; Mourão, P.A.S. Structure and anticoagulant activity of sulfated fucans. Comparison between the regular, repetitive, and linear fucans from echinoderms with the more heterogeneous and branched polymers from brown algae. *J. Biol. Chem.* **1999**, *274*, 7656–7667. [CrossRef]

19. Carlucci, M.J.; Pujol, C.A.; Ciancia, M.; Noseda, M.D.; Matulewicz, M.C.; Damonte, E.B.; Cerezo, A.S. Antiherpetic and anticoagulant properties of carrageenans from the red seaweed *Gigartina skottsbergii* and their cyclized derivatives: correlation between structure and biological activity. *Int. J. Biol. Macromol.* **1997**, *20*, 97–105. [CrossRef]

20. Chevolot, L.; Mulloy, B.; Ratiskol, J.; Foucault, A.; Colliec-Jouault, S. A disaccharide repeat unit is the major structure in fucoidans from two species of brown algae. *Carbohydr. Res.* **2001**, *330*, 529–535. [CrossRef]

21. Percival, E.; McDowell, R.H. *Chemistry and Enzymology of Marine Algal Polysaccharides*; Academic Press: London, UK; New York, NY, USA, 1967; ISBN 9780125506502.

22. Lahaye, M.; Robic, A. Structure and functional properties of ulvan, a polysaccharide from green seaweeds. *Biomacromolecules* **2007**, *8*, 1765–1774. [CrossRef]

23. Sommers, C.D.; Ye, H.; Kolinski, R.E.; Nasr, M.; Buhse, L.F.; Al-Hakim, A.; Keire, D.A. Characterization of currently marketed heparin products: analysis of molecular weight and heparinase-I digest patterns. *Anal. Bioanal. Chem.* **2011**, *401*, 2445–2454. [CrossRef]

24. Mulloy, B.; Hogwood, J.; Gray, E. Assays and reference materials for current and future applications of heparins. *Biologicals* **2010**, *38*, 459–466. [CrossRef]

25. Warkentin, T.E.; Levine, M.N.; Hirsh, J.; Horsewood, P.; Roberts, R.S.; Gent, M.; Kelton, J.G. Heparin-Induced Thrombocytopenia in Patients Treated with Low-Molecular-Weight Heparin or Unfractionated Heparin. *N. Engl. J. Med.* **1995**, *332*, 1330–1336. [CrossRef]

26. Da Silva, M.S.; Sobel, M. Anticoagulants: To bleed or not to bleed, that is the question. *Semin. Vasc. Surg.* **2002**, *15*, 256–267. [CrossRef]

27. Emanuele, R.M.; Fareed, J. The effect of molecular weight on the bioavailability of heparin. *Thromb. Res.* **1987**, *48*, 591–596. [CrossRef]

28. Hao, L.; Zhang, Q.; Yu, T.; Cheng, Y.; Ji, S. Antagonistic effects of ultra-low-molecular-weight heparin on Aβ25–35-induced apoptosis in cultured rat cortical neurons. *Brain Res.* **2011**, *1368*, 1–10. [CrossRef]

29. Sattari, M.; Lowenthal, D.T. Novel oral anticoagulants in development: Dabigatran, Rivaroxaban, and Apixaban. *Am. J. Ther.* **2011**, *18*, 332–338. [CrossRef] [PubMed]

30. van der Wall, S.J.; Klok, F.A.; den Exter, P.L.; Barrios, D.; Morillo, R.; Cannegieter, S.C.; Jimenez, D.; Huisman, M.V. Continuation of low-molecular-weight heparin treatment for cancer-related venous thromboembolism: a prospective cohort study in daily clinical practice. *J. Thromb. Haemost.* **2017**, *15*, 74–79. [CrossRef]

31. Jurd, K.M.; Rogers, D.J.; Blunden, G.; McLellan, D.S. Anticoagulant properties of sulphated polysaccharides and a proteoglycan from *Codium fragile* ssp. *atlanticum*. *J. Appl. Phycol.* **1995**, *7*, 339–345. [CrossRef]

32. Siddhanta, A.K.; Shanmugam, M.; Mody, K.H.; Goswami, A.M.; Ramavat, B.K. Sulphated polysaccharides of *Codium dwarkense* Boergs. from the west coast of India: chemical composition and blood anticoagulant activity. *Int. J. Biol. Macromol.* **1999**, *26*, 151–154. [CrossRef]

33. Matsubara, K.; Matsuura, Y.; Bacic, A.; Liao, M.-L.; Hori, K.; Miyazawa, K. Anticoagulant properties of a sulfated galactan preparation from a marine green alga, *Codium cylindricum*. *Int. J. Biol. Macromol.* **2001**, *28*, 395–399. [CrossRef]

34. Shanmugam, M.; Mody, K.H.; Siddhanta, A.K. Blood anticoagulant sulphated polysaccharides of the marine green algae *Codium dwarkense* (Boergs.) and *C. tomentosum* (Huds.) Stackh. *Indian J. Exp. Biol.* **2001**, *39*, 365–370.

35. Hayakawa, Y.; Hayashi, T.; Lee, J.-B.; Srisomporn, P.; Maeda, M.; Ozawa, T.; Sakuragawa, N. Inhibition of thrombin by sulfated polysaccharides isolated from green algae. *Biochim. Biophys. Acta BBA - Protein Struct. Mol. Enzymol.* **2000**, *1543*, 86–94. [CrossRef]

36. Ciancia, M.; Quintana, I.; Vizcargüénaga, M.I.; Kasulin, L.; de Dios, A.; Estevez, J.M.; Cerezo, A.S. Polysaccharides from the green seaweeds *Codium fragile* and *C. vermilara* with controversial effects on hemostasis. *Int. J. Biol. Macromol.* **2007**, *41*, 641–649. [CrossRef]

37. Maeda, M.; Uehara, T.; Harada, N.; Sekiguchi, M.; Hiraoka, A. Heparinoid-active sulphated polysaccharides from *Monostroma nitidum* and their distribution in the chlorophyta. *Phytochemistry* **1991**, *30*, 3611–3614. [CrossRef]

38. Mao, W.; Li, H.; Li, Y.; Zhang, H.; Qi, X.; Sun, H.; Chen, Y.; Guo, S. Chemical characteristic and anticoagulant activity of the sulfated polysaccharide isolated from *Monostroma latissimum* (Chlorophyta). *Int. J. Biol. Macromol.* **2009**, *44*, 70–74. [CrossRef]

39. Mao, W.-J.; Fang, F.; Li, H.-Y.; Qi, X.-H.; Sun, H.-H.; Chen, Y.; Guo, S.-D. Heparinoid-active two sulfated polysaccharides isolated from marine green algae *Monostroma nitidum*. *Carbohydr. Polym.* **2008**, *74*, 834–839. [CrossRef]

40. Zhang, H.; Mao, W.; Fang, F.; Li, H.; Sun, H.; Chen, Y.; Qi, X. Chemical characteristics and anticoagulant activities of a sulfated polysaccharide and its fragments from *Monostroma latissimum*. *Carbohydr. Polym.* **2008**, *71*, 428–434. [CrossRef]

41. Li, H.; Mao, W.; Zhang, X.; Qi, X.; Chen, Y.; Chen, Y.; Xu, J.; Zhao, C.; Hou, Y.; Yang, Y.; et al. Structural characterization of an anticoagulant-active sulfated polysaccharide isolated from green alga *Monostroma latissimum*. *Carbohydr. Polym.* **2011**, *85*, 394–400. [CrossRef]

42. Li, H.; Mao, W.; Hou, Y.; Gao, Y.; Qi, X.; Zhao, C.; Chen, Y.; Chen, Y.; Li, N.; Wang, C. Preparation, structure and anticoagulant activity of a low molecular weight fraction produced by mild acid hydrolysis of sulfated rhamnan from *Monostroma latissimum*. *Bioresour. Technol.* **2012**, *114*, 414–418. [CrossRef] [PubMed]

43. Mao, W.; Zang, X.; Li, Y.; Zhang, H. Sulfated polysaccharides from marine green algae *Ulva conglobata* and their anticoagulant activity. *J. Appl. Phycol.* **2006**, *18*, 9–14. [CrossRef]

44. Minh Thu, Q.T. Effect of sulfation on the structure and anticoagulant activity of ulvan extracted from green seaweed *Ulva reticulata*. *Vietnam J. Sci. Technol.* **2018**, *54*, 373. [CrossRef]

45. Barros Gomes Camara, R.; Silva Costa, L.; Pereira Fidelis, G.; Duarte Barreto Nobre, L.T.; Dantas-Santos, N.; Lima Cordeiro, S.; Santana Santos Pereira Costa, M.; Guimaraes Alves, L.; Oliveira Rocha, H.A. Heterofucans from the Brown Seaweed *Canistrocarpus cervicornis* with Anticoagulant and Antioxidant Activities. *Mar. Drugs* **2011**, *9*, 124–138. [CrossRef]

46. Li, B.; Lu, F.; Wei, X.; Zhao, R. Fucoidan: Structure and Bioactivity. *Molecules* **2008**, *13*, 1671–1695. [CrossRef] [PubMed]

47. Mestechkina, N.M.; Shcherbukhin, V.D. Sulfated polysaccharides and their anticoagulant activity: A review. *Appl. Biochem. Microbiol.* **2010**, *46*, 267–273. [CrossRef]

48. Adrien, A.; Bonnet, A.; Dufour, D.; Baudouin, S.; Maugard, T.; Bridiau, N. Pilot production of ulvans from *Ulva* sp. and their effects on hyaluronan and collagen production in cultured dermal fibroblasts. *Carbohydr. Polym.* **2017**, *157*, 1306–1314. [CrossRef] [PubMed]

49. Lahaye, M.; Alvarez-Cabal Cimadevilla, E.; Kuhlenkamp, R.; Quemener, B.; Lognoné, V.; Dion, P. Chemical composition and 13C NMR spectroscopic characterisation of ulvans from *Ulva* (Ulvales, Chlorophyta). *J. Appl. Phycol.* **1999**, *11*, 1. [CrossRef]

50. Nishino, T.; Aizu, Y.; Nagumo, T. Antithrombin activity of a fucan sulfate from the brown seaweed *Ecklonia kurome*. *Thromb. Res.* **1991**, *62*, 765–773. [CrossRef]

51. Costa, M.S.S.P.; Costa, L.S.; Cordeiro, S.L.; Almeida-Lima, J.; Dantas-Santos, N.; Magalhães, K.D.; Sabry, D.A.; Albuquerque, I.R.L.; Pereira, M.R.; Leite, E.L.; et al. Evaluating the possible anticoagulant and antioxidant effects of sulfated polysaccharides from the tropical green alga *Caulerpa cupressoides* var. *flabellata*. *J. Appl. Phycol.* **2012**, *24*, 1159–1167. [CrossRef]

52. Ceroni, A.; Dell, A.; Haslam, S.M. The GlycanBuilder: a fast, intuitive and flexible software tool for building and displaying glycan structures. *Source Code Biol. Med.* **2007**, *2*, 3. [CrossRef]

53. Saad, O.M.; Leary, J.A. Delineating mechanisms of dissociation for isomeric heparin disaccharides using isotope labeling and ion trap tandem mass spectrometry. *J. Am. Soc. Mass Spectrom.* **2004**, *15*, 1274–1286. [CrossRef] [PubMed]

54. Qi, X.; Mao, W.; Gao, Y.; Chen, Y.; Chen, Y.; Zhao, C.; Li, N.; Wang, C.; Yan, M.; Lin, C.; et al. Chemical characteristic of an anticoagulant-active sulfated polysaccharide from *Enteromorpha clathrata*. *Carbohydr. Polym.* **2012**, *90*, 1804–1810. [CrossRef]

55. Wang, X.; Zhang, Z.; Yao, Z.; Zhao, M.; Qi, H. Sulfation, anticoagulant and antioxidant activities of polysaccharide from green algae *Enteromorpha linza*. *Int. J. Biol. Macromol.* **2013**, *58*, 225–230. [CrossRef]

56. Rodrigues, J.A.G.; Queiroz, I.N.L.D.; Quinderé, A.L.G.; Vairo, B.C.; Mourão, P.A.D.S.; Benevides, N.M.B. An antithrombin-dependent sulfated polysaccharide isolated from the green alga *Caulerpa cupressoides* has in vivo anti- and prothrombotic effects. *Ciênc. Rural* **2011**, *41*, 634–639. [CrossRef]

57. Li, N.; Mao, W.; Yan, M.; Liu, X.; Xia, Z.; Wang, S.; Xiao, B.; Chen, C.; Zhang, L.; Cao, S. Structural characterization and anticoagulant activity of a sulfated polysaccharide from the green alga *Codium divaricatum*. *Carbohydr. Polym.* **2015**, *121*, 175–182. [CrossRef]

58. Majdoub, H.; Mansour, M.B.; Chaubet, F.; Roudesli, M.S.; Maaroufi, R.M. Anticoagulant activity of a sulfated polysaccharide from the green alga *Arthrospira platensis*. *Biochim. Biophys. Acta BBA - Gen. Subj.* **2009**, *1790*, 1377–1381. [CrossRef]

59. Synytsya, A.; Choi, D.J.; Pohl, R.; Na, Y.S.; Capek, P.; Lattová, E.; Taubner, T.; Choi, J.W.; Lee, C.W.; Park, J.K.; et al. Structural Features and Anti-coagulant Activity of the Sulphated Polysaccharide SPS-CF from a Green Alga *Capsosiphon fulvescens*. *Mar. Biotechnol.* **2015**, *17*, 718–735. [CrossRef] [PubMed]

60. Melo, F.R.; Pereira, M.S.; Foguel, D.; Mourão, P.A.S. Antithrombin-mediated Anticoagulant Activity of Sulfated Polysaccharides: Different mechanisms for heparin and sulfated galactans. *J. Biol. Chem.* **2004**, *279*, 20824–20835. [CrossRef]

61. Mourão, P.A.S. Use of sulfated fucans as anticoagulant and antithrombotic agents: future perspectives. *Curr. Pharm. Des.* **2004**, *10*, 967–981. [CrossRef]

62. Becker, C.F.; Guimarães, J.A.; Mourão, P.A.S.; Verli, H. Conformation of sulfated galactan and sulfated fucan in aqueous solutions: Implications to their anticoagulant activities. *J. Mol. Graph. Model.* **2007**, *26*, 391–399. [CrossRef]

63. De Araújo, C.A.; Noseda, M.D.; Cipriani, T.R.; Gonçalves, A.G.; Duarte, M.E.R.; Ducatti, D.R.B. Selective sulfation of carrageenans and the influence of sulfate regiochemistry on anticoagulant properties. *Carbohydr. Polym.* **2013**, *91*, 483–491. [CrossRef]

64. Ménard, R.; Alban, S.; de Ruffray, P.; Jamois, F.; Franz, G.; Fritig, B.; Yvin, J.-C.; Kauffmann, S. Beta-1,3 glucan sulfate, but not beta-1,3 glucan, induces the salicylic acid signaling pathway in tobacco and *Arabidopsis*. *Plant Cell* **2004**, *16*, 3020–3032. [CrossRef] [PubMed]

65. Dubois, M.; Gilles, K.; Hamilton, J.K.; Rebers, P.A.; Smith, F. A Colorimetric Method for the Determination of Sugars. *Nature* **1951**, *168*, 167. [CrossRef]

66. Bitter, T.; Muir, H.M. A modified uronic acid carbazole reaction. *Anal. Biochem.* **1962**, *4*, 330–334. [CrossRef]

67. Bradford, M.M. A rapid and sensitive method for the quantitation of microgram quantities of protein utilizing the principle of protein-dye binding. *Anal. Biochem.* **1976**, *72*, 248–254. [CrossRef]

68. Jaques, L.B.; Balueux, R.E.; Dietrich, C.P.; Kavanagh, L.W. A microelectrophoresis method for heparin. *Can. J. Physiol. Pharmacol.* **1968**, *46*, 351–360. [CrossRef] [PubMed]

69. Gao, G.; Jiao, Q.; Ding, Y.; Chen, L. Study on quantitative assay of chondroitin sulfate with a spectrophotometric method of azure A. *Guang Pu Xue Yu Guang Pu Fen Xi* **2003**, *23*, 600–602.

70. Singleton, V.L.; Rossi, J.A. Colorimetry of Total Phenolics with Phosphomolybdic-Phosphotungstic Acid Reagents. *Am. J. Enol. Vitic.* **1965**, *16*, 144–158.

71. Ye, F.; Kuang, Y.; Chen, S.; Zhang, C.; Chen, Y.; Xing, X.-H. Characteristics of low molecular weight heparin production by an ultrafiltration membrane bioreactor using maltose binding protein fused heparinase I. *Biochem. Eng. J.* **2009**, *46*, 193–198. [CrossRef]

72. Miller, G.L. Use of Dinitrosalicylic Acid Reagent for Determination of Reducing Sugar. *Anal. Chem.* **1959**, *31*, 426–428. [CrossRef]

73. Mosmann, T. Rapid colorimetric assay for cellular growth and survival: Application to proliferation and cytotoxicity assays. *J. Immunol. Methods* **1983**, *65*, 55–63. [CrossRef]

 *marine drugs*

Article

Antitumour Potential of *Gigartina pistillata* Carrageenans against Colorectal Cancer Stem Cell-Enriched Tumourspheres

João Cotas [1,†], Vanda Marques [2,†], Marta B. Afonso [2], Cecília M. P. Rodrigues [2] and Leonel Pereira [1,3,*]

1   MARE—Marine and Environmental Sciences Centre, Faculty of Science and Technology, University of Coimbra, 3001-456 Coimbra, Portugal; jcotas@gmail.com
2   Research Institute for Medicines (iMed.ULisboa), Faculty of Pharmacy, University of Lisboa, 1649-003 Lisboa, Portugal; vismsmarques@ff.ulisboa.pt (V.M.); mbafonso@ff.ulisboa.pt (M.B.A.); cmprodrigues@ff.ulisboa.pt (C.M.P.R.)
3   Department of Life Sciences, University of Coimbra, 3000-456 Coimbra, Portugal
*   Correspondence: leonel@bot.uc.pt; Tel.: +351-239-855-229
†   These authors contributed equally to this work.

Received: 26 November 2019; Accepted: 10 January 2020; Published: 12 January 2020

Abstract: *Gigartina pistillata* is a red seaweed common in Figueira da Foz, Portugal. Here, the antitumour potential of *G. pistillata* carrageenan, with a known variable of the life cycle, the female gametophyte (FG) and tetrasporophyte (T) was evaluated against colorectal cancer stem cell (CSC) -enriched tumourspheres. FTIR-ATR analysis of *G. pistillata* carrageenan extracts indicated differences between life cycle phases, being FG a κ/ι hybrid carrageenan and T a λ/ξ hybrid. Both carrageenan extracts presented IC_{50} values inferior to 1 µg/mL in HT29-derived CSC-enriched tumourspheres, as well as reduced tumoursphere area. The two extracts were also effective at reducing cellular viability in SW620- and SW480-derived tumourspheres. These results indicate that carrageenans extracted from two *G. pistillata* life cycle phases have antitumour potential against colorectal cancer stem-like cells, specially the T carrageenan.

Keywords: antitumour activity; carrageenan; colorectal cancer; cancer stem cells

1. Introduction

Seaweeds are turning into one of the most attractive natural resources of compounds that can replace chemically synthetized compounds, such as plastic and other petroleum-based products, or those from animal origin, such as collagen (gelatin). Seaweeds main structural component is a polysaccharide-based type; its main function is to maintain the structure of the cell wall, with alginic acid being the main polysaccharide produced in brown algae (Ochrophyta, Phaeophyceae), and agar (agarophytes) and carrageenan (carrageenophytes) in red algae (Rhodophyta).

The use of seaweed biopolymers is increasing in the food industry as natural gelling and emulsifier agents [1]. Carrageenans are one of the main natural texturizing agents in food applications (dairy products, jellies, pet foods, sauces) being considered safe food additives [2]. They are also used in pharmacological formulations, in cosmetics, as biomedical polymer compounds or as lubricant [2–4]. Overall, carrageenans principal functions in industry are as a gelling, stabilizing, and viscosity-building agent.

Diverse types of carrageenan are acquired from different species of the Gigartinales (Rhodophyta). Kappa (κ)-carrageenan is mostly acquired by extraction from the tropical seaweed *Kappaphycus alvarezii* (identified as "cottonii", in the seaweed commercial area related to the food industry), while iota (ι)-carrageenan is mainly extracted from *Eucheuma denticulatum* (commercial denomination "spinosum"). In turn, lambda (λ)-carrageenan is acquired from various species from the genera *Gigartina* and *Chondrus* (commercial denomination "Irish moss").

This unique hydrocolloid consists of alternating galactose and 3,6-anhydrogalactose sugars linked by alternating α-1,3 and β-1,4 glycosidic linkages [5,6]. Carrageenans can be classified as λ, κ, and ι according to the number of sulphated groups by galactose unit (Figure 1), where number, chemical location, and arrangement of these groups defines carrageenan function and bioactivity power [2,7].

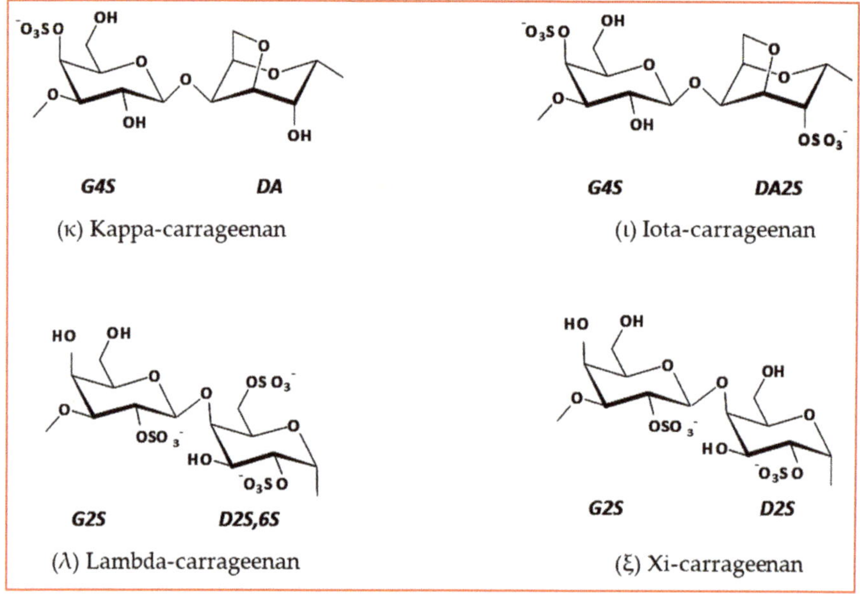

Figure 1. Idealized structure of the chemical units of different types of carrageenan [8].

The Xi (ξ)-carrageenan, a non-commercial carrageenan, has two sulphated ester groups, minus one that is the λ-carrageenan but identical to ι-carrageenan. ξ-carrageenan is a viscous type carrageenan due to the number of sulphated esters present in the structure [8]. The k-carrageenan only has a sulphated ester group by the galactose unit.

Commonly, seaweeds do not produce these flawless and clean carrageenans, but instead a full variety of hybrid configurations of carrageenan within the species life cycle [8]. Intrinsic carrageenans always show a complex hybrid chemical conformation and are frequently a combination of galactans composed of different carrabiose types, where proportions and structures change within the species, ecological, physiological, and developmental conditions [9,10]. Hence, carrageenans are a group of polysaccharides with molecular weight varying between 30 and 5000 kDa but the average molecular weight of extracted carrageenans is between 200 and 800 kDa. [5,11]. These huge polysaccharides are nonnutritive and after ingestion have an analogous effect to dietary fibers, being extremely resistant in the digestive tract [5].

Both the Food and Drugs Administration (FDA) and the European Food Safety Agency (EFSA) have approved the commercial forms of λ-, κ-, and ι-carrageenans for the food industry [5,11]. In the European Union, the carrageenan application is regulated by the Commission Regulation (EU) No 231/2012, which defines that the commercial carrageenan (E 407) comprises essentially potassium, sodium, magnesium, and calcium sulphate esters of galactose and 3,6-anhydrogalactose polysaccharides [11]. Sub-chronic toxicity studies performed in rats showed that a dose of E 407 between 3400–3900 mg/kg body weight per day had no-observed-adverse-effect, fulfilling the EU specification for food level carrageenan food level [11]. On the other hand, the poligeenan, a degraded ι-carrageenan with average molecular weight of 10–20 kDa, is not authorized in food applications within the European Union area. In fact, degraded carrageenans, also known as artificial products derived from carrageenan, are associated with adverse effects [5,12].

Carrageenans are water soluble polymers, their solubility being determined by the type of carrageenan derivation, temperature, pH, and the counter ion in the dissolving solution, where κ-carrageenan is the most soluble. The sodium salt of κ-carrageenan is soluble in cold water, but the potassium salt is soluble only by heating. ι-carrageenan has an intermediate solubility [2]. All types of carrageenan are insoluble in organic solvents including alcohols and ketones [13].

The carrageenan extraction method plays a major role in purity control. As shown by EFSA, when carrageenans are produced by alcohol procedure, they contain approximately 90% anhydrous carrageenan, 8% moisture, and 2% inorganic salts (mainly chlorides), while those manufactured by gel press procedure contain about 77% anhydrous carrageenan, 8% moisture, and up to 15% inorganic salts [11].

Gigartina pistillata (Figure 2) is an edible red seaweed found in both Northeast and Southeast Atlantic and Southeast Asia, and a carrageenan resource for extraction industry. Its morphology is described as the type species of the genus *Gigartina* and their thalli are erect, up to 20 cm tall, dark-red or red-brown, cartilaginous, elastic, dichotomously branched, attached to the substratum through a small disk [14]. *G. pistillata* can show a rare presence of the heterosporic thalli (i.e., producing tetraspores and carpospores in the thalli of one specimen), despite although having an isomorphic triphasic life cycle [6]. In Gigartinaceae, the life cycle phase strongly disturbs carrageenans configuration. Gametophytic life cycle yields a κ/ι-type carrageenans, while tetrasporophytic life cycle yields a λ-type carrageenans, as shown by Pereira et al. [6,9,15,16].

Figure 2. *G. pistillata* in nature, fixed in a rocky platform in Praia do Cabo Mondego, Figueira da Foz, Portugal.

The maximum carrageenan content extracted in *G. pistillata* in the work of Pereira [6,17] was obtained from female gametophyte (FG) samples, with 59.7% of dry weight in late spring; a sample of heterosporic thalli had the minimum value in late autumn, with 22.7% of dry weight.

In the last decades, the biological potential of carrageenans has been explored. Since the 1960s, carrageenan anticoagulant and antithrombotic activity has been studied with λ-carrageenan showing higher anticoagulant potential than κ-carrageenan [18]. Carrageenans have also been shown to selectively inhibit many enveloped viruses [19] and to have antioxidant properties [20]. Carrageenan antitumour potential has only been studied more recently, with several in vivo and in vitro models proposing an antiproliferative action against tumour cells [21], being compounds from marine bio-sources increasingly relevant in the quest to find new biomolecules with antitumour potential.

Cancer is undoubtedly a major cause of worldwide morbidity and mortality, with 18.1 million new cases and 9.6 million deaths estimated in 2018, and with an increasing social and economic impact. Cancer in the gastrointestinal tract plays an important role in these estimates with colorectal cancer being the third most incident (10.2%) and second most mortal (9.2%) [22]. Several studies have highlighted the role of cancer stem cells (CSCs) in colorectal cancer development, progression, and resistance to therapy [23–27].

The increasing evidence of the existence of CSCs has challenged the classical stochastic model of cancer dynamics and has laid the foundation for the hierarchical model of tumourigenesis, where the tumour mass is viewed as a hierarchical and unidirectional system, with CSCs in a foundational position giving rise to heterogeneous cell populations. CSCs display high self-renewal capacities, plasticity potential, high resistance to tumour microenvironment stress factors and quiescence. These properties are believed to be responsible for CSC resistance to chemotherapy, cancer relapse, and metastization [27–30]. The key role of CSCs in cancer development and recurrence opens new venues for novel therapeutic strategies based on the specific and effective targeting of this cell subpopulation. For instance, drugs that selectively inhibit CSC growth, induce cell death, force differentiation, or modulate microenvironment are promising therapeutic strategies [27,30]. By combining CSC-targeting drugs and conventional anti-cancer therapies, one can target both bulk cancer cells and CSCs, being more effective and faster in eradicating the tumour. However, it is important that noncancerous somatic stem cells are spared, particularly in tissues with high cellular turnover rates such as colon [30]. Great efforts have been made in this direction, trying to find novel drugs that specifically target CSC, particularly bioactive compounds from plant sources, where compounds such as resveratrol or curcumin show promising results [29]. In this matter, compounds from marine biosources have been less explored. Nevertheless, compounds isolated from red algae have shown some potential against CSC in an in vitro model of breast cancer [31].

Here, *G. pistillata* carrageenans from two life cycle phases, FG and tetrasporophyte (T) were analyzed by Fourier transform infrared attenuated total reflectance (FTIR-ATR) and their potential against colorectal CSC-enriched tumourspheres was explored.

2. Results

2.1. Carrageenan Extraction Yields

Through the alkali extraction method, from 1 g of dried seaweed, it was possible to obtain 17% ± 0.1 of FG carrageenan and 31% ± 3.2 of T carrageenan.

2.2. FTIR-ATR

Carrageeneenan extracts from *G. pistillata* FG and T life cycles were obtained and analyzed by FTIR-ATR. This technique allowed carrageenan characterization in a fast, nondestructive manner, requiring small amounts of the sample [16]. The obtained spectra were reviewed with bibliographic support [6,8] (Figure 3 and Table 1).

In the *G. pistillata*'s FG FTIR-ATR spectrum (Figure 3a), the first band to emerge is the low intensity 802 band, corresponding to DA2S bond, indicating the presence of ι-carrageenan, while the bands at 845 and 927 cm^{-1}, G4S and DA bonds, respectively, with very close intensities, indicate the presence of κ-carrageenan. A strong percentage of κ-carrageenan is also supported by the 1066 cm^{-1} band (DA). The 970 cm^{-1} band is related to the presence of galactose units.

In the *G. pistillata* T spectrum (Figure 3b) the first band appears at 830 cm^{-1}, corresponding to the G/D2S, this indicates the presence of λ-carrageenan. The following band at 930 cm^{-1}, appears as a low intensity shoulder, in opposition to the strong band observed in the FG carrageenan, the same band (the 927 cm^{-1}, belonging to the same type of bond) in the spectrum presents a more noted band.

Bands at 1007 and 1214 cm^{-1} correspond to sulphated esters groups which dominate the spectrum with their very intense and broad profile, indicating a noted presence of sulphated esters groups in the T carrageenan, because T forms a nongelling carrageenan. It is noted that the 1214 cm^{-1} can be derived from Xi (ξ) carrageenan.

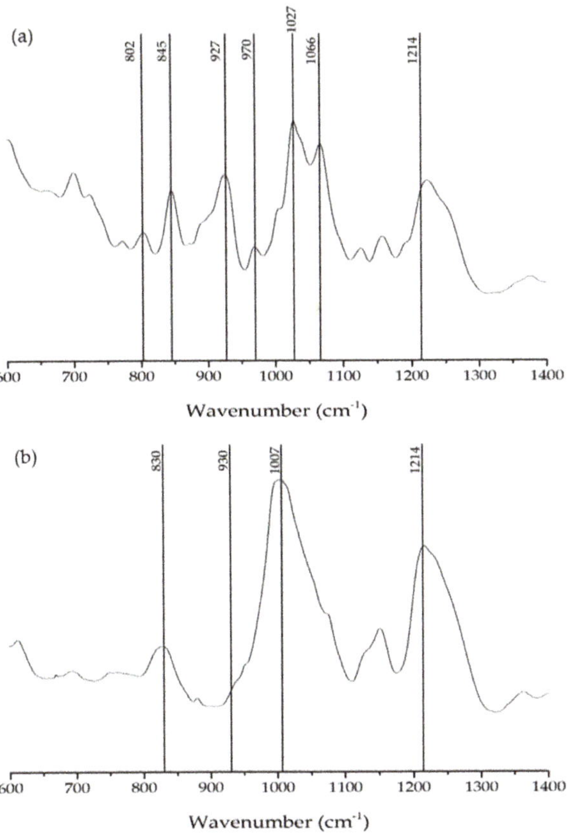

Figure 3. FTIR-ATR spectra of the carrageenan extracted from FG (Female gametophyte) (**a**) and T (tetrasporophyte) (**b**) *G. pistillata* life cycle.

Table 1. FTIR-ATR bands identification and characterization of the *G. pistillata* FG and T life cycle.

Wavelength Numbers (cm^{-1})	Chemical Structure	Bonds/Assignments	Life Cycle Presence
802	C–O–SO$_3$ on C$_2$ of 3,6-anhydrogalactose	DA2S	FG
830	C–O–SO$_3$ on C$_2$ of galactose	G/D2S	T
845	C–O–SO$_3$ on C$_4$ of galactose	G4S	FG
927	C–O of 3,6-anydrogalactose	DA	FG
930	C–O of 3,6-anydrogalactose	DA	T
970	Galactose	G/D	FG
1007	S=O	sulphated esters	T
1027	S=O	sulphated esters	FG
1066	C–O of 3,6-anydrogalactose	DA	FG
1214	S=O	sulphated esters	FG/T

2.3. Anti-CSC Potential of Carrageenan Extracts

The tumoursphere forming assay [32] allows the establishment of 3D spheroids enriched in CSCs from originally 2D adherent colorectal cancer cell cultures (Figure 4a). This cellular model was applied to evaluate the potential of T and FG carrageenan extracts against CSC-enriched tumourspheres derived from colorectal cancer cell lines. In CSC-enriched tumourspheres derived from HT29 cell line, both extracts presented IC_{50} values inferior to 1 µg/mL (0.6572 and 0.7050 µg/mL for FG and T extract, respectively), which are similar to salinomycin (0.1450 µg/mL), a known CSC-targeting agent (Figure 4b) [33].

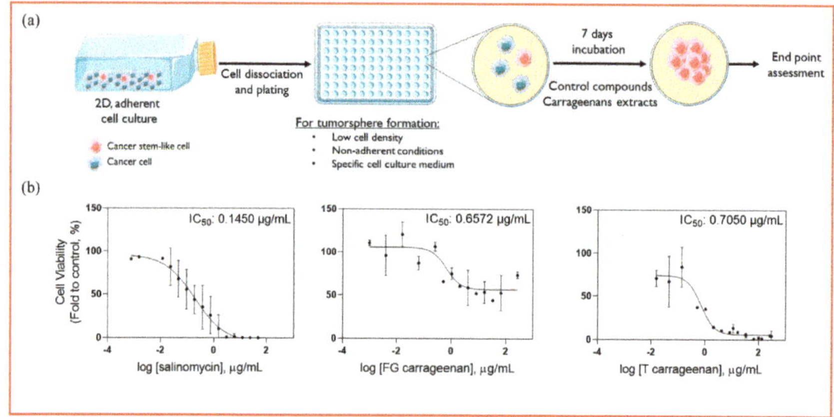

Figure 4. Carrageenan extracts reduce cellular viability in cancer stem cell (CSC)-enriched tumourspheres derived from HT29 colorectal cancer cell line. (**a**) Strategy for CSC-enriched tumoursphere formation; (**b**) IC_{50} determination for salinomycin, T and FG carrageenans in HT29 CSC-enriched tumourspheres. Cell viability was determined by ATP (adenosine triphosphate) metabolism assays. Results are depicted as mean ± SD, $n = 3$.

To assess if this effect was cell line dependent, FG and T carrageenans were incubated at two different concentrations (33 and 100 µg/mL) with CSC-enriched tumourspheres derived from three different human colorectal cell lines: SW620, SW480, and HCT116 cells. T carrageenan extract markedly reduced cellular viability in SW620- and SW480-derived tumourspheres in a dose-independent manner (< 20% for both cell lines), while FG carrageenan was more effective in SW480-derived tumourspheres (< 30% for SW480 and ~50% for SW620) (Figure 5). Reduction in cellular viability was less evident in HTC116-derived tumourspheres (~75% for both T and FG carrageenans) (Figure 5).

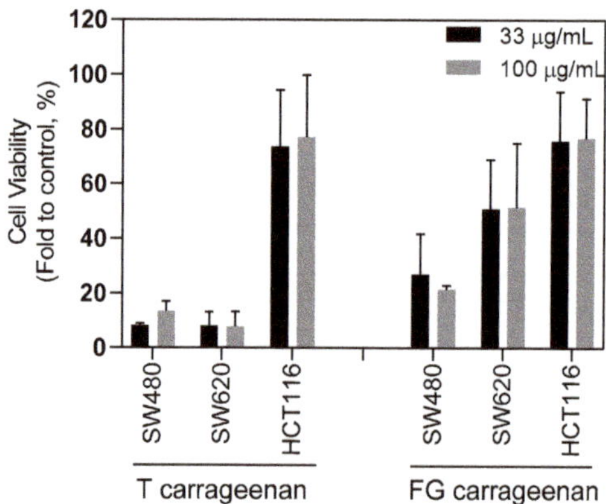

Figure 5. Carrageenan extracts have broad anti-CSC action. Both T and FG carrageenans impact at different extents in CSC-enriched tumourspheres derived from SW620, SW480, and HCT116 cell lines, in a dose-independent manner. Cell viability was determined by ATP metabolism assays. Results are depicted as mean ± SEM, $n = 3$.

Moreover, in HT29-derived tumourspheres, both T and FG extracts impacted on tumoursphere formation, by significantly reducing the tumoursphere area when incubated at their IC_{50} dose ($p = 0.0017$ and $p = 0.0056$ for T and FG, respectively) (Figure 6), similarly to salinomycin control ($p = 0.0007$). On the other hand, tumoursphere numbers slightly increased, particularly for FG carrageenan, although without statistical significance (Figure 6).

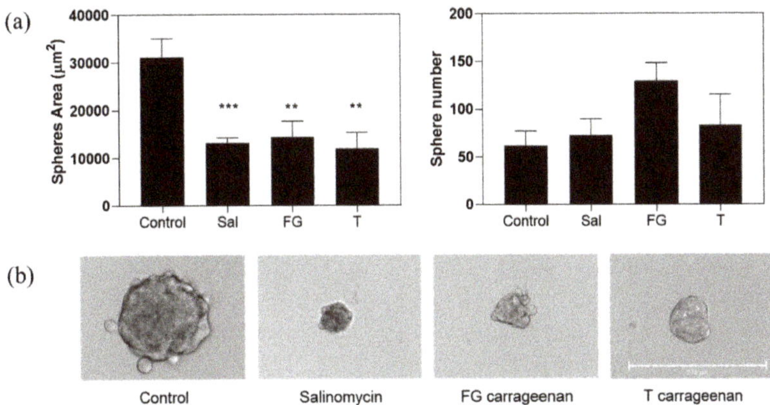

Figure 6. Carrageenan extracts impact on HT29-derived CSC-enriched tumoursphere formation when incubated at their respective IC_{50} dose (FG carrageenan: 0.6572 µg/mL; T carrageenan: 0.7050 µg/mL; salinomycin: 0.1450 µg/mL). (**a**) A significant reduction on tumoursphere size is observed, however tumoursphere number is not significantly altered. Results are depicted as mean ± SEM, $n = 3$. ** $p < 0.005$; *** $p < 0.001$. Sal—salinomycin. (**b**) Representative images of seven-day tumourspheres at 10× magnification. Scale bar, 275 µm.

3. Discussion

Seaweed collection sites and season confer variability of the dominant generation (between gametophytes and tetrasporophytes), and reproductive cycle stage specially affects the carrageenans structure [10,17]. Here, the FG carrageenan yield (17%) was lower when compared with previous reports (59.7%) [17]. This could be explained by differences on the collection season, since in this study sampling was performed in late August and not in the Spring, which is considered the best season [17]. Concerning T carrageenan, the yield (31%) was comparable to that reported before by Amimi et al. (37%) [10].

The extracellular matrix of carrageenans resides mainly in the hydrophilic sulphated polygalactans. The existence of D-galactose and anhydrous D-galactose differentiates the highly sulphated carrageenans from the fewer sulphated agar (anhydrous-L-galactose) [34]. Analyzing the FTIR-ATR spectra of FG *G. pistillata*, strong absorption bands can be observed at ~927 cm^{-1} which is characteristic of carrageenan κ, ι, beta (β), and theta (θ), indicating the presence of C–O of 3,6-anydrogalactose (DA). Bands in the 845 cm^{-1} region (characteristic of the ι and κ carrageenan spectra, and mu (µ) and nu (ν) biological precursors of κ-carrageenan), indicate the presence of C–O–SO$_3$ on C$_4$ of galactose (G4S), and bands in the 970 cm^{-1} area indicate the presence of galactose (G/D), which is characteristic of κ, ι, β, and θ carrageenans spectra. However, the FG spectrum presented absorbance in the 802 cm^{-1} region indicating the presence of C–O–SO$_3$ on C$_2$ of 3,6-anhydrogalactose (DA2S), which is only observed in the ι and θ carrageenan spectra, meaning the presence of small amounts of ι carrageenan [8,16]. These results indicate that the FG carrageenan spectrum belongs to a κ/ι hybrid carrageenan, which is corroborated by previous results [16]. In turn, FTIR-ATR spectra of *G. pistillata* T carrageenan showed a broad band in the 830 cm^{-1} spectral region, indicating the presence of C–O–SO$_3$ on C$_2$ of galactose (G/D2S), which is observed in the λ-carrageenan spectra and λ-carrageenan precursors. In these spectra, bands in the 930 cm^{-1} region indicate the presence of 3,6-anhydrogalactose (DA), characteristic of the κ, ι, β and θ carrageenan spectra [8,16]. Therefore, this spectrum belongs to a hybrid carrageenan λ/ξ spectrum, mostly λ type. Both T and FG carrageenan spectra share strong absorption bands in the 1000–1100 and 1210–1260 cm^{-1} regions, corresponding to sulphated esters groups (S=O), which are always present in sulphated polysaccharide samples, in contrast to other vibrational bands that are characteristic of carrageenans [8]. These results are corroborated by the studies realized by Amimi [9] and Pereira [16] that proved the production of a hybrid λ/ξ carrageenan by *G. pistillata* T and a hybrid κ/ι carrageenan by the *G. pistillata* FG.

Sphere forming assays have been extensively used to evaluate the potential of conventional and novel molecules to target CSCs, since it has been shown that culturing 3D spheres in specific experimental conditions promotes CSC growth and propagation [29,35–37]. The anti-CSC potential of T and FG carrageenan extracts from *G. pistillata* was evaluated using a recently described CSC-enriched tumoursphere forming assay [32], evidencing their bioactivity particularly for T extract. Both T and FG extracts presented IC$_{50}$ values inferior to 1 μg/mL in HT29-derived CSC-enriched tumourspheres. However, when tested in tumourspheres derived from other colorectal cancer cell lines, T extract, with λ/ξ hybrid carrageenans, was more effective in reducing cellular viability, particularly in SW480 and SW620 cell lines, but not in HCT116. This may be since HT29, SW620, and SW480 cell lines share genetic features such as *TP53* and *APC* mutations or chromosomal instability, which are not present in the HCT116 cell line, suggesting potential mechanisms of action. In addition, by impacting on tumoursphere formation through reducing the tumoursphere area, the potential anti-CSC action of T and FG extracts may be due to antiproliferative activity. Conversely, the increase of tumoursphere number could suggest an increase of stemness properties; however, these changes are not significant when compared to the control. Carrageenan extract from *Kappaphycus alvarezii* inhibits cell growth in several cancer cell lines, including HT29 colon cancer cell line with an IC$_{50}$ value of 73.87 μg/mL [38]. Further, it has been demonstrated that carrageenans can delay cell cycle progression in HeLa cells, where λ carrageenans prolonged cell cycle by stalling in G1 and G2/M phases, therefore presenting a strong antiproliferative activity by preventing cellular division [39].

Interestingly, FITR-ATR analysis showed a presence of higher sulphate ester groups in T carrageenan compared to FG carrageenan. The quantity of sulphated esters in the carrageenan extracts may possibly be correlated with the observed antitumour effects. However, previous reports are conflicting. While Murad et al. showed that sulphated carrageenans from the red algae *Palisada perforata* (formerly *Laurencia papillosa*) induced apoptosis in a breast cancer in vitro model [40], Calvo et al. [41] reported that the presence of sulphate groups in native carrageenans diminished their cytotoxic activity in a murine mammary adenocarcinoma cell line.

Yet, several factors must be taken into consideration concerning the use of carrageenans. Although without major toxic effects [21], it has been described that λ carrageenans activate the Wnt/β-Catenin pathway in normal human colonocytes which may induce the development of intestinal polyps [42]. Moreover, one must be aware of a putative carrageenans proinflammatory potential, where reports are again conflicting. While in the past, carrageenans have been shown to induce the formation of ulcerative lesions in a guinea pig model [21] and have been used to induce paw edema in animal models [43], more recent data suggest that carrageenans do not induce the expression on proinflammatory proteins in vitro models [5]. Nevertheless, the immunomodulating effect of λ-carrageenan was explored before, suggesting that this carrageenan could be an efficient adjuvant in cancer immunotherapy by not only inhibiting cell growth but also by enhancing tumour immune response [44].

Overall, carrageenans seem to be a potential new therapeutic strategy against cancer, including therapy against CSC. Further studies are required to evaluate T and FG carrageenans toxicity, CSC-specificity, and potential mechanisms of action.

4. Materials and Methods

4.1. Reagents

Methanol was purchased from José Manuel Gomes dos Santos, Lda., Odivelas, Portugal and acetone from the Ceamed, Lda., Funchal, Portugal. Ethanol was obtained from Valente e Ribeiro. Lda., Belas, Portugal and sodium hydroxide from Sigma-Aldrich GmbH, Steinheim, Germany.

4.2. Seaweed Collection

The specimens of *G. pistillata* (S.G. Gmelin) Stackhouse were collected in Praia do Cabo Mondego, Buarcos, Figueira da Foz (40° 10′ 18.6″ N, 8° 53′ 44.4″ W), Portugal. Sampling was conducted in August 2018 from the sites with well-established *G. pistillata* patches and without epiphytes or degradation visible at eyesight. Once harvested, samples were stored in plastic bags for transport to the laboratory, in a cool box. All samples were washed thoroughly with filtered seawater to remove sand, and epiphytes. Specimens of *G. pistillata* were separated according to the different life cycle phases—T and FG, using a magnifying glass, washed briefly with distilled water to remove salts and dried in an air force oven (Raypa DAF-135, R. Espinar S.L., Barcelona, Spain) at 40 °C, 48 h. The dried algae were finely ground with a commercial mill (Taurus aromatic, Oliana, Spain) ($\leq$ 1 mm) in order to render the samples uniform, and then, stored in a dark room, in a box with silica gel to reduce the humidity, at ambient temperature ($\pm$ 24 °C).

4.3. Carrageenan Refined Alkali Extraction

Alkali extraction was preformed according to the method described by Pereira et al. [15]. The milled seaweed was weighed in a scale (Radwag WLC 1/A2, Radwag, Radom, Poland) and 1 g samples of the FG and T phases was used ($n = 3$). Before extraction, the milled seaweed material (1 g) was resuspended and pretreated with an acetone:methanol (1:1) solution in a final concentration of 1% (m/v) for 16 h, at 4 °C, to eliminate the organic-soluble fraction. The liquid solution was decanted, and the seaweed residues obtained were dried in an air force oven (Raypa DAF-135, R. Espinar S.L., Barcelona, Spain) at 40 °C before the extraction method. The samples were placed in 150 mL of NaOH (1 M) (1 g of initial seaweed: 150 mL of NaOH solution) in a hot water bath system (GFL 1003, GFL, Burgwedel, Germany), at 85–90 °C, for 3 h. The solutions were hot filtered, twice, under vacuum, through a cloth filter supported in a Buchner funnel and a Kitasato flask. The extract was evaporated (rotary evaporator model: 2600000, Witeg, Germany) under vacuum to one-third of the initial volume. The carrageenan was precipitated by adding the warm solution to twice its volume of 96% ethanol. The precipitated carrageenan was washed with ethanol, 48 h at 4 °C, before drying in an air force oven. After drying, the carrageenan was maintained in a close air sample flask stored in a dark and no humidity local, at room temperature before the antitumour and FTIR-ATR assay. For biological assays, the dried carrageenans extracts were grinded and dissolved in ultra-pure, sterile water at a final concentration of 5 mg/mL.

4.4. FTIR-ATR Assay

For FTIR-ATR analysis, the T and FG dried carrageenan extracts were milled using a commercial mill (Taurus aromatic, Spain) to obtain a fine powder, which was subjected to direct analysis. FTIR-ATR spectra were recorded on an IFS 55 spectrometer, using a Golden Gate single reflection diamond ATR system, with no need for sample preparation, since these assays only required dried samples, according to Pereira et al. [6]. All spectra are the average of two independent measurements with 128 scans, each at a resolution of 2 cm^{-1}.

4.5. Cell Culture

Human colorectal carcinoma cell lines HT29, HCT116, SW620, and SW480 were obtained from ECACC (Porton Down, UK). Cell lines were cultured under adherent conditions in RPMI (HT29), McCoy's 5A (HCT116) and Dulbecco's modified Eagle's medium (DMEM; SW620 and SW480), all supplemented with 10% (*v/v*) heat-inactivated fetal bovine serum (FBS) and 1% (*v/v*) antibiotic/antimycotic solution (all from Gibco, Thermo Fisher Scientific, Paisley, UK). All cell cultures were maintained at 37 °C under a humidified atmosphere of 5% CO_2 (HeraCell 150i CO2 incubator, Thermo Fisher Scientific, Paisley, UK).

4.6. Tumoursphere Forming Assay

For the generation of tumourspheres, cell lines were grown as previously described [32]. Briefly, cells were plated at low density (100 and 500 cells/well in 96- and 24-well plates, respectively), in ultra-low attachment plates in a serum-free DMEM/F12 medium supplemented with 1% penicillin-streptavidin, 1% nonessential amino acids, 1% sodium pyruvate, 2% B27 supplement, 1% N2 supplement, 40 ng/mL recombinant human epidermal growth factor (all from Gibco), 4 µg/mL heparin (Biochrom GmbH, Berlin, Germany), and 20 ng/mL recombinant human basic fibroblast growth factor (Peprotech, London, UK).

4.7. IC$_{50}$ Determination

To quantitatively assess the carrageenan-extract inhibitory potency in CSC-enriched tumourspheres, their half maximal inhibitory concentration (IC$_{50}$) was determined through a 10-point dose-response curve. Briefly, HT29 cells were plated in 96-well plates as described above and treated with carrageenan extracts (concentrations ranging from 300 to 0.015 µg/mL), positive control (salinomycin (Sigma-Aldrich, St. Louis, MI, USA); concentrations ranging from 13 to 0.023 µg/mL), or vehicle control (DMSO). Following seven days of incubation, cell viability was assessed based on measurement of ATP metabolism using CellTiter-Glo™ luminescent cell viability assay (Promega, Madison, WI, USA). Luminescence signal was recorded using a GloMax®-MultiDetection System (Promega). IC$_{50}$ determination was preformed using GraphPad Prism 8.0.2. software.

4.8. Validation of CSC-Targeting Action

To assess whether the carrageenan extracts have a generic effect on colorectal CSCs or whether this is a cell line dependent effect, they were tested on CSC-enriched tumourspheres from other colorectal cell lines. Hence, HCT116, SW480, and SW620 cell lines were cultured in 96-well plates as described above to promote CSC-enriched tumoursphere formation and treated with carrageenan extracts (100 and 33.3 µg/mL), positive control (salinomycin, at 0.751 µg/mL), or vehicle control (DMSO). Following seven days of incubation, cell viability was assessed based on measurement of ATP as described before.

4.9. Impact on Tumoursphere Formation

The anti-CSC effect of the carrageenan extracts was further validated by assessing their impact on tumoursphere formation. Hence, HT29 cells were plated in 24-well plates as described before and treated with hit carrageenan extracts, positive control (salinomycin) all at their IC$_{50}$ dose, and vehicle control (DMSO). After seven days of incubation, tumoursphere images were acquired through brightfield microscopy with Invitrogen EVOS™ FL Auto2 imaging system (Invitrogen, Thermo Fisher Scientific). Tumourspheres number and area were determined using ImageJ analysis software (version 1.52a, NIH, USA). Data analysis was preformed using GraphPad Prism 8.0.2. software. A *p*-value inferior to 0.05 was considered significant.

5. Conclusions

This work shows that the carrageenans extracted from the two *G. pistillata* life cycle phases, particularly the T carrageenan, have potential against colorectal cancer CSC-like cells. This could be explained by the higher sulphated ethers content in T carrageenan (λ/ξ) comparatively with the FG carrageenan (κ/ι) with support by the identification via FTIR-ATR. Little is known about the interaction between CSCs and carrageenan, and how this can affect human colorectal cell function. Hence, further studies are needed to understand the carrageenan anti-CSC mechanism of action, particularly for the T carrageenan.

Author Contributions: For Harvest, Seaweed Identification, Alkali Extraction, and Writing, J.C.; Biological Assays and Writing, V.M.; Biological Assays, M.B.A.; FTIR-ATR Analysis, L.P.; Supervision and Manuscript Revision, C.M.P.R. and L.P. All authors have read and agreed to the published version of the manuscript.

Funding: This work has the funding from European Structural & Investment Funds through the COMPETE Programme and from National Funds through Fundação para a Ciência e a Tecnologia (FCT) under the Programme grant SAICTPAC/0019/2015 LISBOA-01-0145-FEDER-016405, and through the strategic projects UID/MAR/04292/2019 (MARE) and UID/DTP/04138/2019 (iMed.ULisboa).

Acknowledgments: The authors thank Paulo Ribeiro Claro (CICECO, University of Aveiro, Portugal) for obtaining the FTIR-ATR spectra.

Conflicts of Interest: The authors declare no conflict of interest.

References

1. Weiner, M.L. Food additive carrageenan: Part II: A critical review of carrageenan in vivo safety studies. *Crit. Rev. Toxicol.* **2014**, *44*, 244–269. [CrossRef]
2. van de Velde, F.; De Ruiter, D.G.A. De Carrageenan. In *Biopolymers Online*; Vandamme, E.J., De Baets, S., Steinbüchel, A., Eds.; Wiley-VCH Verlag GmbH & Co. KGaA: Weinheim, Germany, 2005.
3. Pereira, L. Identification of phycocolloids by vibrational spectroscopy. In *World Seaweed Resources—An Authoritative Reference System*; Critchley, A.T., Ohno, M., Largo, D.B., Eds.; ETI Information Services Ltd.: Amsterdam, The Netherlands, 1998.
4. Imeson, A. Carrageenan. In *Handbook of Hydrocolloids*; Phillips, G.O., Williams, P.A., Eds.; CRC Press: Boca Raton, FL, USA, 2000; pp. 87–102. ISBN 9781845694142.
5. McKim, J.M.; Baas, H.; Rice, G.P.; Willoughby, J.A.; Weiner, M.L.; Blakemore, W. Effects of carrageenan on cell permeability, cytotoxicity, and cytokine gene expression in human intestinal and hepatic cell lines. *Food Chem. Toxicol.* **2016**, *96*, 1–10. [CrossRef] [PubMed]
6. Pereira, L.; Gheda, S.F.; Ribeiro-claro, P.J. Analysis by Vibrational Spectroscopy of Seaweed Polysaccharides with Potential Use in Food, Pharmaceutical, and Cosmetic Industries. *Int. J. Carbohydr. Chem.* **2013**, *2013*, 1–7. [CrossRef]
7. Blakemore, W.R.; Harpell, A.R. Carrageenan. In *Food Stabilisers, Thickeners and Gelling Agents*; Imeson, A., Ed.; Wiley-Blackwell: Oxford, UK, 2009; ISBN 9781444314724.
8. Pereira, L.; Amado, A.M.; Critchley, A.T.; van de Velde, F.; Ribeiro-Claro, P.J. Identification of selected seaweed polysaccharides (phycocolloids) by vibrational spectroscopy (FTIR-ATR and FT-Raman). *Food Hydrocoll.* **2009**, *23*, 1903–1909. [CrossRef]
9. Amimi, A.; Mouradi, A.; Givernaud, T.; Chiadmi, N.; Lahaye, M. Structural analysis of *Gigartina pistillata* carrageenans (Gigartinaceae, Rhodophyta). *Carbohydr. Res.* **2001**, *333*, 271–279. [CrossRef]
10. Amimi, A.; Mouradi, A.; Bennasser, L.; Givernaud, T. Seasonal variations in thalli and carrageenan composition of *Gigartina pistillata* (Gmelin) Stackhouse (Rhodophyta, Gigartinales) harvested along the Atlantic coast of Morocco. *Phycol. Res.* **2007**, *55*, 143–149. [CrossRef]
11. Younes, M.; Aggett, P.; Aguilar, F.; Crebelli, R.; Filipič, M.; Frutos, M.J.; Galtier, P.; Gott, D.; Gundert-Remy, U.; Kuhnle, G.G.; et al. Re-evaluation of carrageenan (E 407) and processed Eucheuma seaweed (E 407a) as food additives. *EFSA J.* **2018**, *16*. [CrossRef]
12. Cohen, S.M.; Ito, N. A Critical Review of the Toxicological Effects of Carrageenan and Processed Eucheuma Seaweed on the Gastrointestinal Tract. *Crit. Rev. Toxicol.* **2002**, *32*, 413–444. [CrossRef]

13. Marburger, A. Alginate und Carrageenane—Eigenschaften, Gewinnung und Anwendungen in Schule und Hochschule. Ph.D. Thesis, Philipps-Universität Marburg, Marburg, Germany, 2003.

14. Pereira, L. *Edible Seaweeds of the World*; CRC Press: Boca Raton, FL, USA, 2016; ISBN 9780429154041.

15. Pereira, L.; Van De Velde, F. Portuguese carrageenophytes: Carrageenan composition and geographic distribution of eight species (Gigartinales, Rhodophyta). *Carbohydr. Polym.* **2011**, *84*, 614–623. [CrossRef]

16. Pereira, L.; Mesquita, J.F. Carrageenophytes of occidental Portuguese coast: 1-spectroscopic analysis in eight carrageenophytes from Buarcos bay. *Biomol. Eng.* **2003**, *20*, 217–222. [CrossRef]

17. Pereira, L. Population studies and carrageenan properties in eight Gigartinales (Rhodophyta) from Western Coast of Portugal. *Sci. World J.* **2013**, *2013*, 939830. [CrossRef] [PubMed]

18. Shanmugam, M.; Mody, K.H. Heparinoid-active sulphated polysaccharides from marine algae as potential blood anticoagulant agents. *Curr. Sci.* **2000**, *79*, 1672–1683.

19. Wang, W.; Wang, S.-X.; Guan, H.-S. The Antiviral Activities and Mechanisms of Marine Polysaccharides: An Overview. *Mar. Drugs* **2012**, *10*, 2795–2816. [CrossRef]

20. Rocha de Souza, M.C.; Marques, C.T.; Guerra Dore, C.M.; Ferreira da Silva, F.R.; Oliveira Rocha, H.A.; Leite, E.L. Antioxidant activities of sulfated polysaccharides from brown and red seaweeds. *J. Appl. Phycol.* **2007**, *19*, 153–160. [CrossRef]

21. Necas, J.; Bartosikova, L. Carrageenan: A review. *Vet. Med.* **2013**, *58*, 187–205. [CrossRef]

22. Bray, F.; Ferlay, J.; Soerjomataram, I.; Siegel, R.L.; Torre, L.A.; Jemal, A. Global cancer statistics 2018: GLOBOCAN estimates of incidence and mortality worldwide for 36 cancers in 185 countries. *CA Cancer J. Clin.* **2018**, *68*, 394–424. [CrossRef]

23. Ricci-Vitiani, L.; Lombardi, D.G.; Pilozzi, E.; Biffoni, M.; Todaro, M.; Peschle, C.; De Maria, R. Identification and expansion of human colon-cancer-initiating cells. *Nature* **2007**, *445*, 111–115. [CrossRef]

24. O'Brien, C.A.; Pollett, A.; Gallinger, S.; Dick, J.E. A human colon cancer cell capable of initiating tumour growth in immunodeficient mice. *Nature* **2007**, *445*, 106–110. [CrossRef]

25. Fabrizi, E.; di Martino, S.; Pelacchi, F.; Ricci-Vitiani, L. Therapeutic implications of colon cancer stem cells. *World J. Gastroenterol.* **2010**, *16*, 3871. [CrossRef]

26. Butler, S.J.; Richardson, L.; Farias, N.; Morrison, J.; Coomber, B.L. Characterization of cancer stem cell drug resistance in the human colorectal cancer cell lines HCT116 and SW480. *Biochem. Biophys. Res. Commun.* **2017**, *490*, 29–35. [CrossRef] [PubMed]

27. Kozovska, Z.; Gabrisova, V.; Kucerova, L. Colon cancer: Cancer stem cells markers, drug resistance and treatment. *Biomed. Pharmacother.* **2014**, *68*, 911–916. [CrossRef] [PubMed]

28. Aponte, P.M.; Caicedo, A. Stemness in Cancer: Stem Cells, Cancer Stem Cells, and Their Microenvironment. *Stem Cells Int.* **2017**, 1–17. [CrossRef] [PubMed]

29. Cianciosi, D.; Varela-Lopez, A.; Forbes-Hernandez, T.Y.; Gasparrini, M.; Afrin, S.; Reboredo-Rodriguez, P.; Zhang, J.; Quiles, J.L.; Nabavi, S.F.; Battino, M.; et al. Targeting molecular pathways in cancer stem cells by natural bioactive compounds. *Pharmacol. Res.* **2018**, *135*, 150–165. [CrossRef]

30. Bouvard, C.; Barefield, C.; Zhu, S. Cancer stem cells as a target population for drug discovery. *Future Med. Chem.* **2014**, *6*, 1567–1585. [CrossRef]

31. de la Mare, J.-A.; Sterrenberg, J.N.; Sukhthankar, M.G.; Chiwakata, M.T.; Beukes, D.R.; Blatch, G.L.; Edkins, A.L. Assessment of potential anti-cancer stem cell activity of marine algal compounds using an in vitro mammosphere assay. *Cancer Cell Int.* **2013**, *13*, 39. [CrossRef]

32. Pereira, D.M.; Gomes, S.E.; Borralho, P.M.; Rodrigues, C.M.P. MEK5/ERK5 activation regulates colon cancer stem-like cell properties. *Cell Death Discov.* **2019**, *5*, 1–13. [CrossRef]

33. Dewangan, J.; Srivastava, S.; Rath, S.K. Salinomycin: A new paradigm in cancer therapy. *Tumor Biol.* **2017**, *39*, 101042831769503. [CrossRef]

34. Chopin, T.; Wagey, B.T. Factorial study of the effects of phosphorus and nitrogen enrichments on nutrient and carrageenan content in *Chondrus crispus* (Rhodophyceae) and on residual nutrient concentration in seawater. *Bot. Mar.* **1999**, *42*, 23–31. [CrossRef]

35. Bahmad, H.F.; Cheaito, K.; Chalhoub, R.M.; Hadadeh, O.; Monzer, A.; Ballout, F.; El-Hajj, A.; Mukherji, D.; Liu, Y.-N.; Daoud, G.; et al. Sphere-Formation Assay: Three-Dimensional in vitro Culturing of Prostate Cancer Stem/Progenitor Sphere-Forming Cells. *Front. Oncol.* **2018**, *8*, 347. [CrossRef]

36. Morrison, B.J.; Steel, J.C.; Morris, J.C. Sphere Culture of Murine Lung Cancer Cell Lines Are Enriched with Cancer Initiating Cells. *PLoS ONE* **2012**, *7*, e49752. [CrossRef] [PubMed]

37. Shaheen, S.; Ahmed, M.; Lorenzi, F.; Nateri, A.S. Spheroid-Formation (Colonosphere) Assay for in Vitro Assessment and Expansion of Stem Cells in Colon Cancer. *Stem Cell Rev. Rep.* **2016**, *12*, 492–499. [CrossRef] [PubMed]
38. Suganya, A.M.; Sanjivkumar, M.; Chandran, M.N.; Palavesam, A.; Immanuel, G. Pharmacological importance of sulphated polysaccharide carrageenan from red seaweed *Kappaphycus alvarezii* in comparison with commercial carrageenan. *Biomed. Pharmacother.* **2016**, *84*, 1300–1312. [CrossRef] [PubMed]
39. Prasedya, E.S.; Miyake, M.; Kobayashi, D.; Hazama, A. Carrageenan delays cell cycle progression in human cancer cells in vitro demonstrated by FUCCI imaging. *BMC Complement. Altern. Med.* **2016**, *16*, 270. [CrossRef]
40. Murad, H.; Ghannam, A.; Al-Ktaifani, M.; Abbas, A.; Hawat, M. Algal sulfated carrageenan inhibits proliferation of MDA-MB-231 cells via apoptosis regulatory genes. *Mol. Med. Rep.* **2015**, *11*, 2153–2158. [CrossRef]
41. Calvo, G.H.; Cosenza, V.A.; Sáenz, D.A.; Navarro, D.A.; Stortz, C.A.; Céspedes, M.A.; Mamone, L.A.; Casas, A.G.; Di Venosa, G.M. Disaccharides obtained from carrageenans as potential antitumor agents. *Sci. Rep.* **2019**, *9*, 6654. [CrossRef]
42. Bhattacharyya, S.; Borthakur, A.; Dudeja, P.K.; Tobacman, J.K. Carrageenan Reduces Bone Morphogenetic Protein-4 (BMP4) and Activates the Wnt/β-Catenin Pathway in Normal Human Colonocytes. *Dig. Dis. Sci.* **2007**, *52*, 2766–2774. [CrossRef]
43. Huang, G.-J.; Pan, C.-H.; Wu, C.-H. Sclareol Exhibits Anti-inflammatory Activity in Both Lipopolysaccharide-Stimulated Macrophages and the λ-Carrageenan-Induced Paw Edema Model. *J. Nat. Prod.* **2012**, *75*, 54–59. [CrossRef]
44. Luo, M.; Shao, B.; Nie, W.; Wei, X.W.; Li, Y.L.; Wang, B.L.; He, Z.Y.; Liang, X.; Ye, T.H.; Wei, Y.Q. Antitumor and Adjuvant Activity of λ-carrageenan by Stimulating Immune Response in Cancer Immunotherapy. *Sci. Rep.* **2015**, *5*, 1–12. [CrossRef]

 *marine drugs*

Article

Effect of Carrageenans on Vegetable Jelly in Humans with Hypercholesterolemia

Ana Valado [1,2,*], Maria Pereira [1], Armando Caseiro [1,3], João P. Figueiredo [4], Helena Loureiro [5], Carla Almeida [6], João Cotas [2] and Leonel Pereira [2,7]

1   Polytechnic Institute of Coimbra, ESTeSC-Coimbra Health School, Department of Biomedical Laboratory Sciences, Rua 5 de Outubro, S. Martinho do Bispo, Apart. 7006, 3046-854 Coimbra, Portugal; fpereiramaria@hotmail.com (M.P.); armandocaseiro@estescoimbra.pt (A.C.)
2   Marine and Environmental Sciences Centre (MARE), Faculty of Sciences and Technology, University of Coimbra, 3001-456 Coimbra, Portugal; jcotas@uc.pt (J.C.); leonel.pereira@uc.pt (L.P.)
3   Unidade I&D Química-Física Molecular, Faculdade de Ciências e Tecnologia, Universidade de Coimbra, Rua Larga, 3004-535 Coimbra, Portugal
4   Polytechnic Institute of Coimbra, ESTeSC-Coimbra Health School, Department of Complementary Sciences, Rua 5 de Outubro, S. Martinho do Bispo, Apart. 7006, 3046-854 Coimbra, Portugal; jpfigueiredo@estescoimbra.pt
5   Polytechnic Institute of Coimbra, ESTeSC-Coimbra Health School, Department of Dietetics and Nutrition, Rua 5 de Outubro, S. Martinho do Bispo, Apart. 7006, 3046-854 Coimbra, Portugal; maria.loureiro@estescoimbra.pt
6   Condi Alimentar, Quinta Palmares Armazém, Rua do Ferro, 2685-459 Camarate, Portugal; carla.almeida@condi.pt
7   Department of Life Sciences, Faculty of Sciences and Technology, University of Coimbra, 3000-456 Coimbra, Portugal
*   Correspondence: valado@estescoimbra.pt; Tel.: +351-23-98-02430

Received: 24 November 2019; Accepted: 20 December 2019; Published: 24 December 2019

Abstract: Changes in lipid profile constitute the main risk factor for cardiovascular diseases. Algae extracted carrageenans are long-chain polysaccharides and their ability to form gels provides for the formation of vegetable jelly. The objective was to evaluate the bioactive potential of carrageenan (E407) in the lipid profile, after ingestion of jelly. A total of 30 volunteers of both sexes, aged 20–64 years and with total cholesterol (TC) values ≥200 mg/dL, who ingested 100 mL/day of jelly for 60 days, were studied. All had two venous blood collections: before starting the jelly intake and after 60 days. At both times, TC, high density lipoprotein cholesterol (HDL-C), low density lipoprotein cholesterol (LDL-C) and triglycerides (TG), were evaluated using commercial kits and spectrophotometer. The statistics were performed using the SPSS 25.0 software and $p < 0.05$ were considered statistically significant. Serum values after 60 days of jelly intake revealed a statistically significant decrease in TC levels (5.3%; $p = 0.001$) and LDL-C concentration (5.4%; $p = 0.048$) in females. The daily intake of vegetable jelly for 60 days showed a reduction in serum TC and LDL-C levels in women, allowing us to conclude that carrageenan has bioactive potential in reducing TC concentration.

Keywords: carrageenan; TC; HDL-C; LDL-C; TC reduction; TG

---

1. Introduction

Today, profound socioeconomic changes have led to changes in lifestyles. A sedentary lifestyle and a poor diet, low in vegetables, sometimes associated with smoking habits, are a strong inducer of alterations in lipid profile. This is evaluated in the laboratory by several parameters: total cholesterol (TC), high-density lipoprotein cholesterol (HDL-C), low-density lipoprotein cholesterol (LDL-C) and triglycerides (TG) levels. Modifications in lipid profile constitute the main risk factor in the development of cardiovascular diseases (CVDs), particularly the increase in TC and LDL-C levels [1].

Despite the description of the main risk factors and availability of pharmacological treatment, 17.9 million people die each year due to CVDs, proving to be the leading cause of death worldwide and presenting itself as one of the biggest epidemiological problems today [2]. A similar trend is observed in Portugal, with a record of 31% of deaths from CVDs [3].

According to Shimazu and collaborators, 2007, epidemiological studies have revealed a low incidence of CVDs and higher longevity in Japan, a region known for the wide introduction of algae in the diet [4]. Based on this principle, experimental and observational descriptive studies were performed on the benefits and applications of macroalgae. These studies demonstrate its richness in minerals, vitamins, dietary fiber and lipid poverty, which probably explains the epidemiological evidence [5,6].

In the diet, carrageenans are linear sulphated polysaccharides, constituents of the cell walls and intercellular spaces of *Chondrus crispus* (Rhodophyta, Gigartinales). They are extracted by aqueous or alkaline processes [7,8]. Industrially they are divided into three groups, Kappa (K), Lambda (λ) and Iota (ι) [7–9], represented on food product labels with the acronym E407 [10]. They are classified as hydrocolloids as they easily incorporate an aqueous solution leading to the formation of gels without changing the taste or color of the mixture. This feature enhances its use in the food, cosmetic and pharmaceutical industries (chocolate milk, gelatin, shampoo, and anti-inflammatories, among others) [10,11].

Carrageenans represent one of the major texturizing agents in the food industry. They are natural ingredients, which have been used for decades in food applications and regarded as safe [12]. The dairy sector accounts for a large part of the carrageenan applications in food products, such as frozen desserts, chocolate milk, cottage cheese, and whipped cream. In addition to this, carrageenans are used in various non-dairy food products, such as instant products, jellies, pet foods, and sauces, and non-food products, such as pharmaceutical formulations, cosmetics, and oil well drilling fluid [13,14]. In general, carrageenan serves as a gelling, stabilizing, and viscosity-building agent.

The literature reports multiple benefits regarding carrageenan consumption, such as, antioxidants, anticancer, antilipidemic, anticoagulants, immunomodifiers, antifungals, antivirals, and digestive health support [8,15]. In agreement with the description of the consumption of macroalgae and/or its components, such as carrageenan, its intake comes up as an alternative to the use of pharmaceutical products.

Some studies have shown an efficient reduction of serum TC, TG and LDL-C levels and an increase of HDL-C in the peripheral blood by the action of carrageenans, as a bioactive principle, mainly of Kappa subtype, justified by the interaction of its constitution (alternating groups of β (1-3)-D-galactose-4-sulfate and α (1-4)-3,6 anhydro-D-galactose) on the digestion [16–18]. The investigation was carried out in vitro [17] and in vivo, the latter performed mainly on Wistar rats [18,19] and population samples, where carrageenans were present as supplement [20] or in enriched foods [21].

The new concept of functional food aims to reconcile food and health benefits with the ease and speed of cooking [22,23]. In the health sciences, the possible substitution of pharmacological therapy for foods rich in bioactive and natural compounds represents a healthier and more appealing way to promote the normal functioning of the organism [24].

Carrageenans are widely used in the food industry as a thickener, gelling agent, stabilizer, and protein suspending agent [25]. Successive research and experiments on colloids led to the formation of gels with an optimal texture resulting from the interaction of carrageenan subtypes (K and ι) leading to the emergence of 100% vegetable origin jelly's with a faster solidification [12,26,27]. The bioactive effect of the carrageenan mixture present in the vegetable jelly makes its ingestion a health benefit by reducing TC. Thus, the consumption of jellies of vegetable origin can be a healthy alternative contributing to the prevention and reduction of CVDs.

In Europe, the adherence to the Mediterranean Food Standard has been much studied to understand if the practices of individuals come close to the recommended ones. In this context, we highlight the PREDIMED instrument [28], used in a multicenter, randomized, nutritional intervention study for the primary prevention of CVDs, developed in Spain as part of the study "PREvención with DIeta MEDiterránea", with the objective to study the effectiveness of this diet in the primary prevention of CVD. It is user-friendly in clinical setting [29], allowing rapid evaluation and immediate feedback in intervention studies.

The present investigation aimed to evaluate the impact of vegetable jelly consumption on lipid profile parameters (TC, HDL-C, LDL-C and TG) in individuals with TC levels equal to or higher than 200 mg/dL and to relate them with the degree of adherence to the Mediterranean diet.

2. Results

2.1. Sample Characterization

The experimental group of 60 days (EG-60 days), with jelly ingestion, consisted of 30 participants, 80% females and 20% males, with an average age of 48.68 ± 10.63 years. The degree of Mediterranean diet adherence was high in 50% of the individuals of this experimental group. The experimental group of 30 days (EG-30 days), with jelly ingestion, involved 12 participants with a predominance of females, 66.6% women and 33.3% men, and an average age of 44.55 ± 11.61 years. Regarding the degree of adherence to the Mediterranean diet, 8 of the 12 participants had a good adherence (Table 1).

Table 1. Sample characterization.

Ingestion Period		30 Days $n = 12$	60 Days $n = 30$
Age (years) (min–max)	–	44.55 ± 11.61 (23–64)	48.68 ± 10.63 (20–64)
Sex	F	8	24
	M	4	6
MeDiet	≥10	4	15
	<10	8	15

F: Female; M: Male; MeDiet: Mediterranean diet.

2.2. FTIR-ATR Spectroscopy

The different types of jelly samples were analyzed by vibrational spectroscopy in order to confirm the existence and characterize the carrageenan type. Figure 1 shows that the four FTIR-ATR spectra of all types of jelly present the same type of spectra, with an evident particular peak of 1065 cm^{-1}. This indicates a gelling type of carrageenan (DA) [9], verified in the kappa and iota carrageenans, which have inherent gelling properties [30].

All the spectra showed a band at 849 cm^{-1}, which is characteristic of D-galactose-4-sulfate (G4S) present in the kappa and iota carrageenan types.

The peak at 908 cm^{-1} indicates the presence of 3,6-anydrogalactose-2-sulfate (DA2S), a type of linkage that belongs to the iota carrageenan type. The peak at 943 cm^{-1} indicates the C–O–C of 3,6-anhydrogalactose (DA) linkage, which is related with the presence kappa or iota carrageenan in the sample. The band at 987 cm^{-1} shows the presence of Galactose (G/D) typically present in the kappa and/or iota types of carrageenan. Nonetheless, the peaks at 1005 cm^{-1}, 1049 cm^{-1}, 1115 cm^{-1}, 1126 cm^{-1}, and 1134 cm^{-1} indicate the presence of sulfate groups, such as sulfate esters and sulfated galactans.

Concluding, vegetal jellies present the same spectra for four different flavors indicating that all flavors have the same colloid, a hybrid kappa/iota carrageenan which is typically present in jellies from vegetal origin (Figure 1) [31].

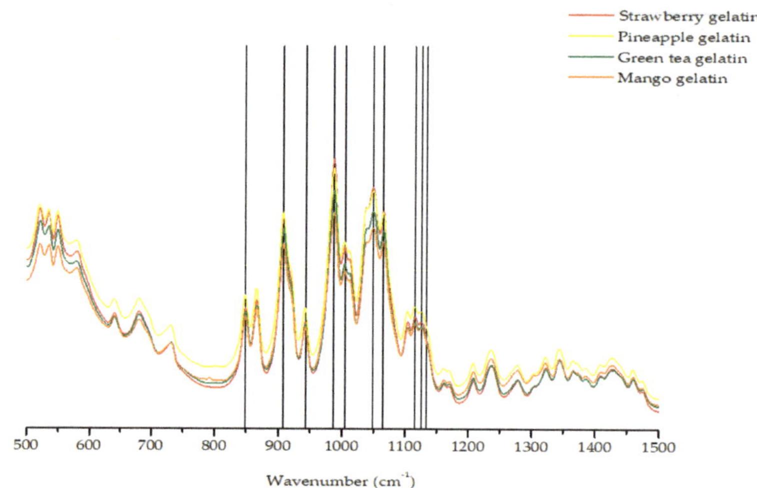

Figure 1. The four FTIR-ATR spectra of different jelly flavors present the same type of spectra with an evident peak of 1065 cm^{-1} that indicates kappa (κ) and iota (ι) hybrid carrageenan [9]. Vegetal jellies present the same spectra indicating they have the same colloid, a hybrid kappa/iota carrageenan.

2.3. Total Cholesterol

The total cholesterol parameter, in the group that ingested carrageenan for 30 days (EG-30 days) at T0 had a concentration of 244.86 ± 31.94 mg/dL and at T1 of 226.18 ± 40.99 mg/dL, representing a statistically significant decrease of 7.63% (18.67 mg/dL), p = 0.021 (Figure 2). The group of individuals evaluated after 60 days of ingestion (EG-60 days) presented TC concentration at T0 of 236.92 ± 30.48 mg/dL and at T2 of 224.40 ± 23.47 mg/dL, revealing a statistically significant decrease of 5.3% (12.52 mg/dL), p = 0.001 (Figure 3). Analysis by gender showed a significant decrease (p = 0.004) in women at T2, comparatively to T0 (Table 2).

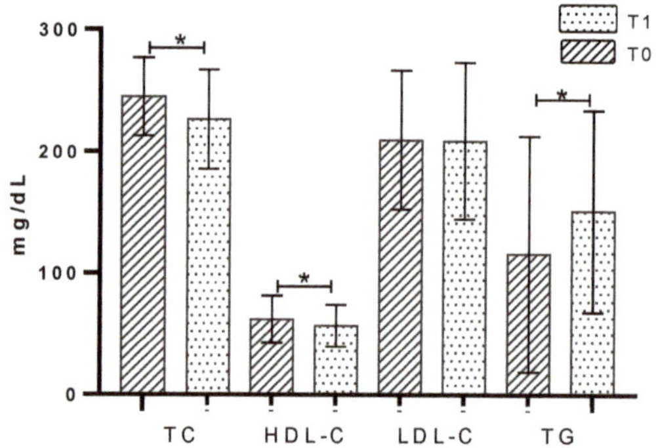

Figure 2. Lipid profile evaluation in experimental group with vegetable jelly ingestion for 30 days. T0: before consumption of jelly; T1: after 30 days of consumption of jelly; TC: Total cholesterol; HDL-C: High-density lipoprotein cholesterol; LDL-C: Low-density lipoprotein cholesterol; TG: Triglycerides; * p < 0.05.

Experimental group-60 days

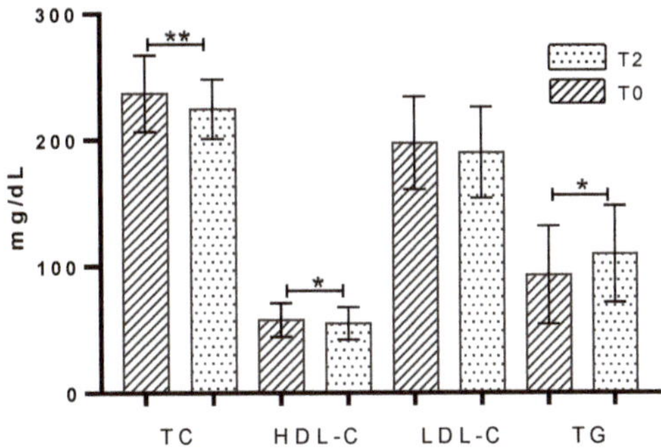

Figure 3. Lipid profile evaluation in experimental group with vegetable jelly ingestion for 60 days. T0: before consumption of jelly; T2: after 60 days of consumption of jelly; TC: Total cholesterol; HDL-C: High-density lipoprotein cholesterol; LDL-C: Low-density lipoprotein cholesterol; TG: Triglycerides; * $p < 0.05$; ** $p < 0.01$.

Table 2. Characterization of lipid profile analytical parameters by sex on EG-60 days.

	Sex	
	Female	**Male**
TC (T0)	234.51 ± 31.06	244.95 ± 29.62
TC (T2)	220.43 ± 21.16 **	237.76 ± 27.82
HDL-C (T0)	60.85 ± 13.15	46.78 ± 6.65
HDL-C (T2)	58.41 ± 12.10	43.46 ± 7.28
LDL-C (T0)	190.58 ± 33.85	220.12 ± 38.77
LDL-C (T2)	180.35 ± 29.07 *	221.48 ± 40.11
TG (T0)	87.88 ± 38.06	109.75 ± 40.18
TG (T2)	101.65 ± 31.16	135.87 ± 50.83 *

The results are presented in mg/dL. TC: Total cholesterol; HDL-C: High-density lipoprotein cholesterol; LDL-C: Low-density lipoprotein cholesterol; TG: Triglycerides. T0: before consumption of jelly; T2: after 60 days of consumption of jelly; * $p < 0.05$; ** $p < 0.01$ versus T0.

2.4. HDL-Cholesterol

The EG-30 days group, before ingestion of the jelly (T0), had an average concentration of 62.20 ± 19.30 mg/dL. After 30 days (T1), it presented 56.88 ± 17.39 mg/dL, which represents a statistically significant decrease of 8.6% (5.32 mg/dL), $p = 0.008$ (Figure 2). The concentration of HDL-C in EG-60 days presented at T0 a concentration of 57.60 ± 13.30 mg/dL and at T2 of 54.97 ± 12.78 mg/dL, with a decrease of 4.6% (2.63 mg/dL) in the parameter, with statistical significance, $p = 0.037$ (see Figure 3). The distribution by gender revealed no statistically significant differences and is shown in Table 2.

2.5. LDL-Cholesterol

The LDL-C parameter concentration at EG-30 days presented mean values of 209.54 ± 57.13 mg/dL at T0 and at T1 of 208.87 ± 64.36 mg/dL, showing no differences (Figure 2). The individuals belonging to the EG-60 days presented a mean T0 concentration of 197.39 ± 36.5 mg/dL and T2 of 189.83 ± 33.73 mg/dL, showing a 3.8% decrease (7.56 mg/dL; $p = 0.062$) (Figure 3). Considering an evaluation by gender of the individuals in the statistical analysis of the parameter, a statistically significant reduction of 5.4% (10.28 mg/dL) was obtained for females, $p = 0.048$ (Table 2).

2.6. Triglycerides

The EG-30 days exhibited at T0 a concentration of 116.09 ± 30.02 mg/dL and at T1 a concentration of 151.28 ± 31.8 mg/dL, representing a 30.3% increase (35.19 mg/dL), statistically significant, $p = 0.038$ (Figure 2). The EG-60 days group presented at T0 the serum concentration of 92.93 ± 38.88 mg/dL and at 60 days (T2) the average concentration was 109.55 ± 38.35 mg/dL, representing a statistically significant increase of 17.8% (16 mg/dL), $p = 0.014$ (Figure 3). Analysis by gender revealed a significant increase associated with males at time T2, versus T0 (Table 2).

2.7. MeDiet-Adherence to Mediterranean Diet

The analysis of the results considering the MeDiet questionnaire evaluation led to the conclusion that the adhesion or not of a Mediterranean diet does not influence the action of carrageenan, present in jelly, in the lipid profile.

3. Discussion

Changes in lipid profile constitute the main risk factor in the development of CVDs, considered the leading cause of death worldwide [1,2]. The need to reduce the mortality rate has led to epidemiological studies showing that seaweed ingestion induces an antilipidemic effect [4,5]. Intake of fiber-rich foods has been described as a protective factor for numerous pathologies, such as hypertension, obesity, diabetes, and many others [32]. Algae, rich in water-soluble fibers such as carrageenan, play an important role as a source of fiber in the diet [13,22]. Introduction of seaweed constituents in food products is nowadays frequently and widely used. As example, the inclusion of Kappa–Iota hybrid carrageenan, identified by food additive code E407, does not alter the taste or color of mixtures and optimizes their consistency, texture, and water retention [7,8]. Due to this characteristic, carrageenan is widely used as a gelling agent in the food industry in the production of chocolate milk, ice cream, beer, and jelly [10]. In addition to the antilipidemic effect, numerous benefits to human health are attributed to carrageenan as antioxidant, immunomodulatory, antiviral, and digestive health support, thus revealing its high bioactive potential [9,14]. However, for cultural reasons it is difficult to estimate the impact of a carrageenan-rich diet associated with individuals not inserted in populations of Asian culture and to evaluate their lipid profile, giving the present study a pioneer status. Thus, two time periods of 30 and 60 days of daily ingestion of vegetable origin jelly were tested. In the absence of studies with similar methodology developed in Caucasian populations, the time periods were based on studies by Sokolova et al. [20] and by Panlasigui et al. [21].

According to Panlasigui and coauthors, participants ingested the mixed carrageenans in their own food for eight weeks [21]. Also, Sokolova's research applied carrageenan as a dietary supplement to individuals previously diagnosed with CVDs [20]. Participants ingested 250 mL per day of a supplement consisting of a hybrid K–λ carrageenan (3:1) for 20 days [20]. Our results showed a statistically significant decrease in TC and HDL-C concentration in both periods (60 and/or 30 days) after daily ingestion of 100 mL of vegetable jelly. Also, a reduction in TC and LDL-C levels in females when assessed by gender distribution and EG-60 days was noted. In contrast, males recorded an increase in TG concentration.

The TC levels decreased by 7.6% after 30 days and 5.3% after 60 days. The decrease was corroborated by Sokolova et al. [20], but with a smaller percentage, which may be due to the different study design involving cardiovascular patients with supplemental carrageenan consumption resulting in a 16.5% decrease in TC [20]. Also, Panlasigui and coauthors [21], in their research on eating carrageenan-rich foods at the three most important meals (breakfast, lunch and dinner) showed a reduction in TC levels by 33%. Concerning HDL-C concentration, our results decreased by 8.6% and 4.6% after 30 and 60 days respectively, as opposed to the work of other authors [20,21] which showed an increase. For the concentration of the LDL-C fraction after 60 days, it showed a tendency to decrease, evidenced in relation to females showing a significant decrease, which is in accordance with the results of Panlasigui et al. and Matthan et al. [21,33]. The increase in TG concentration was 30.3% for 30 days and 17.8% after 60 days of jelly ingestion. It should be noted that it has not reached the maximum reference values [34]. For TG, the literature presents ambiguous results, since Panlasigui and coauthors showed a 32% reduction [21], while Sokolova and collaborators showed an increase [20]. However, low cholesterol absorption is known to be associated with increased TG and decreased HDL in both sexes [33]. Despite the multiple benefits to human health, the mechanism of action of carrageenans is not yet completely understood and only their interaction with the digestive process is documented, being classified as long-chain carbohydrates [35,36]. When carrageenans reach the intestinal tract, the volume and viscosity of the intestinal content increases and leads to increased digestion time and reduced efficiency, since the availability of products for enzymatic action is compromised [37].

On the other hand, it is capable of capturing cholesterol and bile salts, inhibiting lipase, causing a decrease in cholesterol absorption directly and indirectly, since the emulsifying action performed by bile salts and lipase becomes deficient, disabling micelle formation [38–40]. However, the action of carrageenans is not restricted to the partial effect of gastrointestinal barrier, but also has endogenous repercussions. Excretion of bile salts precludes their reuse and, therefore, synthesis is required once more [20,21,40,41]. The mobilization of TG occurs from energy reserves into the bloodstream because cholesterol is required for synthesis and it is in low circulating concentrations [42].

The bioactive potential of carrageenan comes from its ability to decrease cholesterol absorption, leading to a decreased cholesterol absorption rate and consequent increase in endogenous cholesterol synthesis rate, maintaining the balance of values, a fundamental process to maintaining the proper functioning of the body [33]. According to some researchers, the relationship between cholesterol absorption markers and serum parameters has shown that when cholesterol absorption markers decrease, serum triglyceride concentration increases and HDL-C levels decrease [33,43]. As a complement, Matthan et al. [33] verified that when high endogenous cholesterol level marker values were detected, the serum LDL-C fraction values decreased. An inverse relationship between expression markers and LDL-C concentration is justified by the increased expression of the LDL-C receptor in the cell, which causes an increase in cholesterol internalization rate, thereby decreasing 3-hydroxy-3-methylglutaryl-coenzyme A (HMG-CoA) reductase activity and, therefore, decreasing of endogenous cholesterol synthesis [33,43]. Thus, carrageenan ingestion leads to decreased cholesterol absorption rate, leading to increased serum TG concentration and reduced HDL-C. On the other hand, the need for cholesterol for bile salt synthesis, as a result of carrageenan-induced excretion, leads to the promotion of endogenous cholesterol synthesis, leading to a decrease in serum LDL-C levels [20,40,41]. Also, the gender difference leads to metabolic changes where the hypolipidemic properties of estrogens induce a protective role in women [44]. The low activity of the enzyme HMG-CoA reductase in women indicates that the neo-synthesis of cholesterol is physiologically lower in females, until menopause, compared with males [33,44]. Also, adding the principle that the presence of estrogens increases intestinal cholesterol absorption in humans, it seems to justify the significant reduction observed in LDL-C and TC levels in women as a result of decreased absorption efficiency of induced cholesterol molecules by vegetable jelly [41,44].

In conclusion, carrageenans constitute a bioactive potential in reducing total cholesterol levels in the body. Regular consumption of vegetable origin jelly is beneficial, contributing to lower total cholesterol levels and also to decreased LDL-C levels in females. Ingestion of vegetable jelly can therefore be considered a good practice as a protection against the development of cardiovascular diseases.

Although these results are considered relevant, the need to increase the number of participants in this type of study in order to consolidate the presented results in future complementary works must be emphasized.

4. Materials and Methods

4.1. Sample Characterization

The study comprised 30 volunteers of both sexes, aged between 20 and 64 years. In the selection of the sample, the following inclusion criteria were applied: over 18 years old, with previous indication of elevated TC levels (equal to or higher than 200 mg/dL), and apparently healthy. No restriction on participants' eating habits was indicated. As exclusion criteria, pharmacological therapy for hypercholesterolemia was considered.

The study participation involved the ingestion of 100 mL of vegetable jelly per day, preferably after dinner, for 60 days. This feature allowed the formation of two experimental groups. The experimental group with jelly ingestion for 60 days (EG-60 days) was composed of 30 participants. Of these, 12 were harvested after 30 days, constituting the experimental group with jelly ingestion for 30 days (EG-30 days). In the jelly composition, the presence of a hybrid Kappa–Iota carrageenan polysaccharide, integrated in the gelling agent, was equal in the various flavours (pineapple, green tea, mango, strawberry). The analysis of the carrageenan composition present in the jellies was performed by attenuated total reflectance Fourier transform infrared spectroscopy (FTIR-ATR) spectroscopy on the IFS 55 equipment using the Golden Gate ATR single reflection diamond system. The jelly with vegetable origin is available on any commercial food surface. In its composition we find the following ingredients: sugar, gelling agent (carrageenan, potassium citrates and sucrose), acidity regulators (citric acid and sodium citrates), aromas (sulfites), antioxidant (ascorbic acid, vitamin C) and dyes (anthocyanins and beta carotene). For the composition and energy distribution of a portion of 92 g, equivalent to 100 mL of jelly ingested, the following mean values are given: energy 266 kJ/63 kcal (3% DR) where DR is the reference dose for an average adult (8400 kJ/2000 kcal) per serving; lipids, including saturated 0 g; carbohydrates 16 g (6% DR), of which 15 g (17% DR) are sugars; fiber 0 g; proteins 0 g; salt 0.08 g (1% DR); Vitamin C: 29 mg (36% NRV) where NRV is the nutrient reference value. The preparation of jelly involves the addition of a 90 g sachet of powdered preparation to 500 mL of boiling water, followed by distribution into five portions.

According to the inclusion criteria, venous blood was collected at three different times: (T0) before starting the consumption of vegetable jelly, (T1) after 30 days and (T2) 60 days after the first jelly. Posteriorly, the analytical parameters TC, HDL-C, LDL-C and TG were performed at various times and the adherence to the Mediterranean diet was evaluated.

4.2. Sample Collection and Preparation

Venous blood samples were collected (about 5 mL) into a vacuum-dried gel tube and centrifuged at 1800× g, 4 °C for 10 min. Serum was aliquoted and stored at −20 °C until adjustments were made. The TC was quantified by the Liquick cor-CHOL KIT, from Cormay, Poland, according to the enzymatic colorimetric glycerophosphate oxidase method. The determination of TG was performed by the Liquick cor-TG KIT, from Cormay, Poland, using the enzymatic colorimetric method with cholesterol esterase and cholesterol oxidase. The HDL-C quantification was performed with the Prestige 24i HDL-DIRECT KIT, Cormay, Poland with help of the enzymatic direct colorimetric method using cholesterol esterase and cholesterol oxidase. The concentrations were evaluated by spectrophotometry according to the protocol and the Prestige 24i equipment, *Tokyo Boeki Medical System* Ltd., Tokyo, Japan.

The determination of LDL-C was calculated by Friedewald's formula (LDL-C = [TC]–[HDL-C]–([TG]/5)) using the TC, HDL-C and TG concentrations previously quantified. The formula is applied only at TG concentrations <400 mg/dL [34,45].

The adherence to the Mediterranean diet was evaluated by applying the MeDiet questionnaire, integrated in PREDIMED [28]. Demographic characterization was also obtained by the questionnaire. The PREDIMED questionnaire presents 14 questions from which the respondent is categorized as having good or poor adherence to Mediterranean diet. The response to each of the 14 items is scored 1 if it meets the criteria defined as characteristic of this type of feed (possible range 0–14 points).

A final score of 10 points or higher represents good adherence to the standard. This questionnaire uses predefined targets for food consumption [46].

4.3. Statistical Analysis

The statistical treatment was performed using SPSS 25.0 Software, IBM®, United States and the graphs using GraphPad Prism version 8.1.2. Descriptive analysis of the variables was performed with mean, maximum, minimum, and standard deviation calculations. The Wilcoxon's t-test was used for comparison of paired quantitative variables. The results are presented as mean ± standard deviation and considered statistically significant for p-value < 0.05.

4.4. Confidentiality and Data Protection

The study was conducted in accordance with the principles of the Declaration of Helsinki, ensuring maximum protection and confidentiality of data, having been previously approved by the Ethics Committee of the Polytechnic Institute of Coimbra. All subjects gave their informed consent for inclusion before they participated in the study. The study was conducted in accordance with the Declaration of Helsinki, and the protocol was approved by the Ethics Committee of Polytechnic Institute of Coimbra (n° 44/2018).

Author Contributions: A.V. was responsible for conceptualization, resources, blood collection, writing review and research supervision; M.P. was responsible for study design, data realization and analysis, and writing of the original draft; A.V. and M.P. contributed to the elaboration of the paper; A.C. participated in the methodology, performance and validation of laboratory analysis; J.P.F. was responsible for statistical treatment and results analysis; H.L. helped in the methodology; C.A. participated in methodology and resources; J.C. held the spectra; L.P. was responsible for curating the data and reviewing the manuscript. All authors have read and agreed to the published version of the manuscript.

Funding: The authors acknowledge financial support from the Portuguese Foundation for Science and Technology (FCT) through the strategic project UID/MAR/04292/2019 granted to MARE.

Acknowledgments: To all participants and to Condi Alimentar, who made available the commercial jelly for the study.

Conflicts of Interest: The authors declare no conflict of interest.

References

1. Klop, B.; Elte, J.W.; Cabezas, M.C. Dyslipidemia in obesity: Mechanisms and potential targets. *Nutrients* **2013**, *5*, 1218–1240. [CrossRef] [PubMed]
2. WHO. Cardiovascular Diseases CVDs 2018. Available online: https://www.who.int/cardiovascular_diseases/en/ (accessed on 19 September 2019).
3. INE. *Anuário Estatístico*, 109rd ed.; Instituto Nacional de Estatística: Lisboa, Portugal, 2017; p. 254.
4. Shimazu, T.; Kuriyama, S.; Hozawa, A.; Ohmori, K.; Sato, Y.; Nakaya, N.; Nishino, Y.; Tsubono, Y.; Tsuji, I. Dietary patterns and cardiovascular disease mortality in Japan: A prospective cohort study. *Int. J. Epidemiol.* **2007**, *36*, 600–609. [CrossRef] [PubMed]
5. Burtin, P. Nutritional value of seaweeds. *Electron. J. Environ. Agric. Food Chem.* **2003**, *2*, 498–503.
6. Lordan, S.; Ross, R.P.; Stanton, C. Marine bioactives as functional food ingredients: Potential to reduce the incidence of chronic diseases. *Mar. Drugs* **2011**, *9*, 1056–1100. [CrossRef]
7. Younes, M.; Aggett, P.; Aguilar, F.; Crebelli, R.; Filipic, M.; Jose Frutos, M.; Galtier, P.; Gott, D.; Gundert-Remy, U.; Kuhnle, G.G.; et al. Re-evaluation of carrageenan (E 407) and processed Eucheuma seaweed (E 407a) as food additives. *EFSA J.* **2018**, *16*, e05238.
8. Kenn, H. *Genu Carrageenan Book*; CP Kelco Inc.: Lille Skensved, Denmark, 2002; p. 28.
9. Pereira, L.; Amado, A.M.; Critchley, A.T.; van de Velde, F.; Ribeiro-Claro, P.J.A. Identification of selected seaweed polysaccharides (phycocolloids) by vibrational spectroscopy (FTIR-ATR and FT-Raman). *Food Hydrocoll.* **2009**, *23*, 1903–1909. [CrossRef]
10. Pereira, L. *Algae, Uses in Agriculture, Gastronomy and Food Industry*; Câmara Municipal de Viana do Castelo: Viana do Castelo, Portugal, 2010; p. 67.

11. García, I.A.F.; Castrovielo, R.A.; Neira, C.D. *Las Algas en Galicia—Alimentacion y Otros Usos. Xunta de Galicia*; Consellería de Pesca; Marisqueo e Acuicultura: Galicia, Spain, 1993; p. 229.

12. Soares, P.A.; de Seixas, J.R.; Albuquerque, P.B.; Santos, G.R.; Mourão, P.A.; Barros, W., Jr.; Correia, M.T.; Carneiro-da-Cunha, M.G. Development and characterization of a new hydrogel based on galactomannan and κ-carrageenan. *Carbohydr. Polym.* **2015**, *134*, 673–679. [CrossRef]

13. van de Velde, F.; de Ruiter, G.A. Carrageenan. In *Biopolymers Polysaccharides II, Polysaccharides from Eukaryotes*; Vandame, S.D., Baets, S.D., Steinbèuchej, A., Eds.; Wiley-VCH: Weinheim, Germany, 2002; Volume 6, pp. 245–274.

14. Pereira, L. Biological and therapeutic properties of the seaweed polysaccharides. *Int. Biol. Rev.* **2018**, *2*, 1–50.

15. Silva, F.R.F.; Dore, C.M.P.G.; Marques, C.T.; Nascimento, M.S.; Benevides, N.M.B.; Rocha, H.A.O.; Chavante, S.F.; Leite, E.L. Anticoagulant activity, paw edema and pleurisy induced carrageenan: Action of major types of commercial carrageenans. *Carbohydr. Polym.* **2010**, *79*, 26–33. [CrossRef]

16. Blackwood, A.; Salter, J.; Dettmar, P.; Chaplin, M. Dietary fibre, physicochemical properties and their relationship to health. *J. R. Soc. Promot. Health* **2000**, *120*, 242–247. [CrossRef]

17. Chen, F.; Deng, Z.; Zhang, Z.; Zhang, R.; Xu, Q.; Fan, G.; Luo, T.; Julian, D.; Clements, Mc. Controlling lipid digestion profiles using mixtures of different types of microgel: Alginate beads and carrageenan beads. *J. Food Eng.* **2018**, *238*, 156–163. [CrossRef]

18. Amano, H.; Kakinuma, M.; Coury, D.; Ohno, H.; Hara, T. Effect of a seaweed mixture on serum lipid level and platelet aggregation in rats. *Fish. Sci.* **2005**, *71*, 1160–1166. [CrossRef]

19. Gomez, E.; Jimenez, A.; Ruperez, P. Effect of the red seaweed *Mastocarpus stellatus* intake on lipid metabolism and antioxidant status in healthy Wistar rats. *Food Chem.* **2012**, *135*, 806–811. [CrossRef] [PubMed]

20. Sokolova, E.; Bogdanovich, L.; Ivanova, T.; Byankina, A.; Kryzhanovskiy, S.; Yermak, I. Effect of carrageenan food supplement on patients with cardiovascular disease results in normalization of lipid profile and moderate modulation of immunity system markers. *PharmaNutrition* **2014**, *2*, 33–37. [CrossRef]

21. Panlasigui, L.; Baello, O.; Dimatangal, J.; Dumelod, B. Blood cholesterol and lipid-lowering effects of carrageenan on human volunteers. *Asia Pac. J. Clin. Nutr.* **2003**, *12*, 209–214.

22. Cardoso, S.; Pereira, O.; Seca, A.; Pinto, D.; Silva, A. Seaweeds as preventive agents for cardiovascular diseases: From nutrients to functional foods. *Mar. Drugs* **2015**, *13*, 6838–6865. [CrossRef]

23. Williams, M.; Ragasa, C. Functional foods: Opportunities and challenges for developing countries. *World Bank* **2006**, *19*, 37683.

24. Siro, I.; Kápolna, E.; Kapolna, B.; Lugasi, A. Functional food. Product development, marketing and consumer acceptance—A review. *Appetite* **2008**, *51*, 456–467. [CrossRef]

25. Dickson-Spillman, M.; Siegrist, M.; Keller, C. Attitudes toward chemicals are associated with preference for natural food. *Food Qual. Pref.* **2011**, *22*, 149–156. [CrossRef]

26. Campo, V.; Kawano, D.; Silva, D.; Carvalho, I. Carrageenans: Biological properties, chemical modifications and structural analysis—A review. *Carbohydr. Polym.* **2009**, *77*, 167–180. [CrossRef]

27. Bui, V.; Nguyen, B.; Renou, F.; Nicolai, T. Rheology and microstructure of mixtures of iota and kappa-carrageenan. *Food Hydrocoll.* **2019**, *89*, 180–187. [CrossRef]

28. Martinez-Gonzalez, M.A.; Corella, D.; Salas-Salvado, J.; Ros, E.; Covas, M.I.; Fiol, M.; Wärnberg, J.; Arós, F.; Ruíz-Gutiérrez, V.; Lamuela-Raventós, R.M.; et al. PREDIMED Study Investigators. Cohort profile: Design and methods of the PREDIMED study. *Int. J. Epidemiol.* **2012**, *41*, 377–385. [CrossRef] [PubMed]

29. Schroder, H.; Fito, M.; Estruch, R.; Martínez-González, M.A.; Corella, D.; Salas-Salvadó, J.; Lamuela-Raventós, R.; Ros, E.; Salaverría, I.; Fiol, M.; et al. A short screener is valid for assessing Mediterranean diet adherence among older Spanish men and women. *J. Nutr.* **2011**, *141*, 1140–1145. [CrossRef] [PubMed]

30. Du, L.; Brenner, T.; Xie, J.; Liu, Z.; Wang, S.; Matsukawa, S. Gelation of iota/kappa Carrageenan Mixtures. In *Gums and Stabilisers for the Food Industry 18: Hydrocolloid Functionality for Affordable and Sustainable Global Food Solutions*, 1st ed.; Williams, P., Philips, G.R.S.C., Eds.; Royal Society of Chemistry: Cambridge, UK, 2016; pp. 47–55.

31. Pereira, L.; Gheda, S.; Ribeiro-Claro, P. Analysis by vibrational spectroscopy of seaweed polysaccharides with potential use in food, pharmaceutical, and cosmetic industries. *Int. J. Carbohydr. Chem.* **2013**, *2013*. [CrossRef]

32. Anderson, J.; Baird, P.; Davis, R.; Ferreri, S.; Knudtson, M.; Koraym, A.; Waters, V.; Williams, C.L. Health benefits of dietary fiber. *Nutr. Rev.* **2009**, *67*, 188–205. [CrossRef]

33. Matthan, N.; Zhu, L.; Pencina, M.; D'Agostino, R.; Schaefer, E.; Lichtenstein, A. Sex-specific differences in the predictive value of cholesterol homeostasis markers and 10-year cardiovascular disease event rate in Framingham Offspring Study participants. *Am. Heart J.* **2013**, *2*, e005066. [CrossRef]

34. Piepoli, M.F.; Hoes, A.W.; Agewall, S.; Albus, C.; Brotons, C.; Catapano, A.L.; Cooney, M.T.; Corrà, U.; Cosyns, B.; Deaton, C.; et al. ESC Scientific Document Group. European guidelines on cardiovascular disease prevention in clinical practice: The sixth joint task force of the European society of cardiology and other societies on cardiovascular disease prevention in clinical practice (constituted by representatives of 10 societies and by invited experts) developed with the special contribution of the European Association for Cardiovascular Prevention & Rehabilitation (EACPR). *Elsevier* **2016**, *252*, 207–274.

35. McKim, J. Food additive carrageenan: Part I: A critical review of carrageenan in vitro studies, potential pitfalls, and implications for human health and safety. *Crit. Rev. Toxicol.* **2014**, *44*, 211–243. [CrossRef]

36. Weiner, M.; Nuber, D.; Blakemore, W.; Harriman, J.; Cohen, S. A 90-day dietary study on kappa carrageenan with emphasis on the gastrointestinal tract. *Food Chem. Toxicol.* **2007**, *45*, 98–106. [CrossRef]

37. Awang, A.; Ng, J.; Matanjun, P.; Sulaiman, M.; Tan, T.; Ooi, Y. Anti-obesity property of the brown seaweed, Sargassum polycystum using an in vivo animal model. *J. Appl. Phycol.* **2014**, *26*, 1043–1048. [CrossRef]

38. Weiner, M. Food additive carrageenan: Part II: A critical review of carrageenan in vivo safety studies. *Crit. Rev. Toxicol.* **2014**, *44*, 244–269. [CrossRef] [PubMed]

39. Chater, P.I.; Wilcox, M.; Cherry, P.; Herford, A.; Mustar, S.; Wheater, H.; Wheater, H.; Brownlee, I.; Seal, C.; Pearson, J. Inhibitory activity of extracts of Hebridean brown seaweeds on lipase activity. *J. Appl. Phycol.* **2016**, *28*, 1303–1313. [CrossRef] [PubMed]

40. Wang, W.; Onnagawa, M.; Yoshie, Y.; Suzuki, T. Binding of bile salts to soluble and insoluble dietary fibers of seaweeds. *Fish. Sci.* **2001**, *67*, 1169–1173. [CrossRef]

41. Sokolova, E.; Kravchenko, A.; Sergeeva, N.; Davydova, V.; Bogdanovich, L.; Yermak, I. Effect of carrageenans on some lipid metabolism components in vitro. *Carbohydr. Polym.* **2019**. [CrossRef]

42. McIntosh, M.; Miller, C. A diet containing food rich in soluble and insoluble fiber improves glycemic control and reduces hyperlipidemia among patients with type 2 diabetes mellitus. *Nutr. Rev.* **2001**, *59*, 52–55. [CrossRef] [PubMed]

43. Andrade, I.; Santos, L.; Ramos, F. Cholesterol absorption and synthesis markers in Portuguese hypercholesterolemic adults: A cross-sectional study. *Eur. J. Int. Med.* **2016**, *28*, 85–90. [CrossRef] [PubMed]

44. De Marinis, E.; Matini, C.; Trentalance, A.; Pallottini, V. Sex differences in hepatic regulation of cholesterol homeostasis. *J. Endocrinol.* **2008**, *198*, 635–643. [CrossRef]

45. Mehta, R.; Reyes-Rodriguez, E.; Yaxmehen, B.-C.O.; Guerrero-Díaz, A.C.; Vargas-Vázquez, A.; Cruz-Bautista, I.; AAguilar-Salinas, C. Performance of LDL-C calculated with Martin's formula compared to the Friedewald equation in familial combined hyperlipidemia. *Atherosclerosis* **2018**, *277*, 204–210. [CrossRef]

46. Schulze, M.B.; Hoffmann, K. Methodological approaches to study dietary patterns in relation to risk of coronary heart disease and stroke. *Br. J. Nutr.* **2006**, *95*, 860–869. [CrossRef]

MDPI

St. Alban-Anlage 66

4052 Basel

Switzerland

Tel. +41 61 683 77 34

Fax +41 61 302 89 18

www.mdpi.com

Marine Drugs Editorial Office

E-mail: marinedrugs@mdpi.com

www.mdpi.com/journal/marinedrugs